THE MATHEMATICAL

AND OTHER WRITINGS

OF

ROBERT LESLIE ELLIS, M.A.

Cambridge:
PRINTED BY C. J. CLAY, M.A.
AT THE UNIVERSITY PRESS.

Ever affectionately yrs
R. L. Ellis

Published by Deighton, Bell & Co. Cambridge.

THE MATHEMATICAL AND OTHER WRITINGS

OF

ROBERT LESLIE ELLIS, M.A.

LATE FELLOW OF TRINITY COLLEGE, CAMBRIDGE.

EDITED BY

WILLIAM WALTON, M.A.

TRINITY COLLEGE; MATHEMATICAL LECTURER AT
MAGDALENE COLLEGE, CAMBRIDGE.

With a Biographical Memoir

BY

THE VERY REVEREND HARVEY GOODWIN, D.D.

DEAN OF ELY.

CAMBRIDGE:
DEIGHTON, BELL, AND CO.
LONDON: BELL AND DALDY.
1863.

DEDICATORY LETTER.

Dear Lady Affleck,

Having undertaken with your sanction the publication of the writings of your lamented brother in a collected form, I may be permitted to address to you a few observations on the principles by which I have been guided in the fulfilment of an act of piety to the memory of a friend of many years, whom I shall ever remember with affection and veneration.

The greater part of the writings contained in this volume had previously appeared in Scientific Journals, and Reports, some few in Educational Treatises. I have had no scruple whatever in reprinting these works, since their publication had already taken place by the author's own act or permission.

The question of the publication of manuscripts presented to my mind greater difficulty. During his long years of suffering he was in the habit of dictating to his friends various speculations in adaptation to their different tastes and pursuits, evidently for the most part not intended for the press. A great many mathematical investigations, which may be described as interesting problems, communicated to me on numerous occasions either by dictation or by letter, are in my possession. Out of these manuscripts I have ventured to publish only two, the one on the Retardation of Sunrise, and the other a new Solution of a Problem in the First Book of Newton's Principia. All the manuscripts entrusted to me by you, classical, philological, botanical, and mathematical, works of a more elaborate character, I have not hesitated to include in this volume, from a conviction that they will be interesting to many readers, and from an impression, grounded on internal evidence, that the probability of their ultimate publication may have been contemplated by their author.

The difficulty under which an amanuensis labours, in transferring accurately to paper the words of one afflicted by severe illness, is at all times considerable. In the manuscripts placed in my hands the errors were necessarily, by reason of the peculiar nature of the subjects, very numerous. The obstacles, arising from this source, in the way of preparing for the press some portions of the work with proper emendations, I should have regarded as not a little formidable, had it not been for the zealous assistance of Mr Munro, Fellow of Trinity College, to whom I am indebted for the corrections of the text in all the philological and classical writings which had not previously been published.

I may mention also that to the Dean of Ely, who at my request and with your entire approbation undertook most heartily the composition of the Biographical Memoir, I am under obligation for occasional advice, and for much kindly interest in the progress of the work through the press.

The engraving has been taken from an admirable portrait by Samuel Lawrence, in the possession of Professor Grote, by his kind permission.

Hoping that I have adequately discharged my duties as Editor of this Collection of your brother's writings, and at any rate conscious that I have done my best, I beg to dedicate to you this volume, and to subscribe myself,

Your faithful servant,

WILLIAM WALTON.

CHESTERTON, *Oct.* 26, 1863.

CONTENTS.

ERRATUM.

p. 1, 4, 5. *For* Bernouilli, *read* Bernoulli.

BIOGRAPHICAL MEMOIR

OF

ROBERT LESLIE ELLIS, M.A.

LATE FELLOW OF TRINITY COLLEGE, CAMBRIDGE.

BY

HARVEY GOODWIN, D.D.

DEAN OF ELY.

"Il avait une promptitude infinie à tout saisir, une mémoire prodigieuse et une faculté méthodique et rectifiante pour tirer, comme par une chimie naturelle, quelque chose de précieux de tout ce qui s'offrait à lui, soit dans la conversation, soit dans la lecture. Tout sujet d'entretien lui était bon; il acceptait volontiers celui qu'on mettait sur le tapis, et il étonnait les indifférents par les trésors qu'il tirait à l'instant de la mine qu'ils lui avaient offerte sans y songer. Son esprit était comme une bibliothèque encyclopédique bien ordonnée, qu'il suffisait d'ouvrir à la lettre qu'on voulait, pour en faire sortir des richesses."

BIOGRAPHICAL MEMOIR

OF

ROBERT LESLIE ELLIS.

THE publication in a collected form of the papers which this volume contains is to be regarded, to a certain extent, as a tribute of affection. From this point of view the work would hardly seem to be complete without some notice of the life and character of the author. That he was a man of no ordinary attainments, in at least one field of knowledge, will be sufficiently evident to those who know no more of him than they can gather from this portion of his writings; that he had powers distinct from those of mathematical research, is evident from what he has done as the editor of the philosophical works of Bacon; but even these published records of his intellect will perhaps fail to convey to readers that impression of remarkable and various ability which was made, I believe, upon all those who were brought into personal contact with him. It may therefore be interesting to the general reader, besides completing the memorial character of the volume, if an attempt be made by one of his contemporaries to give some account of what he was.

The mere facts of the life are few and simple. The external picture of it may be very easily drawn. It was short, quiet, uneventful, but very full of suffering. The plan which I shall adopt in the following memoir will be this: I shall first give the story of the life in as compact a form as may be possible, and then endeavour to lay before the reader some estimate of the mind and character.

ROBERT LESLIE ELLIS was born at Bath, August 25, 1817, being the youngest of a family consisting of three sons and three daughters. His mother's health was not good, and from her he appears to have inherited that highly nervous constitution, which became, during a considerable portion of his life, as we shall see hereafter, the medium of great suffering. His father was a man of cheerful disposition, of active and well cultivated intellect, fond of speculative inquiry, and in worldly circumstances independent. His character and his mode of dealing with Robert, as a child, had a great influence upon him throughout his life: he became his father's companion from a very early age, and the affection with which he referred in later life to his father's care and to the happy days of his boyhood, could not fail to strike those who had the pleasure of knowing him intimately.

I do not find that as a child he exhibited any extraordinary symptoms of precocity[1], though it is manifest, from records of his boyish doings made by himself, that he was very forward in his studies, and that he took an interest in his work, and

[1] With reference to what is said in the text, and possibly the reader may think in contradiction to it, I insert here a memorandum, which I find, amongst the papers intrusted to me, and which appears to be in his father's hand.

"The following numerical theorem, if not curious in itself, may perhaps be esteemed so, as coming from a boy of eight years old, who was not far advanced in the ordinary rules of arithmetic.

"If any number be added to its equal, subtracted from its equal, multiplied by its equal, and divided by its equal, then the *sum*, the *difference*, the *product*, and the *quotient* of these equal numbers, added together, will equal the square of the next higher number."

That is to say, if n be the number,

$$(n+n)+(n-n)+n\times n+n\div n=n^2+2n+1=(n+1)^2.$$

exerted his mind upon the subjects to which it was directed in a manner by no means usual.

He was never at school, but had the advantage of two tutors at Bath, one in classics, the other in mathematics[1]. He worked for them with great earnestness, and I find from his own memoranda, that in the year 1827, when he was about ten years old, he was doing equations, and reading Xenophon and Virgil, besides giving some attention to French and drawing. These same memoranda shew that at this time, in addition to his ordinary work with his tutors, he was reading books not usually read by boys at such an age, Cuvier's *Theory of the Earth*, *The Edinburgh Journal of Science*, *The Edinburgh Review*, &c.

One remark which is suggested by the boyish records left behind him is, that it is clear that from an early age Ellis had an extreme delight in knowledge for its own sake: he had not the ordinary stimulus of school emulation, indeed he was singularly free from the influence of competition until his college days: but it is manifest, from his own account, that his progress in knowledge, and perhaps especially in mathematical knowledge, was a source of very keen delight.

In 1829, that is, when twelve years old, he began to read Mechanics. In the early part of 1830 he commenced the Differential Calculus; from which he rapidly proceeded to the Integral Calculus; and towards the middle of the year he speaks of being engaged with his tutor in finding the lengths and areas of curves.

Meanwhile his general reading, for which he was dependent upon his father's library and upon that of the Bath Institution, was most multifarious; but each particular subject seems to have been carefully studied, and an opinion formed upon it.

Thus his education proceeded quietly and also rapidly under his father and private tutors for several years.

[1] His mathematical tutor was Mr T. S. Davies, afterwards of Woolwich; his classical, Mr H. A. S. Johnstone.

This home education had, I think, a perceptible effect upon his future character. The effect was not bad in the sense in which that epithet is generally believed to be applicable to home education; but there might be observed in him a kind of elderly sobriety of manner, not amounting to stiffness, but conveying the impression that he had been accustomed to converse with those older than himself, and standing out in marked contrast with that lively boyish freedom and gaiety which is especially the characteristic of young men educated at the great public schools.

In October, 1834, he became the pupil of the Rev. James Challis, then Rector of Papworth St Everard, in Cambridgeshire, who soon after was appointed and still remains Plumian Professor of Astronomy in the University of Cambridge. His residence at Papworth was, however, very short; his health gave way, and at the end of six weeks he was compelled to return home. Here he remained for about two years, not coming up to the University in 1835, as originally intended, but postponing the event, on the ground of health, to the following year.

He came into residence as a Pensioner of Trinity College, in October 1836, being entered as a pupil of the Rev. G. Peacock, afterwards Lowndean Professor and Dean of Ely. During his undergraduate career his health was not strong, but I think he was never compelled by illness to desist from his course of study. He was very much in advance of the men of his year in mathematical acquirement, and had already read most of the subjects which usually occupy an undergraduate's time. He was himself much amused at the surprise expressed by his tutor, Mr Peacock, when at an early stage of his College life in answer to the question, "What are you chiefly reading now?" he replied, "Woodhouse's Isoperimetrical Problems." He read mathematics chiefly without the aid of a private tutor, but in his third year and his last term had the advantage of

Mr Hopkins's direction[1]. I was myself a pupil of Mr Hopkins' at the time; but Ellis never read with the class of which I was one; in fact, he did not need the kind of lecture which was adapted to myself and others; he required only that his reading should be arranged, and put in a form suitable for the Cambridge examinations.

The only occasion upon which I was brought into contact with him as a fellow-student was in attending Professor Peacock's lectures on Plane Astronomy. I remember well the astonishment with which I witnessed his demeanour during the lectures: he made no note, he asked no question; but he quietly remarked as we left the lecture-room together one day, "It saves one the trouble of reading these things up."

It was in fact a great advantage to him to be able to substitute the use of his ears for that of his eyes. His sight was very tender, and during the latter part of his undergraduate career he regularly employed a person to read to him high mathematical subjects. He once mentioned to me incidentally that the theory of the Earth's Figure, as given in Pratt's Mechanical Philosophy, was in this manner read to him; which instance I here record as an indication of a power of mental effort possible to very few, and the magnitude of which mathematicians will appreciate.

During his undergraduate career I was not intimately acquainted with him: probably he had no desire to increase his circle of friends beyond that which was naturally brought round him in his own college: and his manner was not such as to encourage rapid intimacy. I do not think that at this period the number of his intimate friends was large even within Trinity College, and sometimes a feeling of desolation and want of

[1] In a note to me Mr Hopkins says, with reference to his recollections of Ellis as a pupil, "On one point he always seemed to puzzle me. The extent and definiteness of his acquirement, and his maturity of thought, were so great, so entirely pertaining to the *man*, that I could hardly conceive when he could have been a boy."

sympathy oppressed him painfully. He once described to me the forcible manner in which he was affected in College Chapel by those words of the Psalm, "I had no place to flee unto, and no man cared for my soul." This melancholy feeling, which sometimes assumed a very painful intensity, was no doubt connected with the weak state of his bodily health, and the highly nervous temperament which naturally belonged to him: it was the source of much suffering, I fear, even in that period which preceded the most distressing portion of his life.

Ellis was to be seen sometimes at the debates of the Union Society, but he seldom took any part in them. On one or two occasions, however, when domestic troubles had arisen, and squabbles of a somewhat personal kind ran high, he stood up as a pacificator; he was heard with marked respect, and his suggestions were readily adopted.

In January, 1840, he passed his examination for B.A. degree. In consequence of complaints which had been made of the coldness of the Schools, in which the Candidates for Mathematical Honours were then examined, the examination took place for two or three years in the Lecture Rooms of Trinity College. The difficulty was afterwards solved by the proper warming of the Senate-House. When we visited the rooms on the day before the Examination to inspect our places, I found that the alphabetical arrangement of names combined with the conditions imposed by the length of the tables had brought Ellis and myself almost immediately opposite to each other, and I was rather pleased with the thought of seeing him in actual work. He however made a special request that his seat might be changed, (I do not exactly know why,) and was allowed to be placed in a different room; so that I saw nothing of him during the examination.

Those who knew anything of the relative powers of the men of the year had no doubt as to which place Ellis must occupy, if only his health should enable him to do himself justice in the examination. His health of course introduced an element of uncertainty; but when the examination was concluded, and it was found that he had been able to take every paper, the result was quite sure. He was Senior Wrangler; and I can truly say, that for myself I had almost as much satisfaction in seeing his name at the top of the list, as in seeing my own next to it, for I felt convinced that it was his rightful position, and that nothing but the accident of ill health could have put any of his competitors above or even near him.

His appearance in the Senate-House when he took his degree was very striking. He looked very pale and ill, but this perhaps enhanced the intellectual beauty of his countenance. A person who was present remarked to me very pithily, "If I had seen him before, I could have told you you could not beat him."

In October, 1840, he was elected Fellow of Trinity College. He retained his fellowship until the year 1849, that is, for seven years after the degree of M.A., when as a layman he ceased to be a fellow in due course.

His intention after taking his degree was to read for the bar, and at one time there was a notion of his entering upon political life by becoming a candidate for his native city Bath. His name was publicly discussed with reference to the election, but the design was given up on the ground of the weakness of his health. Had he been a candidate, it would have been on Whig principles; he was not a very earnest politician, but always professed himself a Whig, a profession which was probably strengthened by his intimacy with Sir William Napier, to whom he always expressed himself as much attached.

With regard to the bar, he was duly called, but did not study long with the intention of practising. The fact is that his worldly position was unexpectedly altered. Both of his elder brothers died, and he thus became heir in expectation of considerable property, and soon by the death of his father heir in possession. He was thus deprived of the chief inducement to labour as a lawyer; and had it been otherwise, it is clear that his health would never have enabled him to undergo the necessary drudgery. Nor indeed would the actual practice of law-courts have been very congenial to his feelings and tastes: law in the abstract he loved exceedingly, as we shall see presently, but law as it is concerned with the actual strifes and quarrels of mankind would have been eminently distasteful to him.

As a Fellow of Trinity he made his College his home, except for a short period after his election. Here he continued his mathematical reading, but not with any very definite purpose. He became very intimate with the late D. F. Gregory, who was then Fellow of Trinity College, and who did good service to mathematics by the establishment of the *Cambridge Mathematical Journal;* when Gregory resigned the editorship shortly before his death, Ellis took the office, and edited part of the third and fourth volumes of the journal, in the latter of which he inserted a short biographical memoir of his friend[1].

In January, 1844, Ellis was Moderator. On this occasion, being myself one of the Examiners, I was thrown into closer relations with him than before, and commenced that real intimacy which lasted as long as his life. His problem paper on this occasion was singularly elegant, but perhaps too refined for its purpose. His fellow moderator, O'Brien of Caius Col-

[1] Ellis's name appears as editor on the title-page of the fourth volume of the Journal. I may take this opportunity of observing that the parallel drawn between Gregory and Ellis in a very kind and warmhearted obituary notice of the latter, which was inserted in the *Athenæum* of Feb. 11, 1860, seems to me not justified. They resembled each other, no doubt, in the fact that both were good mathematicians and both real philosophers and lovers of truth: but beyond this very general resemblance the parallel does not hold. They were much attached, and Ellis felt the loss of his friend keenly: but neither in mind nor in manner was there much likeness between them.

lege, and he, published their problems with their own solutions, soon after the Examination. He bore the labour of the Examination better than could have been expected; he was a very pleasant workfellow, being always ready to fill up the intervals of work with that rich and varied conversation which his friends remember so well. Shall I be pardoned if I mention that his fees as Moderator were transferred to Addenbrooke's Hospital?

He worked for the Senate-House again in 1845. It is customary for the Moderator of one year to act as Examiner the next[1], but he was desirous of escaping the labour, and had declined to serve. There was, however, a difficulty in finding a substitute; I was myself one of the Moderators and felt anxious that he should serve, which at my earnest entreaty he at length consented to do. It was in this year that Professor W. Thomson took his degree; great expectations had been excited concerning him, and I remember Ellis remarking to me with a smile, "You and I are just about fit to mend his pens." He again got through the Examination much better than could have been expected; in fact, the effort seemed to do him good; when he had consented to act he said, "I feel all the better for having done something plucky."

It has already been mentioned that the study of the principles of Law was very agreeable to him. At this period of his life he devoted much time to the study of the Civil Law, and he has left behind him several volumes of notes made in the course of his reading. The only appointment concerning which I ever heard him express any strong wish was that of the Professor-

[1] It may be mentioned, for the benefit of readers not acquainted with Cambridge customs, that the distinction between Moderator and Examiner is practically merely this, that the papers of original problems are set wholly by the Moderators. Constitutionally the difference is, that the office of Moderator is an old statutable office, whereas the Examiners were added by Grace of the Senate, in consequence of the increase of the number of candidates for mathematical honours.

ship of Civil Law. He acknowledged to me that he would have felt gratified by the tenure of this office, which is the more remarkable when taken in connection with the fact that on the occasion of one of the Mathematical Professorships being vacant he expressed no desire to be appointed, but, on the contrary, declared that he would not consent to be nominated as a candidate. Indeed it is a mistake to suppose that Ellis was in any exclusive or even preponderating degree devoted to mathematics: his mathematical power was no doubt very great, but I think not greater than several other powers, and certainly his taste by no means exclusively leaned in this direction, as his intimate friends very well knew. But of this more hereafter.

He did not give himself in any degree to tuition during his Cambridge residence. So far as I know he never had a private pupil; he gave a few College Lectures upon high mathematical subjects, but he did this only as locum-tenens for friends upon whom the task devolved. Probably his health would have interfered with any regular occupation of this kind; but besides this, he had not, I think, any taste or any special fitness for imparting knowledge to average minds; his remarks were always suggestive, and he could throw light upon almost any subject which could be brought forward, but he usually assumed a considerable amount of knowledge on the part of those with whom he conversed, and sometimes (as it seemed to me) he was obscure, in consequence perhaps of the neatness and conciseness which were so remarkable in his conversation.

At the request of the British Association, which held its annual meeting at Cambridge in 1845, he undertook a Report upon the progress of certain branches of pure mathematics. This Report is reprinted in the present volume. It represents a great amount of labour and research, and I have no doubt that the preparation of it was a source of pleasure to him, as it refers to a department of mathematics which was with Ellis a special favourite.

It was during his residence as a Fellow in Trinity College that he undertook, in conjunction with Mr James Spedding and Mr Douglas Denon Heath, to edit the works of Bacon. The philosophical section of the works was the share allotted to Ellis. No literary occupation could have been more congenial to his taste, and the prefaces to the several treatises which he was able more or less to complete, especially the "General Preface to the Philosophical Works," are perhaps the most valuable thing which he has left behind him. He was engaged upon the preface to the *Novum Organum*, when he was stopped by illness; and so complete and sudden was the break in his health that he never completed it. The last sentence that he wrote will be found on page 100 of the first volume of his edition of Bacon's Works. It is an affecting monument, and as such I here produce it. "Again he affirms that he does not inculcate, as some might suppose, a —;" to which Mr Spedding has appended a note, "Mr Ellis had written thus far when the fever seized him."

The mention of Bacon has led me to anticipate the course of events. In the years 1847 and 1848 he visited Malvern for the benefit of his health, still making Trinity College his head-quarters, and he certainly appeared to be strengthened by the course of treatment to which he was submitted. He had, I believe, always intended, at the expiration of the tenure of his fellowship, to go abroad; partly perhaps for the general advantages of travel, and partly with the belief that his health would be improved by residence in a warmer climate. He desired to settle himself in some place possessing a good library, where he might complete his work for the edition of Bacon. Accordingly in the autumn of 1849 he went to Nice. After remaining there some little time he started, not well in health, for the journey by post along the Riviera. The first night he slept at Mentone, and, as he believed, in a damp bed. The next day he arrived early at S. Remo, but feeling indisposed determined to proceed no further that day. In the evening he took the last walk which he was ever able to take, otherwise than as a cripple; he described afterwards to one of his friends the profound effect produced upon his mind, possibly rendered more sensitive by approaching illness, by the loveliness of the scene. That night, which was one of horrors to him, he was seized with a rheumatic fever, which for several days put his life in great danger. A physician, who was called in, seems to have exerted himself with great kindness and to the utmost of his skill, to do all that could be done. Ellis always retained an affectionate remembrance of him. He ordered his patient to be bled extensively, and after a few days the imminent danger was passed. Rheumatism however remained fixed hopelessly upon him; he was ever after in constant pain, with very little use of any part of his body; and the rest of his life, ten years, may be described as a long process of gradual dissolution.

After a residence of nearly three months at S. Remo, he was brought home by easy stages. He visited several places after his return, London, Brighton, Bath, Malvern, Tunbridge Wells, consulting various physicians, with no apparent result. At length giving up all hope of amendment, he fixed himself in 1853 at Trumpington, a village two miles from Cambridge, of which his friend Professor Grote was Vicar, for the sake of being near the University and his old friends. When he arrived he was unable to walk, but could drive out in a carriage, and in the house he could move from one place to another on the same floor by means of a chair set upon wheels: after some time however he became entirely confined to the house; then to his bed, where he remained in a sadly suffering condition till the day of his death.

He wrote me a letter shortly before his arrival at Trumpington, telling me that he had taken Anstey Hall (the name of his house[1] at Trumpington), and adding that he was coming

[1] When Ellis first came into residence at Cambridge his father had some

to leave his bones amongst us, and he trusted we should "give him a little earth for charity." During his lingering illness I saw him not unfrequently, and am able to record, from personal recollection, some few things which may be pleasant to those who knew him, and not unprofitable perhaps to the general readers of this volume. In the early part of his residence in Anstey Hall, he was well enough to enjoy the society of his friends to a very considerable extent; he sat in his invalid chair, with sometimes two or three persons present, pouring forth his varied stores of knowledge as in olden days; in fact, it should be stated here once for all, that during the whole of his lingering sickness his mental powers never appeared to be in the smallest degree impaired; it was a wonder to note the perfect action of the mind, at a time when the body was a mere distorted and attenuated heap of skin and bones. But this brightness of intellect doubtless made the suffering more acute. His life for several years was a constant looking of death in the face, with scarcely an interval of ease or obliviousness. By degrees the resource of the society of his friends began to be diminished; frequently we called and found him unable to see us; and even relatives staying in the house could not be admitted into his chamber for days together.

In the earlier part of his illness he was able to give some attention, but not much, to Bacon; some of the notes which have been since printed were dictated at this time. The thought of leaving his work imperfect could not fail to be painful to him, and the pain would be increased by the very high standard of excellence which he set up for himself in all matters which he undertook. Latterly he could not bear the subject of Bacon to be alluded to: if it happened to be introduced, he would

intention of engaging Anstey Hall, in order that he might be near his son, whose health even then was, as we have seen, not strong. This circumstance had made Ellis always feel an interest in the house.

say, "We don't talk about it in this room." He amused himself also with mathematical investigations, which he was able to carry on in a remarkable manner, without paper or figure, "in his head," as the common phrase is. It was in this way that he discovered for himself what he believed to be a new view of Napier's rules for the solution of right-angled spherical triangles. His discovery involves so curious a piece of history that I shall venture for a moment to dwell upon it. Ellis found out in his illness, that Napier's rules, instead of being, as they have been stated to be in Cambridge books, from Professor Woodhouse downwards, a mere *memoria technica*, were all capable of being deduced from one geometrical construction. He sent me a paper which he dictated on the subject, and which I requested him to allow me to communicate to the Cambridge Philosophical Society. Having gained his permission, I thought it well to examine the literature of the subject, and above all to see what Napier had himself said. On turning to Napier's famous tract, *Mirifici Logarithmorum Canonis Descriptio*, I found that Ellis had in fact rediscovered Napier's own original conception of the problem.

It was during his illness that he dictated his remarks concerning the construction of bees' cells; I believe also that he thought out at this time his demonstration of the tautochronous quality of the cycloid. Indeed he had usually some mathematical question running in his head, which served him for recreation during his easier moments. Nor were other subjects excluded; it was in this season of extreme bodily weakness that he corresponded with the late Dr Gilly on the Romaunce language, discussed the date of the "Noble Lesson," and criticized Dr Gilly's edition of the Vaudois Gospel of S. John. I have also before me a considerable number of letters dictated to friends, dealing with subjects so different, according to the tastes of the persons to whom they were sent, that it seems difficult to believe them to be the production of the same person. He dictated also papers on Vegetable spirals, on Comparative Metrology, and on various points of Etymology.

But the most remarkable effort of his illness was the dictation of a pamphlet (reprinted in this volume) on the subject of a Chinese Dictionary, and the best mode of constructing such a work. This pamphlet was in the form of a letter to the Rev. J. Power, the Librarian of the University, who had kindly assisted him in his literary researches, and had supplied him with the most recent literature bearing upon the Chinese language. I can give no opinion of its value, but can hardly be wrong in regarding it as a marvellous exertion of mental vigour under very depressing conditions.

In truth his taste for language was as marked as that for mathematics. I have just now remarked incidentally upon his study of the Romaunce language; Gothic also appears from his letters as having received considerable attention; he has left a paper on Sanscrit; and amongst the modern languages, besides the usual acquirements of French and German, I perceive that he was well versed in Italian, and that he had given attention to Danish and Spanish. He thoroughly enjoyed the study of a language, and I remember very well the pleasure which he expressed at having had the courage to communicate a French memoir to Liouville's *Journal:* it seemed to me that the writing mathematics in a foreign language gave him almost as much satisfaction as the mathematical results themselves[1]. I ought, perhaps, to mention that I found him one day reading the New Testament in Swedish, which he told me he had "picked up" since he had been ill[2].

I have said that through his long illness Ellis retained his

[1] He was very fond of translating. Amongst his papers are some translations of Danish ballads, Spanish ballads, Andersen's Tales, &c.

[2] During his illness his notes to his physician were usually written in Latin.

vigour of mind. It is wonderful that in such suffering, and in the consciousness of approaching dissolution the mind should have been capable of dwelling calmly upon subjects of abstract interest, such as investigations in pure mathematics; he himself felt that there was something strange in the occupation. In a note accompanying a mathematical paper he writes to me: "I have been very miserable all this week. God will mend it, when His will is. It seems strange that my mind still runs at all upon triangles, and I am not at all sure that it is right it should. I need not tell you to think charitably of me in this as in other respects." His mind by no means however dwelt upon triangles to the exclusion of more solemn subjects, as I shall have occasion to shew more fully presently; but it appeared to have a vigour of action and a fulness of matter which no external circumstances could affect, and so far as my observation went he conversed with the same facility and command of his subject during his illness, as in earlier days.

It would be only painful to draw as vivid a picture as might be easily drawn of the protracted sufferings which he had to endure. Medicine could do nothing more for him than mitigate the severity of the disease, which seemed to claim as its own one muscle after another, as it slowly approached the heart. "One twitch *there*," said he to me one day, speaking of his heart, "and then I shall know the great secret." For a considerable period, reading (as might be supposed) was a relief; the weary hours of night, sometimes rendered horrible by the fear of dreams if sleep should come upon him[1], were beguiled with books, a lamp suspended over his head giving him the necessary light. By and bye however this resource also failed: the eyes began to be affected by the complaint, and for about

[1] He one day represented his condition to me in words curiously resembling those of Job: "When I say, My bed shall comfort me, my couch shall ease my complaint; then Thou scarest me with dreams, and terrifiest me through visions." Job vii. 13, 14.

two years before his death he was almost entirely blind[1]. Books were read to him, and he dictated occasionally to an amanuensis, but the loss of sight was a very severe addition to his sufferings.

The following lines, which may be taken as a sample of the love of epigram[2] which belonged to him, were sent to Dr Paget, his physician, when the blindness was gaining upon him.

Contortos artus nunc culcita celat, at olim
  Terra teget melius: sit modo et illa levis.
Et quam vix possunt oculi tolerare dolentes
  Lux fugit, ac tenebris mox adopertus ero.

Mar. 9, 1857.

In the earlier part of his illness, I made the remark one day that he appeared to me to be a little better; he at once said

[1] His eyes were first attacked in April, 1856; he was unable to read in July, 1857.

[2] This love of epigram was very striking. Here is an instance. During his illness an old friend wrote to him asking him for some new conundrums. It so happened that on the day of receiving this request he fancied that he had discovered from Dr Paget, that he was labouring under Bright's disease; he sent the following answer:

Si petis hinc ænigma novum, si ludicra poscis,
  Quod nuper didici scribere cur dubitem?
Morbus, qui clarum fecit qui nomine clarus
  Semper erat, solvet vincula queis teneor.

On the same day he wrote on the same subject in a different style:

DEAR PAGET,

I think it well to thank you for your most kind note. It came a few minutes before my dinner. That over, I told W—* its purport, and desired him to take notice how little it disturbed me.

Of course, no such communication can ever be matter of indifference, and least of all to a person like me, in whom the power of suffering and of being anxious has been but little impaired by years of suffering and of anxiety.

But, to use John Bradford's words, "He who has helped me till now will not leave me when I have most need, for His truth and mercy sake;" and of necessity I am less anxious about many things than I have long been.

Yours,

R. L. ELLIS.

Feb. 19, (1857).

* His servant.

very earnestly, "O pray do not say so!" He recognized from the first the sure character of his disease, and he desired that his sufferings might not be protracted; at the same time he was perfectly patient, even cheerful, and reverently acknowledged in all his afflictions the governing hand of God.

But it is time that I should leave the story of my friend's sickness, upon which affectionate remembrance tempts me to linger, in order that I may endeavour to give such estimate as I can of his mind and character.

Speaking generally I should say that his intellect was the most remarkable that I have known. It was made up of a combination of powers so delicately balanced, and working together in such perfect harmony, that it would be difficult to say that any one predominated over the rest. Popularly in Cambridge he might be regarded as specially a mathematician, because he was Senior Wrangler, but those who knew anything of him were fully aware that mathematics was only one of his acquirements, and that in conversation mathematics by no means presented itself as the chief or even the favourite subject of his thoughts. Indeed, I think, it would be difficult to say that there was any one subject in which his mental powers were manifested more decidedly than in others; and the only marked deficiency which I ever detected was in respect of music, for which he had no special taste. I do not mean that he had no sense of melody: this was far from being the case: but his conversation never ran upon the great masters of music and their works, as it did upon almost all other subjects.

Doubtless, however, mathematical power belonged to him in a very large degree. With the present volume in the reader's hands it would be superfluous to say much concerning the special departments of mathematical investigation in which his taste impelled him; but I may remark that he seemed most naturally to associate his mathematics with the past; he delighted to discuss the principles of investigations already known, to trace the history of processes, to examine the philosophy of a subject, to hunt up its literature, or to simplify its treatment. His memoir on the Foundations of the Theory of Probabilities and that on the Method of Least Squares, which stand at the opening of this volume, appear to me to represent as well as possible his special taste, so far as he had a special taste, with regard to mathematics. He always seemed to talk on the subject of Probabilities with great pleasure, and as one in which he was thoroughly at home. The remarkable little essay on the Theory of Matter was also one which I think gave him much satisfaction. His taste did not seem to lead him much in the direction of elaborate physical experiment, nor do I remember that on any occasion he worked in this path of investigation. His mind in fact was rather that of the philosopher than the physicist; his impulse was rather to contemplate existing knowledge, than to take up a particular line of physical investigation and press forward knowledge upon that one line.

This characteristic of mind gave a great charm to his conversation: his thoughts were, so to speak, set in a rich historical framework, and they were produced with an ease and readiness which I have never seen equalled. In referring to his conversational powers generally, I may record the singular accuracy of his speech; this was perhaps partly a natural gift, and partly the result of early education; certainly it was very wonderful; his sentences were not only full of thought and of references to literature of all kinds, but they were so remarkably correct in their construction and elegant in diction. He was one of the few men who could have borne a Boswell with great advantage to their reputation[1].

[1] This remark was made to me by one very intimate with him, and I cordially adopt it. In fact, with obviously wide differences, there was a good deal of curious

He was a good scholar, and very fond of the Greek and Roman literature. I believe I am justified in saying, that his knowledge of that literature was really more extensive and thorough than that of many whose reputation as scholars has been much greater. He could enjoy a discussion of a point of classical philology as keenly as one on scientific subjects, and when so engaged no one would have supposed that mathematics was his favourite study. Indeed, as I have already intimated, mathematics could not in any proper sense be so described: Civil Law was certainly as much a favourite: and he seemed to be most happy in conversation, when the subject was one of a philosophical character. I have often felt disposed to compare his mental constitution in many respects with that of Leibnitz. Each was the philosopher quite as emphatically as the mathematician. Leibnitz, I may observe, was one of his favourites, and he mentioned to me one day with some feeling of amusement that a Fellow of Trinity had spoken to him of Leibnitz, under the title "*your* Leibnitz," as though the old feeling of jealousy were still lurking in the College.

His love of philosophy fitted him especially to be the editor of Bacon. It was a work, I believe, which he undertook with all his heart, and relinquished with extreme pain and only under a sense of imperious necessity. He was assisted too by his familiarity with the philosophical speculations of the middle ages: there was something congenial to his own cast of mind in the discussions of the schoolmen, and I think he appreciated Bacon all the more in virtue of his appreciation of those, whose processes of thought and methods of argument it was Bacon's task to supersede.

similarity between Ellis and Dr Johnson. Ellis was an excellent conversationalist: he not only expressed himself with singular precision, but he was a patient listener, and readily caught up and retained in his memory the remarks of those with whom he conversed.

In producing his knowledge he was much assisted by the strength and clearness of his memory. All his knowledge seemed to be as it were *in hand;* it was not merely that he knew where to look for information, although that is a great thing and as much as many clever men are satisfied to accomplish; but information upon the most various subjects, the most trivial and the most important, seemed to be at call upon all occasions. His long painful illness had no apparent effect upon this faculty. On one occasion, some years after he had been confined to the house, he wished to describe a certain picture in a room which he had only once entered, and that merely for a morning call: to my astonishment he mentioned the pictures one after another as they hung on the wall, and so identified that to which he desired to refer.

Ellis was very fond of expressing his thoughts in verse. Some of his notebooks are full of poetical scraps, generally somewhat melancholy in their tone, but expressing (I have no doubt) the feelings of his mind at the time of writing. I will produce two or three of these scraps in this place: it will be seen that they all belong to the period preceding his last long illness.

1.

E'en in the days when life is dear
And we would fain live on for aye,
Then let it not forgotten be
    That death draws near.

2.

And when we fall on sadder hours,
And gladly would lie down and die,
Remember that we live to do
    God's will; not ours.

DINANT, 1846,
    Written *April* 9.

The two following were written, I presume, in Trinity College.

1.

The flower of life is gone—'tis well
We know each flower will have its day.
It were not wise did you repine
Because the spring has passed away.

2.

The flower was cankered in the bud,
It was a sickly faded flower—
What recks it now? however fair
It still had died before this hour.

3.

'Aye, but it died and left no fruit,
And therefore I must needs lament.'
'Lament not for the buried past,
For all in vain your grief is spent.'

4.

''Tis true, and days to come may bring
A sharper woe than all the past,
And every passing year may seem
To me more bitter than the last.'

5.

'Think not of this, but turn to Him
Who bids His wearied follower rest,
About the altars of whose house
The swallow builds itself a nest.

6.

Turn you to Him whose voice can calm
The working of our troubled sea,
Who brings the way-worn traveller
Unto the port where he would be.

7.

To Him who said these solemn words,
"Blessed are they that weep,"
Who now as in the days of old
Gives His beloved sleep.'

Jan. 12, 1848.

I do lament me for the days gone by,
All spent in loving frail mortality:
Else I had been what now I cannot be,
For I had wings wherewith to mount on high.

Thou king of heaven eternal, hear my cry!
Thou knowest all my grief and misery:
Recal, O God, my wandering soul to Thee,
And its shortcomings with Thy grace supply!
O let me die in harbour and at rest,
Though I have lived in tempest and in strife:
So when I journey hence at Thy behest
My death will be less worthless than my life.
Vouchsafe in life and death to succour me;
Thou knowest that I trust in nought but Thee.

Amen and Amen.

Miserere mei Domine et nunc et in horâ mortis.

Feb. 24, 1848.

I have no wish to indulge in any extravagant eulogy of my friend, but I should leave a sad blank in this brief memorial of him if I did not say that his moral qualities were not below his intellectual. His manner might be accounted by some persons cold, and he was certainly not one with whom familiar intercourse and the thorough freedom of friendship were attained rapidly; but those who knew him well knew that he possessed one of the most gentle of hearts, with a delicate consideration for the feelings of others, and a most grateful sense of any kindness shewn to himself. Above all his sense of honour and propriety was perfect; nothing shabby or mean could exist in the same place with Leslie Ellis. This is a description which I am convinced that all who knew anything of him will certify to be correct, and I think I need not add more to it: his intellectual faculties were undoubtedly his most striking characteristics, and the fruits of his intellect, which he was permitted to gather during his brief period of health and activity, remain as an indication of what he might have done; let it suffice to say that those qualities of which his works can tell but little, were such as we delight to contemplate in union with great intellectual gifts.

There is one other point which must not be omitted, but which must be treated with delicacy. It would be unjust not to add a word upon the more definitely religious aspect of his character; I say unjust, because the remarkable keenness of his mind concerning mathematical and other questions, during an illness in which his life was hanging constantly by a thread, may give those who are disposed to do so the occasion of remarking that his mind might very well have been occupied with more solemn thoughts. Let such persons then be satisfied by knowing that more solemn thoughts did occupy his mind. I think that as his sickness advanced and his bodily powers were diminished, his mind gradually found more settled peace and rested more surely upon the love of God and the merits of the Saviour. Certainly there was much to tempt him to murmur, but I never noticed any murmuring propensity or any tendency to do otherwise than bow to God's will and accept with patience a mysterious and painful dispensation. In the early part of his illness, he asked me whether I had ever thought much or heard a sermon upon Habakkuk iii. 17, 18: "Although the fig-tree shall not blossom, neither shall fruit be in the vines; the labour of the olive shall fail, and the fields shall yield no meat; the flock shall be cut off from the fold, and there shall be no herd in the stalls: yet I will rejoice in the Lord, I will joy in the God of my salvation." He made no application of the verses to his own circumstances, only remarking how striking the language was; but it was evident to me that his own case was in his mind.

He always begged me to read prayers with him. I usually introduced the Collect from the Visitation of the Sick; on one occasion I omitted it; he noticed the omission at my next visit, and begged me to use it. Anyone who remembers the substance of that Collect will see the value of this simple anecdote.

His own bodily weakness and utter abstraction from all works of active piety intensified his desire of doing something for the benefit of his fellow-creatures, and made him grieve over his forced indolence. I do not mean that his charitable feelings first germinated in his sick room: this was very far from being the case: but his sense of his own inability to discharge any active duty made him more keenly sensible of the privilege of being permitted to exert ourselves for God and for our brethren. On reading the account of the death of Captain Gardiner in his noble but not wisely-arranged effort to found a mission in Patagonia, he expressed a wish that his own life might have had a similar termination. Indeed a sense of the spiritual needs of mankind appeared to grow upon him as his own bodily weakness brought him nearer to the great realities of existence: I can never forget the earnestness with which he said to me at a late period of his illness, "The thing above all others which strikes me, as I lie here on my bed, is the intense wickedness of mankind." "I feel," he continued, "as if I should be constrained, did God ever raise me up again, to rush in amongst them, as Barnabas and Paul did amongst the people of Lystra, and rend my clothes and say, Sirs! why do ye these things[1]?"

His speculative mind, acting under the peculiar conditions to which it was subjected by his diseased body, could hardly fail to look sometimes anxiously into the future, and guess what might be the nature of the life prepared for him in the world to which he was brought so near: as time went on however the keen discipline of affliction seemed to have taught him that he must "stand and wait," and simply look forward with calm hope.

[1] I ventured to introduce this reminiscence into a sermon preached before the University, March 22, 1863, and since published.

I find a similar anecdote amongst some recollections, which have been kindly put in my hands by the Rev. J. P. Norris. Speaking of children, and his strong disapproval of giving them prizes for mere cleverness, he added, "There is another point connected with children that I feel with an intensity which I would give much to have felt years ago, the sacred duty of keeping them pure. Wounded Arthur, speaking to Sir Bedivere, threatens to arise and slay him with his hands if he fail in his behest; and I feel sometimes as if I could arise from this bed and tear to shreds some of the books that are left in children's way."

In fact it was impossible not to observe that throughout his illness he perceived, that it was to be regarded in the light of a divine discipline, however mysterious such discipline might be. "Aches and pains," he said to a friend, "have been my teachers of late." "Fiat voluntas Tua," he observed to the same friend, taking up his *De Imitatione Christi*, "is after all the only prayer. Domine, modo sum in tribulatione, et non est cordi meo bene, sed multum vexor a præsenti passione......Et nunc inter hæc quid dicam? Domine, fiat voluntas Tua; ego bene merui tribulari et gravari."

His last days were days of peace; and his last words were so striking that I think it right to put them here upon record. It will be remembered that for a considerable period before his death he had been quite blind: just before his departure, which was in a certain sense sudden though so long expected, he exclaimed, "I see a light!" and so expired. Possibly some physical explanation of this exclamation may be given: for myself I would rather look upon it as indicative of something spiritual, and as announcing the arrival of a glorious change for which his imprisoned soul had long waited and earnestly prayed[1].

He was released from his sufferings May 12, 1859, and was buried at Trumpington. His grave is at the South-East corner of the churchyard, and bears the simple inscription—

ROBERT LESLIE ELLIS

BORN 25 AUGUST 1817

DIED 12 MAY 1859.

BLESSED IS THE MAN THAT HATH SET HIS HOPE IN THE LORD.
Ps^{m}. 40. $\bar{v}$. 5.

[1] He once observed that he wondered no one had ever chosen for an epitaph the words of Psalm cxvi. 14, "Thou hast broken my bonds in sunder."

ON THE FOUNDATIONS OF THE THEORY OF PROBABILITIES*.

THE Theory of Probabilities is at once a metaphysical and a mathematical science. The mathematical part of it has been fully developed, while, generally speaking, its metaphysical tendencies have not received much attention.

This is the more remarkable, as they are in direct opposition to the views of the nature of knowledge, generally adopted at present.

2. The theory received its present form during the ascendancy of the school of Condillac. It rejects all reference to *à priori* truths as such, and attempts to establish them as mathematical deductions from the simple notion of probability. Are we prepared to admit, that our confidence in the regularity of nature is merely a corollary from Bernouilli's theorem? That until this theorem was published, mankind could give no account of convictions they had always held, and on which they had always acted? If we are not, what refutation have we to give? For these views are entitled to refutation, from the general reception they have met with, from the authority of the great writers by whom they were propounded, and even from the imposing form of the mathematical demonstration in which they are invested.

* *Transactions of the Cambridge Philosophical Society,* Vol. VIII. [Read Feb. 14, 1842.]

I shall be satisfied if the present essay does no more than call attention to the inconsistency of the theory of probabilities with any other than a *sensational* philosophy.

3. As the first principles of the mathematical theory are familiar to every one, I shall merely recapitulate them.

If on a given trial, there is no reason to expect one event rather than another, they are said to be equally possible.

The probability of an event is the number of equally possible ways in which it may take place, divided by the total number of such ways which may occur on the given trial.

If a_1, b_1, ... m_1, denote equally possible cases which may occur on one trial, a_2b_2......k_2 those which may occur on a second trial, a_3b_3.....p_3 those belonging to a third, &c.: then $a_1b_2a_3$...... $a_1a_2b_3$......&c. &c. are all equally possible complex results.

Hence it follows that on the repetition of the same trial k times, the probability that an event whose simple probability is m will occur p times is

$$\frac{1 \,.\, 2 \ldots k}{1 \,.\, 2 \ldots p \; 1 \,.\, 2 \ldots (k-p)} \, m^p (1-m)^{k-p}:$$

this follows merely by the doctrine of combinations. These are all the propositions to which I shall have occasion to refer.

4. If the probability of a given event be correctly determined, the event will, on a long run of trials, tend to recur with frequency proportional to this probability.

This is generally proved mathematically. It seems to me to be true *à priori.*

When on a single trial we expect one event rather than another, we necessarily believe that on a series of similar trials the former event will occur more frequently than the latter. The connection between these two things seems to me to be an ultimate fact, or rather, for I would not be understood to deny the possibility of farther analysis—to be a fact, the evidence of which must rest upon an appeal to consciousness. Let any one endeavour to frame a case in which he may expect one event on a single trial, and yet believe that on a series of trials another will occur more frequently; or a case in which he may be able to divest himself of the belief that the expected event will occur more frequently than any other.

For myself, after giving a painful degree of attention to the point, I have been unable to sever the judgment that one event is more likely to happen than another, or that it is to be expected in preference to it, from the belief that on the long run it will occur more frequently.

5. It follows as a limiting case, that when we expect two events equally, we believe they will recur equally on the long run. In this belief we may of course be mistaken: if we are, we are wrong in expecting the two events equally, and in thinking them equally possible. Conversely, if the events are truly equally possible, they really will tend to recur equally on a series of trials. But this proves the proposition placed at the head of the section: for if any event can occur in a out of b equally possible ways, its probability is $\frac{a}{b}$: and if all these b cases tend to recur equally on the long run, the event must tend to occur a times out of b; or in the ratio of its probability. Which was to be proved.

6. Let us now examine the mathematical demonstration of this proposition. In entering upon it, we are supposed to have no reason whatever to believe that equally possible events tend to occur with equal frequency.

It is well known that what is called Bernouilli's theorem, relates to the comparative magnitudes of the several terms of the binomial expansion.

The general term of

$$\{m + (1 - m)\}^k \text{ is } \frac{\lfloor k \rfloor}{\lfloor p \rfloor \lfloor k - p \rfloor} m^p (1 - m)^{k-p},$$

which is the probability that an event whose simple probability is m will recur p times on k trials; and hence the connexion between the binomial expansion and the theory of probabilities.

7. A particular example will suffice to illustrate what seems to me to be the essential defect of the mathematical proof of the proposition in question.

A coin is to be thrown 100 times: there are 2^{100} definite sequences of heads and reverses, all equally possible if the coin is fair. One only of these gives an unbroken series of 100 heads. A very large number give 50 heads and 50 reverses; and Bernouilli's theorem shows that an absolute majority of the 2^{100} possible sequences give the difference between the number of heads and reverses less than 5.

If we took 1000 throws, the absolute majority of the 2^{1000} possible sequences give the difference less than 7, which is proportionally smaller than 5. And so on.

Now all this is not only true, but important.

But it is not what we want. We want a reason for believing that on a series of trials, an event tends to occur with frequency proportional to its probability; or in other words, that generally speaking, a group of 100 or 1000 will afford an approximate estimate of this probability.

But, although a series of 100 heads can occur in one way only, and one of 50 heads and 50 reverses in a great many, there is not the shadow of a reason for saying that therefore the former series is a rare and remarkable event, and the latter, comparatively at least, an ordinary one.

Non constat, but the single case producing 100 heads may occur so much oftener than any case which produces 50 only, that a series of 100 heads may be a very common occurrence, and one of 50 heads and 50 reverses may be a curious anomaly.

Increase the number of trials to 1000, or to 10,000. Precisely the same objection applies: namely, that in Bernouilli's theorem, it is merely proved that one event is more probable than another, *i.e.* by the definition can occur in more equally possible ways, and that there is no ground whatever for saying, it will therefore occur oftener, or that it is a more natural occurrence. On the contrary, the event shown to be improbable may occur 10,000 times for once that the probable one is met with.

To deny this, is to admit that if an event can take place in more equally possible ways, it will take place more frequently. But if this is admitted, Bernouilli's theorem is unnecessary. It leaves the matter just where it was before, and introduces no new element into the question.

8. Thus, both by an appeal to consciousness, and by the

impossibility of dispensing with such an admission, we are led to recognize the principle, that when an event is expected rather than another, we believe it will occur more frequently on the long run. And thus we perceive that we are in the habit of forming judgments as to the comparative frequency of recurrence of different possible results of similar trials. These judgments are founded, not on the fortuitous and varying circumstances of each trial, but on those which are permanent—on what is called the nature of the case. They involve the fundamental axiom, that on the long run, the action of fortuitous causes disappears. Associated with this axiom is the idea of an average among discordant results, &c.

I conceive this axiom to be an *à priori* truth, supplied by the mind itself, which is ever endeavouring to introduce order and regularity among the objects of its perceptions.

9. With a view to conciseness, I omit several interesting points which here present themselves—namely, the connection between the axiom just stated, and the inductive principle; the real utility of Bernouilli's theorem; and what seems to me to be the true definition of probability, founded on a reference to the ratios developed on the long run.

I proceed to illustrate what has been said by a few passages from Laplace's "*Essai Philosophique sur les Probabilités.*"

10. It seems obvious that no mathematical deduction from premises which do not relate to laws of nature, can establish such laws. Yet it is beyond doubt that Laplace thought Bernouilli's theorem afforded a demonstration of a general law of nature, extending even to the moral world.

At p. xlii. of the Essay, prefixed as an Introduction to the third edition of the Théorie des Probabilités, after giving some account of the theorem of James Bernouilli, Laplace proceeds: "On peut tirer du théorème précédent cette conséquence qui doit être regardée comme une loi générale, savoir que les rapports des effets de la nature, sont à fort peu près constants, quand ces effets sont considérés en grand nombre....Je n'excepte pas de la loi précédente, les effets dus aux causes morales."

It appears not to have occurred to Laplace, that this theorem is founded on the mental phenomenon of expectation. But it is clear that expectation never could exist, if we did not believe in the general similarity of the past to the future, *i. e.* in the regularity of nature, which is here deduced from it.

A little further on,..."Il suit encore de ce théorème que dans une série d'évènemens indéfiniment prolongée, l'action des causes régulières et constantes doit l'emporter à la longue, sur celle, des causes irrégulières....Ainsi des chances favorables et nombreuses étant constamment attachées à l'observation des principes éternels de raison de justice et d'humanité, qui fondent et qui maintiennent les societés; il y a un grand avantage à se conformer à ces principes, et de graves inconvéniens à s'en écarter. Que l'on consulte les histoires, et sa propre expérience on y verra tous les faits venir à l'appui de ce résultat du calcul." Without disputing the truth of the conclusion, we may doubt whether it is to be considered as a "résultat du calcul."

The same expression occurs immediately afterwards in another passage, in which the writer seems to allude to the history of his own times, and to the ambition of the great chieftain whom he at one time served.

Indeed it would seem as if to Laplace all the lessons of history were merely confirmations of the "résultats du calcul." We are tempted to say with Cicero—"hic ab artificio suo non recessit."

11. The results of the theory of probabilities express the number of ways in which a given event can occur, or the proportional number of times it will occur on the long run: they are not to be taken as the measure of any mental state; nor are we entitled to assume that the theory is applicable wherever a presumption exists in favour of a proposition whose truth is uncertain.

Nevertheless it has been applied to a great variety of inductive results; with what success and in what manner, I shall now attempt to enquire.

12. Our confidence in any inductive result varies with a variety of circumstances; *one* of these is the number of particular cases from which it is deduced. Now the measure of this confidence which the theory professes to give, depends on this number exclusively. Yet no one can deny, that the force of the

induction may vary, while this number remains unchanged. This consideration appears almost to amount to a *reductio ad absurdum.*

13. If, on m occasions, a certain event has been observed, there is a presumption that it will recur on the next occasion. This presumption the theory of probabilities estimates at $\frac{m+1}{m+2}$. But here two questions arise; What shall constitute a "next occasion"? What degree of similarity in the new event to those which have preceded it, entitles it to be considered a recurrence of the same event?

Let me take an example given by a late writer:—

Ten vessels sail up a river. All have flags. The presumption that the next vessel will have a flag is $\frac{11}{12}$. Let us suppose the ten vessels to be Indiamen. Is the passing up of any vessel whatever, from a wherry to a man of war, to be considered as constituting a "next occasion"? or will an Indiaman only satisfy the conditions of the question?

It is clear that in the latter case, the presumption that the next *Indiaman* would have a flag is much stronger, than that, as in the former case, the next *vessel* of any kind would have one. Yet the theory gives $\frac{11}{12}$ as the presumption in both cases. If right in one, it cannot be right in the other. Again, let all the flags be red. Is it $\frac{11}{12}$ that the next vessel will have a red flag, or a flag at all? If the same value be given to the presumption in both cases, a flag of any other colour must be an impossibility.

It is to be noticed, that I only refer to the visible differences among different kinds of vessels, and not to any knowledge we may have about them from previous acquaintance.

14. I turn to a more celebrated application of the theory.

All the movements of the planetary system, known as yet, are from west to east. This undoubtedly affords a strong presumption in favour of some common cause producing motion in that direction. But this presumption depends not merely upon the number of observed movements, but also on the natural affinity which in a greater or less degree appears to exist among them.

This is so natural a reflection, that Lacroix, in calculating the mathematical value of the presumption, omits the rotatory movements, and, I believe, those of the secondary planets, in order, as he expressly says, to include none but similar movements. But in the admission thus by implication made, that regard must be had to the similarity of the movements, too much is conceded for the interests of the theory. For are the retained movements absolutely similar? The planets move in orbits of unequal eccentricity and in different planes: they are themselves bodies of various sizes; some have many satellites and others none. If these points of difference were diminished or removed, the presumption in favour of a common cause determining the direction of their movements would be strengthened; its calculated value would not increase, and *vice versa.*

Again, up to the close of 1811, it appears (Laplace) that 100 comets had been observed, 53 having a direct and 47 a retrograde movement. If these comets were gradually to lose the peculiarities which distinguish them from planets—we should have 64 planets with direct movement, 47 with retrograde. The presumption we are considering would, in such a case, be very much weakened. At present, we unhesitatingly exclude the comets on account of their striking peculiarities: in the case supposed we should with equal confidence include them in the induction. But at what precise point of their transition-state are we abruptly, from giving them no weight at all in the induction, to give them as much as the old planets?

15. It is difficult to acquiesce in a theory which leads to so many conclusions seemingly in opposition to the common sense of mankind.

One of the most singular of them may, perhaps, serve as a key to explain their nature. When any event, whose cause is unknown, occurs, the probability that its *à priori* probability was greater than $\frac{1}{2}$ is $\frac{3}{4}$. Such at least is the received result. But in reality, the *à priori* probability of a given event has no absolute determinate value independent of the point of view in which it is considered. Every judgment of probability involves an analysis of the event contemplated. We toss a die, and an ace is thrown. Here is a complex event. We resolve it into, (1) the tossing of the die; (2) the coming up of the ace. The first constitutes the 'trial,' on which different possible results might have occurred; the second is the particular result which actually did occur. They are in fact related as *genus* and *differentia*. Besides both, there are many circumstances of the event; as how the die was tossed, by whom, at what time, rejected as irrelevant.

This applies in every case of probability. Take the case of a vessel sailing up a river. The vessel has a flag. What was the *à priori* probability of this? Before any answer can by possibility be given to the enquiry, we must know (1) what circumstances the person who makes it rejects as irrelevant. Such as, *e.g.* the colour of which the vessel is painted, whether it is sailing on a wind, &c. &c.; (2) what circumstances constitute in his mind the 'trial;' the experiment which is to lead to the result of flag or no flag; must the vessel have three masts? must it be square rigged? (3) What idea he forms to himself of a flag. Is a pendant a flag? Must the flag have a particular form and colour? Is it matter of indifference whether it is at the peak or the main? Unless all such points were clearly understood, the most perfect acquaintance with the nature of the case would not enable us to say what was the *à priori* probability of the event: for this depends, not only on the event, but also on the mind which contemplates it.

The assertion therefore that $\frac{3}{4}$ is the probability that any observed event had on an *à priori* probability greater than $\frac{1}{2}$, or that three out of four observed events had such an *à priori* probability, seems totally to want precision. *A priori* probability to what mind? In relation to what way of looking at them?

16. Let us see if this will throw any light on the question. Let h be a large number. And suppose we took h trials and that the probability of a certain event from each (considered in a determinate manner) was $\frac{1}{m}$; let us take a second set of h trials for which the same quantity is $\frac{2}{m}$: and so on to $\frac{m-1}{m}$ and 1.

When the trials have taken place, we shall have approximately,

$$h\left(\frac{1}{m}+\frac{2}{m}+\ldots\ldots+\frac{m-1}{m}+1\right)$$

of the sought events. Of these

$$h\left\{\left(\frac{1}{2}+\frac{1}{m}\right)+\left(\frac{1}{2}+\frac{2}{m}\right)+\ldots\ldots+1\right\},$$

had *à priori* a probability greater than $\frac{1}{2}$. Summing these series and dividing the second by the first, we get $\frac{3m+2}{4m+4}$, for the ratio which the latter class of events bears to the total number. The limit of this, when m is infinite, or when we take an infinite number of sets of trials, is $\frac{3}{4}$, which is the received result.

17. Thus, it appears this result is based upon some thing equivalent to the following assumption:—There are an infinity of events whose simple probability *à priori* is x, and another infinite number for which it is x'. These two infinities bear to one another the definite ratio of equality (x and x' may represent *any* quantity from 0 to 1). Now in reality, as we have seen, these numbers are not only infinite, but *in rerum naturâ* indeterminate, and therefore the assumption that they bear to one another a definite ratio is illusory.

And this assumption runs through all the applications of the theory to events whose causes are unknown.

This position could be *directly* proved only by an analysis of the various ways in which this part of the subject has been considered, which would require a good deal of detail. Those who take an interest in the question, may without much difficulty satisfy themselves, whether the view I have taken (which at least avoids the manifest contradictions of the received results) is correct.

18. I will add only one remark. If in (16) instead of taking one event from each of the trials there specified, we had taken p in succession, and kept account only of those sequences of p events each, which contained none but events of the kind sought; we should have had of such sequences

$$h\left(\frac{1}{m^p}+\frac{2^p}{m^p}+\ldots\ldots 1\right),$$

of which

$$h\left\{\left(\frac{1}{2}+\frac{1}{m}\right)^p + \ldots\ldots 1\right\}$$

would have belonged to trials where the simple *à priori* probability was $> \frac{1}{2}$: the ratio of these two expressions is ultimately

$$\frac{\int_{\frac{1}{2}}^{1} x^p dx}{\int_{0}^{1} x^p dx} = 1 - \left(\frac{1}{2}\right)^{p+1}.$$

This is the expression applied to determine the probability of a common cause among similar phenomena, as in the case already mentioned of the planets.

But this application is founded on a *petitio principii:* we *assume* that all the phenomena are allied: that they are the results of repetitions of the same trial, that they have the same simple probability; all that, setting other objections aside, we really determine, is the probability, that this simple probability common to all these allied phenomena is $> \frac{1}{2}$.

But how does this determine the force of the presumption that the phenomena *are* allied, or, to use Condorcet's illustration, that they all come out of the same infinite lottery?

19. The object of this little essay being to call attention to the subject rather than fully to discuss it, I have omitted several questions which entered into my original design.

The principle on which the whole depends, is the necessity of recognizing the tendency of a series of trials towards regularity, as the basis of the theory of probabilities.

I have also attempted to show that the estimates furnished by what is called the theory *à posteriori* of the force of inductive results are illusory.

If these two positions were satisfactorily established, the theory would cease to be, what I cannot avoid thinking it now is, in opposition to a philosophy of science which recognizes ideal elements of knowledge, and which makes the process of induction depend on them.

ON THE METHOD OF LEAST SQUARES*.

THE importance attached to the method of least squares is evident from the attention it has received from some of the most distinguished mathematicians of the present century, and from the variety of ways in which it has been discussed.

Something, however, remains to be done—namely, to bring the different modes in which the subject has been presented into juxta-position, so that the relations which they bear to one another may be clearly apprehended. For there is an essential difference between the way in which the rule of least squares has been demonstrated by Gauss, and that which was pursued by Laplace. The former of these mathematicians has in fact given two different demonstrations of the method, founded on quite distinct principles. The first of these demonstrations is contained in the *Theoria Motûs*, and is that which is followed by Encke in a paper of which a translation appeared in the *Scientific Memoirs*. At a later period Gauss returned to the subject, and subsequently to the publication of Laplace's investigation gave his second demonstration in the *Theoria Combinationis Observationum.*

The subject has been also discussed by Poisson in the *Connaissance des Tems* for 1827, and by several other French writers. Poisson's analysis is founded on the same principle as Laplace's: it is more general, and perhaps simpler. It is not, however, my intention to dwell upon mere differences in the mathematical part of the enquiry.

The consequence of the variety of principles which have been made use of by different writers has naturally been to produce some perplexity as to the true foundation of the method. As the results of all the investigations coincided, it was natural to suppose that the principles on which they were founded were

* *Transactions of the Cambridge Philosophical Society*, Vol. VIII. [Read March 4, 1844.]

essentially the same. Thus Mr Ivory conceived that if Laplace arrived at the same result as Gauss, it was because in the process of approximation he had introduced an assumption which reduced his hypothesis to that on which Gauss proceeded. In this I think Mr Ivory was certainly mistaken; it is at any rate not difficult to show that he had misunderstood some part at least of Laplace's reasoning: but that so good a mathematician could have come to the conclusion to which he was led, shows at once both the difficulty of the analytical part of the inquiry, and also the obscurity of the principles on which it rests. Again, a recent writer on the Theory of Probabilities has adopted Poisson's investigation, which, as I have said, is the development of Laplace's, and which proves in the most general manner the superiority of the rule of least squares, whatever be the law of probability of error, provided equal positive and negative errors are equally probable. But in a subsequent chapter we find that he coincides in Mr Ivory's conclusion, that the method of least squares is not established by the theory of probabilities, unless we assume one particular law of probability of error.

These two results are irreconcilable; either Poisson or Mr Ivory must be wrong. The latter indeed expressed his dissent from all that had been done by the French mathematicians on the subject, and in a series of papers in the *Philosophical Magazine* gave several demonstrations of the method of least squares, which he conceived ought not to be derived from the theory of probabilities. In this conclusion I cannot coincide; nor do I think Mr Ivory's reasoning at all satisfactory.

From this imperfect sketch of the history of the subject, we perceive that the methods which have been pursued may be thus classified.

1. Gauss's method in the *Theoria Motûs*, which is followed and developed by Encke and other German writers.

2. That of Laplace and Poisson.

3. Gauss's second method.

4. Those of Mr Ivory.

I proceed to consider these separately, and in detail.

For the analysis of Laplace and Poisson, I have substituted another, founded on what is generally known as Fourier's theorem, having been first given by him in the *Théorie de la Chaleur.* It will be seen that the mathematical difficulty is greatly diminished by the change.

GAUSS'S FIRST METHOD.

This method is founded on the assumption that in a series of direct observations, of the same quantity or magnitude, the arithmetical mean gives the most *probable* result.. This seems so natural a postulate that no one would at first refuse to assent to it. For it has been the universal practice of mankind to take the arithmetical mean of any series of equally good direct observations, and to employ the result as the approximately true value of the magnitude observed.

The principle of the arithmetical mean seems therefore to be true *à priori.* Undoubtedly the conviction that the effect of fortuitous causes will disappear on a long series of trials, is an immediate consequence of our confidence in the permanence of nature. And this conviction leads to the rule of the arithmetical mean, as giving a result which as the number of observations increases *sine limite*, tends to coincide with the true value of the magnitude observed. For let a be this value, x the observed value, e the error, then we have

$$\begin{aligned} x_1 - a &= e_1 \\ x_2 - a &= e_2 \\ \&c. \quad &= \&c. \end{aligned}$$

And as on the long run the action of fortuitous causes disappears, and there is no permanent cause tending to make the sum of the positive differ from that of the negative errors, $\Sigma e = 0$, and therefore

$$\Sigma (x_1 - a) = 0;$$

$$\text{or,} \qquad a = \frac{1}{n} \Sigma x_1;$$

which expresses the rule of the arithmetical mean, and which is thus seen to be absolutely true ultimately when n increases *sine limite.*

In this sense therefore the rule in question is deducible from *à priori* considerations. But it is to be remarked, that it is not the only rule to which these considerations might lead us. For not only is $\Sigma e = 0$ ultimately, but $\Sigma fe = 0$, where fe is any function such that $fe = -f(-e)$; and therefore we should have

$$\Sigma f(x-a) = 0$$

as an equation which ultimately would give the true value of x when the number of observations increases *sine limite*, and which therefore for a finite number of observations may be looked on in precisely the same way as the equation which expresses the rule of the arithmetical mean. There is no discrepancy between these two results. At the limit they coincide: short of the limit both are approximations to the truth. Indeed, we might form some idea how far the action of fortuitous causes had disappeared from a given series of observations by assigning different forms to f, and comparing the different values thus found for a.

No satisfactory reason can be assigned why, setting aside mere convenience, the rule of the arithmetical mean should be singled out from the other rules which are included in the general equation $\Sigma f(x-a) = 0$.

Let us enquire, therefore, whether there is any sufficient reason for saying that the rule of the arithmetical mean gives the *most probable* value of the unknown magnitude. In the first place, it is only one rule out of many among which it has no prerogative but that of being in practice more convenient than any other: in the second place, if this were not so, it would not follow that in the accurate sense of the words it gave the *most probable* result. This objection I shall defer for a moment, and proceed to consider the manner in which Gauss makes use of the postulate on which his method is founded.

From the first principles of what is called the theory of probabilities *à posteriori*, it appears that the most probable value which can be assigned to the magnitude which our observations are intended to determine, is that which shall make the *à priori* probability of the observed phenomena a maximum. That is to say, if a be the true value sought, x_1 being the value observed at the first observation, x_2 the corresponding quantity for the second, and so on, the errors at the first, second, &c. observation must be $x_1, -a, x_2 - a$, &c., respectively; and if $\phi\epsilon \,.\, d\epsilon$ be the probability of an error ϵ in any observation of the series, the quantity which is to be made a maximum for a is proportional to

$$\phi(x_1 - a)\,\phi(x_2 - a)\ldots\,\phi(x_n - a).$$

Equating to zero the differential of this with respect to a, we find

$$\frac{\phi'(x_1 - a)}{\phi(x_1 - a)} + \&c. + \frac{\phi'(x_n - a)}{\phi(x_n - a)} = 0,$$

as the equation for determining a in x. Let $\frac{\phi'}{\phi} = \psi$, then it becomes

$$\Sigma_1^n\, \psi(x - a) = 0.$$

Now we have assumed that the most probable value of a is given by the equation

$$\Sigma_1^n\,(x - a) = 0:$$

and it is impossible to make these equations generally coincident, without assuming that

$$\psi\epsilon = m\epsilon, \ m \text{ being any constant};$$

$$\text{hence } \frac{\phi'\epsilon}{\phi\epsilon} = m\epsilon,$$

$$\text{and } \phi\epsilon = Ce^{\frac{1}{2}m\epsilon^2}.$$

Now as the error ϵ is necessarily included in the limits $-\infty\ +\infty$, we must have

$$\int_{-\infty}^{+\infty} \phi\epsilon\, d\epsilon = \frac{C\sqrt{2}\sqrt{\pi}}{\sqrt{m}} = 1,$$

$$C = \sqrt{\frac{m}{2\pi}}:$$

or if we adopt the usual notation, and replace m by $2h^2$,

$$C = \frac{h}{\sqrt{\pi}}, \text{ and } \phi\epsilon = \frac{h}{\sqrt{\pi}}\, e^{-h^2\epsilon^2}.$$

Consequently, we are thus led to adopt one particular law of probability of error as alone congruent with the rule of the arithmetical mean.

But, in fact, we are perfectly sure that in different classes of observations the law of probability of error must vary, and we have no direct proof that in any class it coincides with the form assigned to it. Therefore one of two things must be true, either the rule of the arithmetical mean rests on a mere illusory prejudice, or, if it has a valid foundation, the reasoning now stated must be incorrect. Either alternative is opposed to Gauss's investigation. For the reasons already given, we are, I think, led to adopt the latter, and then the question arises, wherein does the incorrectness of the reasoning reside? It resides in the ambiguity of the words *most probable.* For let us consider what they imply in the theory of probabilities *à posteriori.*

Suppose there were m different magnitudes $a_1 a_2 \dots a_m$, and that each of these were observed n times in succession. Let this process be repeated p times, p being a large number which increases *sine limite.* Thus we shall have pm sets of observations each containing n observations.

Of these a certain number K will coincide with the set of observations supposed to be actually under discussion; and we shall have the equation

$$k_1 + k_2 + \dots k_m = K:$$

where k is that portion of K which is derived from observations of a_k.

Then, ultimately, the *most probable* value which the given series of observations leads us to assign to a, is (supposing a is susceptible only of the values $a_1 a_2 \dots a_m$) equal to a_r, r being such that the corresponding quantity k_r is the maximum value of k.

To make the case now stated entirely coincident with the one which we are in the habit of considering, we have only to suppose (making m infinite) that the series of magnitudes $a_1 \dots a_m$ includes all possible magnitudes from $-\infty$ to $+\infty$.

Now from this statement, it is clear there is no reason for supposing that because the arithmetical mean would give the true result if the number of observations were increased *sine limite*, it must give the *most probable* result the number of observations being finite.

The two notions are heterogeneous: the conditions implied by the one may be fulfilled without introducing those required by the other: and we have already seen that by losing sight of this distinction, we are led to the inadmissible conclusion, that a principle recognised as true *à priori* necessarily implies a result, viz. the universal existence of a special law of error, not only not true *à priori*, but not true at all.

Having stated what seem to me to be the objections in point of logical accuracy to this mode of considering the subject, I will briefly point out the manner in which, from the law of error already obtained, the method of least squares is to be deduced.

Let

$$\left.\begin{aligned} \epsilon_1 &= a_1x + b_1y + \&c. - V_1 \\ \epsilon_2 &= a_2x + b_2y + \&c. - V_2 \\ \&c. &= \&c. \\ \epsilon_n &= a_nx + b_ny + \&c. - V_n \end{aligned}\right\} \quad\ldots\ldots\ldots\ldots(\alpha)$$

be the system of equations of condition, which are to be combined together so as to give the values of x, y, &c. The error committed at the first observation is ϵ_1, at the second ϵ_2, and so on; each observation corresponding to an equation of condition.

The probability of the concurrence of all these errors is, (according to the law of error already arrived at) proportional to

$$e^{-h^2[(a_1x+b_1y+\&c.-V_1)^2+(a_2x+b_2y+\&c.-V_2)^2+\&c.]},$$

and it is to be made a maximum by the most probable values of x, y, &c. These values will therefore make

$$(a_1x + b_1y + \&c. - V_1)^2 + (a_2x + b_2y + \ldots - V_2)^2 + \ldots,$$

a minimum: that is to say, they will make the sum of the squares of errors a minimum.

Hence *the method of least squares.* The conditions of the minimum give the linear equations:

$$\left.\begin{aligned} x\Sigma a^2 + y\Sigma ab + \&c. &= \Sigma a V \\ x\Sigma ab + y\Sigma b^2 + \&c. &= \Sigma b V \\ \&c. \qquad\qquad &= \&c. \end{aligned}\right\} \quad\ldots\ldots\ldots\ldots(\beta),$$

in which system there are always the same number of equations as there are unknown quantities to be determined.

The next investigation of the principle of the method of least squares which I shall attempt to analyze is that of Laplace.

LAPLACE'S DEMONSTRATION.

If, in order to determine x from the equations of condition stated in the last paragraph, we multiply the first by μ_1, the second by μ_2, &c., and add: ($\mu_1\mu_2$, &c. fulfilling the conditions

$$\Sigma\mu a = 1,\ \Sigma\mu b = 0,\ \&\text{c.} = 0)$$

we find $$x = \Sigma\mu V - \Sigma\mu\epsilon;$$

and if we assume that $\Sigma\mu\epsilon$ is equal to zero, then the resulting value of x is $\Sigma\mu V$: the error of this determination being the quantity $\Sigma\mu\epsilon$, which we have assumed to be equal to zero, without knowing whether it really is so or not.

Now supposing there are n equations of condition, and p quantities to be determined, and that n is greater than p, then we see that there are n factors μ_1, $\mu_2 \ldots \mu_n$, and p conditions for them to fulfil. They may therefore be subjected to $n-p$ additional conditions.

This being premised, let us consider the probability that the quantity $\Sigma\mu\epsilon$ will not be less than α, or greater than β, α and β being any quantities whatever. The law of probability of error at each observation being given, the question is evidently analogous to the common problem of finding the chance, that with a given set of dice the number of points thrown shall not be less than one given number or greater than another.

We may therefore suppose that the probability in question has been determined: call it P. Suppose also that we have taken $\alpha = -l$ and $\beta = l$, l being any positive quantity.

Then P is a function of l, and of $\mu_1 \ldots \mu_n$.

Let us now so determine $\mu_1 \ldots \mu_n$, (subject to the conditions already specified,) that P may be a maximum. When this is done, it follows that there is a greater probability that the error in our determination of x, viz. $\Sigma\mu\epsilon$, lies within the limits $\pm l$, than if we had made use of any other set of factors whatever.

On this principle Laplace determines what he calls the most advantageous system of factors.

It does not follow that the value thus obtained for x is the most *probable* value that could be assigned for it. But if we consider a large number of sets of observations, (the quantities a, b, &c. being the same for all) then the error which we commit by using Laplace's factors will in a greater proportion of cases lie between $\pm l$, than if we had used any other system of factors.

The investigation has reference merely to the different ways in which by the method of factors a given set of linear equations may be solved.

We now enter on the analysis requisite to determine P.

Let the probability that $\Sigma\mu\epsilon$ will be precisely equal to u, be pdu. Then manifestly

$$P = \int_{-l}^{+l} pdu\,;$$

and we have therefore only to determine p.

Let $\epsilon_1\ \epsilon_2 \ldots \epsilon_n$ be the errors which occur at the first second &c. observation; $\phi_1\,\epsilon_1\,d\,\epsilon_1$, $\phi_2\,\epsilon_2\,d\,\epsilon_2 \ldots \phi_n\,\epsilon_n\,d\,\epsilon_n$ be the probabilities of their occurrence: the form of the function ϕ determining the law of probability of error, which, for greater generality, we suppose different at each observation. The probability of the concurrence of these errors is of course

$$\phi_1\epsilon_1\phi_2\epsilon_2\ldots\phi_n\epsilon_n d\epsilon_1\ldots d\epsilon_n \ldots\ldots\ldots\ldots\ldots\ldots \quad (1),$$

and the first principles of the theory of probabilities show that the value of pdu will be obtained by integrating (1), $\epsilon_1 \ldots \epsilon_n$ being subjected to the condition $\Sigma\mu\epsilon = u$.

Thus

$$pdu = \int \phi_1\epsilon_1\phi_2\epsilon_2\ldots\phi_n\epsilon_n d\epsilon_1\ldots d\epsilon_n \ldots\ldots\ldots\ldots \quad (2)$$

with the relation

$$\mu_1\epsilon_1 + \mu_2\epsilon_2 \ldots + \mu_n\epsilon_n = u.$$

Consequently

$$pdu = d\epsilon_n \int \phi_1\epsilon_1 \ldots \phi_{n-1}\epsilon_{n-1}\phi_n \frac{u - \mu_1\epsilon_1 \ldots - \mu_{n-1}\epsilon_{n-1}}{\mu_n} d\epsilon_1 \ldots d\epsilon_{n-1} \ldots (3).$$

Now by Fourier's theorem

$$\phi_n \frac{u - \mu_1\epsilon_1 - \ldots - \mu_{n-1}\epsilon_{n-1}}{\mu_n}$$

$$= \frac{1}{\pi}\int_0^\infty d\alpha \int_{-\infty}^{+\infty} \phi\epsilon_n \cos\left(\alpha \frac{u - \mu_1\epsilon_1 - \ldots - \mu_{n-1}\epsilon_{n-1} - \mu_n\epsilon_n}{\mu_n}\right) d\epsilon_n,$$

which, replacing $\frac{\alpha}{\mu_n}$ by α, becomes

$$\frac{\mu_n}{\pi}\int_0^\infty d\alpha \int_{-\infty}^{+\infty} \phi\epsilon_n \cos\alpha\,(u - \Sigma\mu\epsilon)\, dc_n.$$

Therefore

$$pdu = \frac{\mu_n d\epsilon_n}{\pi}\int_0^\infty d\alpha \int_{-\infty}^{+\infty} d\epsilon_1 \ldots \int_{-\infty}^{+\infty} d\epsilon_n \phi_1\epsilon_1 \ldots \phi_n\epsilon_n \cos\alpha(u - \Sigma\mu\epsilon) \ldots (4).$$

Now if u and ϵ_n are to vary together

$du = \mu_n d\epsilon_n$, and therefore

$$p = \frac{1}{\pi}\int_0^\infty d\alpha \int_{-\infty}^{+\infty} d\epsilon_1 \ldots \int_{-\infty}^{+\infty} d\epsilon_n \phi_1\epsilon_1 \ldots \phi_n\epsilon_n \cos\alpha\,(u - \Sigma\mu\epsilon) \ldots\ldots (5).$$

And finally,

$$P = \frac{1}{\pi}\int_{-l}^{+l} du \int_0^\infty d\alpha \int_{-\infty}^{+\infty} d\epsilon_1 \ldots \int_{-\infty}^{+\infty} d\epsilon_n \phi_1\epsilon_1 \ldots \phi_n\epsilon_n \cos\alpha\,(u - \Sigma\mu\epsilon) \ldots (6).$$

Now let us suppose that equal positive and negative errors are equally probable. In this case $\phi\epsilon = \phi(-\epsilon)$, and consequently,

$$\int_{-\infty}^{+\infty} \phi\epsilon \sin\alpha\mu\epsilon d\epsilon = 0.$$

Hence (6) will become

$$P = \frac{2}{\pi}\int_0^l du \int_0^\infty \cos\alpha u d\alpha \int_{-\infty}^{+\infty} \phi_1\epsilon_1 \cos\alpha\mu_1\epsilon_1 d\epsilon_1 \ldots \int_{-\infty}^{+\infty} \phi_n\epsilon_n \cos\alpha\mu_n\epsilon_n d\epsilon_n \ldots (7).$$

The next step is to find an approximate value of this expression.

$$\text{When } \alpha = 0 \int_{-\infty}^{+\infty} \phi\epsilon \cos \alpha\mu\epsilon d\epsilon = \int_{-\infty}^{+\infty} \phi\epsilon d\epsilon = 1,$$

as the error ϵ must have some value lying between $\pm\infty$.

It is clear this is the greatest value the integral in question can have, and therefore as n increases *sine limite*, the continued product

$$\int_{-\infty}^{+\infty} \phi_1\epsilon_1 \cos \mu_1\alpha\epsilon_1 d\epsilon_1 \ldots \int_{-\infty}^{+\infty} \phi_n\epsilon_n \cos \mu_n\alpha\epsilon_n d\epsilon_n$$

decreases *sine limite*, (being the product of n factors each less than unity) except for values of α differing infinitesimally from zero.

$$\text{Let } k^2 = \int_0^{\infty} \phi\epsilon \,.\, \epsilon^2 d\epsilon, \quad \kappa^4 = \int_0^{\infty} \phi\epsilon \,.\, \epsilon^4 d\epsilon,$$

and develope each of the cosines in the above-written continued product. It is thus seen to be equal to

$$1 - \alpha^2\Sigma\mu^2k^2 + \alpha^4\left(\frac{1}{12}\Sigma\mu^4\kappa^4 + \Sigma\mu_1^2\mu_2^2k_1^2k_2^2\right) - \&c.$$

Again, n being very large and ultimately infinite, it is evident that $\Sigma\mu^4\kappa^4$ is of the same order of magnitude as n, while $\Sigma\mu_1^2\mu_2^2k_1^2k_2^2$ is of the order of n^2, the former term of the coefficient of α^4 may therefore be neglected in comparison with the latter, which again may be replaced by $\frac{1}{2}(\Sigma\mu^2k^2)^2$, from which it differs by a quantity of the order of n. Similar remarks apply with respect to the higher powers of α.

Thus the continued product may be replaced by

$$1 - \alpha^2\Sigma\mu^2k^2 + \frac{1}{2}\alpha^4(\Sigma\mu^2k^2)^2 - \frac{1}{2\,.\,3}\alpha^6(\Sigma\mu^2k^2)^3 + \&c.$$

or by $e^{-\alpha^2\Sigma\mu^2k^2}$; a function which is coincident with it when α is infinitesimal. When α is finite both are, as we have seen, infinitesimal.

Consequently,

$$P = \frac{2}{\pi}\int_0^l du \int_0^{\infty} \cos \alpha u d\alpha \,.\, e^{-\alpha^2\Sigma\mu^2k^2} \ldots\ldots\ldots\ldots\ldots\ldots\ldots (8),$$

or

$$P=\frac{1}{(\pi\Sigma\mu^2k^2)^{\frac{1}{2}}}\int_0^l e^{-\frac{u^2}{4\Sigma\mu^2k^2}}\,du=\frac{2}{\sqrt{\pi}}\int_0^{\frac{l}{2\sqrt{\Sigma\mu^2k^2}}} e^{-v^2}\,dv\ldots\ldots(9),$$

where we have supposed

$$u=2\,(\Sigma\mu^2k^2)^{\frac{1}{2}}v.$$

It is evident, that whatever l may be, this expression for P is a maximum when

$$\Sigma\mu^2k^2 \text{ is a minimum.}$$

Hence we get the following remarkable conclusion: When the number of observations increases *sine limite* the most advantageous system of factors are those which make

$$\Sigma\mu^2k^2 \text{ a minimum.}$$

It remains to determine μ from the condition of the minimum taken in connexion with those already stated, viz.

$$\Sigma\mu a=1,\ \Sigma\mu b=0,\ \&c.=0.$$

We have

$$\left.\begin{array}{ll}\Sigma k^2\mu d\mu &=0\\ \Sigma a d\mu &=0\\ \Sigma b d\mu &=0\\ \&c. &=0\end{array}\right\}\ldots\ldots\ldots\ldots(A).$$

Let $\lambda_1, \lambda_2 \ldots \lambda_p$ be indeterminate factors, then we may put

$$\left.\begin{array}{l}k_1^2\mu_1=a_1\lambda_1+b_1\lambda_2+\&c.\\ k_2^2\mu_2=a_2\lambda_1+b_2\lambda_2+\&c.\\ \&c.=\&c.\end{array}\right\}\ldots\ldots\ldots\ldots(B).$$

From the n equations (B) we deduce a new system of p equations. To obtain the first of these, we multiply equations (B) by $\frac{a_1}{k_1^2}$, $\frac{a_2}{k_2^2}$, &c. respectively, and add the results. For the second, we employ instead of the factors $\frac{a_1}{k_1^2}$, &c., the factors $\frac{b_1}{k_1^2}$, $\frac{b_2}{k_2^2}$, &c. and then proceed as before. And similarly for the others.

In consequence of the relations

$$\Sigma\mu a = 1,\ \Sigma\mu b = 0,\ \&c. = 0,$$

the new system of equations will be

$$\left.\begin{aligned} 1 &= \lambda_1 \Sigma\frac{a^2}{k^2} + \lambda_2 \Sigma\frac{ab}{k^2} + \&c. \\ 0 &= \lambda_1 \Sigma\frac{ab}{k^2} + \lambda_2 \Sigma\frac{b^2}{k^2} + \&c. \\ 0 &= \&c. \end{aligned}\right\} \cdots\cdots\cdots (C).$$

These p equations determine $\lambda_1, \lambda_2, \ldots \lambda_p$, and thus in virtue of (B) the values of $\mu_1, \mu_2, \ldots \mu_n$ become known. Finally as

$$x = \mu_1 V_1 + \mu_2 V_2 + \ldots + \mu_n V_n,$$

x will be completely determined.

Now let us recur to the original equations of condition stated in the last paragraph.

$$\left.\begin{aligned} \epsilon_1 &= a_1 x + b_1 y + \&c. - V_1 \\ \epsilon_2 &= a_2 x + b_2 y + \&c. - V_2 \\ \&c. &= \&c. \\ \epsilon &= a_n x + b_n y + \&c - V_n \end{aligned}\right\} \cdots\cdots\cdots (\alpha).$$

From this system we deduce a new one, containing p equations. The first of these is got by multiplying equations (α) by $\frac{a_1}{k_1^2}, \frac{a_2}{k_2^2}$, &c., and adding the results: the second by using the factors $\frac{b_1}{k_1^2}, \frac{b_2}{k_2^2}$, &c.: and so on as before. The resulting system will be, neglecting all errors,

$$\left.\begin{aligned} x\Sigma\frac{a^2}{k^2} + y\Sigma\frac{ab}{k^2} + \&c. &= \Sigma\frac{a}{k^2} V \\ x\Sigma\frac{ab}{k^2} + y\Sigma\frac{b^2}{k^2} + \&c. &= \Sigma\frac{b}{k^2} V \\ \&c. \qquad\qquad &= \&c. \end{aligned}\right\} \cdots\cdots\cdots (\beta').$$

The system (β') contains as many equations as there are unknown quantities x, y, &c. I proceed to show that if x be determined from this system, its value will be the same as if it had been obtained from the most advantageous system of factors, namely, that which is determined by means of (B) and (C). In order to prove this, we multiply equations (β') by λ_1, λ_2, &c., and add the results. Then, in virtue of (C)

$$x = \lambda_1 \Sigma \frac{a}{k^2} V + \lambda_2 \Sigma \frac{b}{k^2} V + \&c.$$

Or,

$$x = (\lambda_1 a_1 + \lambda_2 b_1 + \&c.) \frac{V_1}{k_1^2} + (\lambda_1 a_2 + \lambda_2 b_2 + \&c.) \frac{V_2}{k_2^2} + \&c.$$

that is to say, as is seen on referring to (B),

$$x = \mu_1 V_1 + \mu_2 V_2 + \ldots + \mu_n V_n,$$

as before; which proves that the system (β') gives the same value for x as the most advantageous system of factors. Moreover, as (β') is symmetrical in x and a, y and b, &c. it is clear that it will also give the most advantageous values for y and the other unknown quantities.

When the law of probability of error is the same at every observation $k_1 = k_2 =$ &c. and (β') reduces itself to (β) given at p. 18 as the result of the method of least squares. In the general case, it expresses the modification which the method of least squares must undergo, when all the observations are not of the same kind, namely, that instead of making the function $\Sigma\,(ax + by + \&c. - V)^2$ a minimum with respect to xy, &c., we must substitute for it the function $\Sigma \frac{1}{k^2} (ax + by + \&c. - V)^2$, and then proceed as before.

Such, in effect, is Laplace's demonstration, except that he supposed the law of error the same at each observation. The form in which I have presented it is wholly unlike his. The introduction of Fourier's theorem enables us to avoid the theory of combinations, and also the use of imaginary symbols. It must be admitted that there are few mathematical investigations less inviting than the fourth chapter of the *Théorie des Probabilités*, which is that in which the method of least squares is proved.

It may be worth while to recur to the general formula:

$$P=\frac{1}{\pi}\int_{-l}^{+l} du \int_0^{\infty} d\alpha \int_{-\infty}^{+\infty} d\epsilon_1 \ldots \int_{-\infty}^{+\infty} d\epsilon_n \phi_1 \epsilon_1 \ldots \phi_n \epsilon_n \cos \alpha \, (u - \Sigma\mu\epsilon).$$

It is certain that $\Sigma\mu\epsilon$ lies between the limits $\pm \infty$. Therefore when $l = \infty$, P should be equal to unity. I proceed to show that this is the case.

$$P = \frac{1}{\pi}\int_{-\infty}^{+\infty} e^{-m^2u^2} du \int_0^{\infty} d\alpha \int_{-\infty}^{+\infty} d\epsilon_1 \ldots \int_{-\infty}^{+\infty} d\epsilon_n \, \phi_1\epsilon_1 \ldots \phi_n\epsilon_n \cos \alpha \, (u - \Sigma\mu\epsilon)$$

when $m = 0$.

Effecting the integration for u,

$$P_\infty = \frac{1}{m\sqrt{\pi}}\int_0^{\infty} e^{-\frac{\alpha^2}{4m^2}} d\alpha \int_{-\infty}^{+\infty} d\epsilon_1 \ldots \int_{-\infty}^{+\infty} d\epsilon_n \, \phi_1\epsilon_1 \ldots \phi_n\epsilon_n \cos \alpha \, \Sigma\mu\epsilon \;\ldots\ldots\; (10)$$

when $m = 0$,

since

$$\int_{-\infty}^{+\infty} e^{-m^2u^2} \cos \alpha u \, du = \frac{\sqrt{\pi}}{m} e^{-\frac{\alpha^2}{4m^2}}$$

and $\int_{-\infty}^{+\infty} e^{-m^2u^2} \sin \alpha u \, du = 0.$

Integrating for α, we see that when $m = 0$

$$P_\infty = \int_{-\infty}^{+\infty} d\epsilon_1 \ldots \int_{-\infty}^{+\infty} d\epsilon_n \, \phi_1\epsilon_1 \ldots \phi_n \epsilon_n e^{-(m\Sigma\mu\epsilon)^2} \ldots (11).$$

Or,

$$P_\infty = \int_{-\infty}^{+\infty} \phi_1\epsilon_1 \, d\epsilon_1 \ldots \int_{-\infty}^{+\infty} \phi_n\epsilon_n \, d\epsilon_n \;\ldots\ldots\ldots\ldots\; (12).$$

And as each of these integrals is separately equal to unity,

$P = 1$, which was to be proved.

I proceed to show that in a particular case in which the value of P can be accurately determined, Laplace's approximation is correct. It has sometimes been thought that the introduction of the negative exponential involves a *petitio principii*, and is equivalent to assuming a particular law of error. It is therefore desirable, and I am not aware that it has hitherto been done, to verify his result in an individual case.

Let the law of error be the same in all the observations, and such that $\phi\epsilon = \frac{1}{2}\epsilon^{\mp\epsilon}$, the upper sign to be taken when ϵ is positive.

Let $\mu_1 = \mu_2 = \&c. = 1$, then

$$p = \frac{1}{\pi}\int_0^\infty d\alpha \int_{-\infty}^{+\infty} (\tfrac{1}{2}\epsilon^{\mp\epsilon_1})\, d\epsilon_1 \ldots\ldots \int_{-\infty}^{+\infty} (\tfrac{1}{2}e^{\mp\epsilon_n})\, d\epsilon_n \cos\alpha\,(u - \Sigma\epsilon),$$

or,

$$p = \frac{1}{\pi}\int_0^\infty \frac{\cos u\alpha}{(1+\alpha^2)^n}\, d\alpha, \text{ since } \int_0^\infty e^{-\epsilon}\cos\alpha\epsilon d\epsilon = \frac{1}{1+\alpha^2}\,.$$

The value of p is thus given by a known definite integral, which has been discussed by M. Catalan in the fifth volume of *Liouville's Journal.*

It may be developed in a series of powers of u. Up to $u^{2(n-1)}$ no odd power of u can appear in this development, for $\int_0^\infty \frac{\alpha^{2p}}{(1+\alpha^2)^n}\, d\alpha$ is finite while p is less than n, and therefore the integral may be developed by Maclaurin's theorem. For higher powers the method ceases to be applicable, and we must complete the development by other means. But as we suppose n to increase *s. l.* the integral tends to become developable in a series of even powers only of u. Thus

$$\int_0^\infty \frac{\cos u\alpha}{(1+\alpha^2)^n}\, d\alpha = \int_0^\infty \frac{d\alpha}{(1+\alpha^2)^n} - \tfrac{1}{2}u^2\int_0^\infty \frac{\alpha^2}{(1+\alpha^2)^n}\, d\alpha + \&c.$$

Let

$$\int_0^\infty \frac{d\alpha}{(1+\alpha^2)^n} = f(n).$$

Then

$$\int_0^\infty \frac{\alpha^2 d\alpha}{(1+\alpha^2)^n} = -\Delta f(n-1)\,;$$

and generally,

$$\int_0^\infty \frac{a^{2p}da}{(1+a^2)^n} = \pm\, \Delta^p f(n-p).$$

Now

$$f(n) = \frac{\pi}{2} \cdot \frac{1\,.\,3\ldots 2n-3}{2\,.\,4\ldots 2n-2};$$

$$-\Delta f(n-1) = \frac{\pi}{2} \cdot \frac{1\,.\,3\ldots 2n-5}{2\,.\,4\ldots 2n-2}\,.\,1,$$

$$\Delta^2 f(n-2) = \frac{\pi}{2} \cdot \frac{1\,.\,3\ldots 2n-7}{2\,.\,4\ldots 2n-2}\,.\,1\,.\,3;$$

and generally,

$$\pm\,\Delta^p f(n-p) = \frac{\pi}{2} \cdot \frac{1\,.\,3\ldots(2n-3-2p)}{2\,.\,4\ldots 2n-2}\,.\,1\,.\,3\ldots(2p-1)$$

$$= f(n)\,.\,\frac{1\,.\,3\ldots(2p-1)}{(2n-1-2p)\ldots(2n-3)}\,.$$

Thus

$$\int_0^\infty \frac{\cos uada}{(1+a^2)^n} = fn\left\{1 - \frac{1}{2} \cdot \frac{1}{2n-3}\,u^2 + \frac{1}{2\,.\,3\,.\,4}\,\frac{1\,.\,3}{2n-5\,.\,2n-3}\,u^4 - \&c.\right\}$$

The coefficient of u^{2p} is

$$\pm f n\,.\,\frac{1.3\ldots(2p-1)}{2\,.\,3\ldots 2p} \cdot \frac{1}{(2n-1-2p)\ldots(2n-3)}$$

or,

$$\pm f n\;\frac{1}{1\,.\,2\ldots p} \cdot \frac{1}{2^p} \cdot \frac{1}{(2n-1-2p)\ldots(2n-3)}\,.$$

Let n become infinite, this becomes

$$f(n)\;\frac{1}{1\,.\,2\ldots p}\;\frac{1}{(4n)^p};$$

and we have only to determine what $f(n)$ then becomes.

Now by Wallis's theorem

$$\left(\frac{2}{\pi}\right)^{\frac{1}{2}} = \frac{1\,.\,3\ldots(2n-1)^{\frac{1}{2}}}{2\,.\,4\ldots 2n-2}\,.\ \text{ult.}$$

Therefore $fn = \dfrac{1}{2}\left(\dfrac{2\pi}{2n-1}\right)^{\frac{1}{2}}$ when n is infinite,

$$\text{or, } fn = \frac{1}{2}\left(\frac{\pi}{n}\right)^{\frac{1}{2}}.$$

Consequently,

$$\int_0^\infty \frac{\cos u\alpha d\alpha}{(1+\alpha^2)^n} = \frac{1}{2}\left(\frac{\pi}{n}\right)^{\frac{1}{2}}\left\{1 - \frac{u^2}{4n} + \frac{1}{2}\cdot\frac{u^4}{(4n)^2} - \&c.\right\}$$

$$= \frac{1}{2}\left(\frac{\pi}{n}\right)^{\frac{1}{2}} e^{-\frac{u^2}{4n}} \text{ when } n \text{ is infinite.}$$

Therefore,

$$p = \frac{1}{2\sqrt{n\pi}} e^{-\frac{u^2}{4n}},$$

$$\text{and } P = \frac{1}{\sqrt{n\pi}} \int_0^l e^{-\frac{u^2}{4n}} du.$$

Now the value given for P at p. 23 is

$$P = \frac{1}{\sqrt{\pi \Sigma \mu^2 k^2}} \int_0^l e^{-\frac{u^2}{4\Sigma\mu^2 k^2}} du.$$

In the present case $\mu = 1$;

$$k^2 = \tfrac{1}{2}\textstyle\int_0^\infty e^{-\epsilon}\epsilon^2 d\epsilon = 1 \text{; and consequently } \Sigma\mu^2 k^2 = n.$$

Thus

$$P = \frac{1}{\sqrt{n\pi}} \int_0^l e^{-\frac{u^2}{4n}} du, \text{ as before.}$$

Thus Laplace's approximation coincides with the result obtained by an independent method.

This example serves to show distinctly the nature of the approximation in question.

The function p having been developed in a series of powers of u, we take the *principal term* in the coefficient of each power of u; that is, the term divided by the lowest power of n. We neglect for instance every such term as $\frac{1}{n^{p+\delta}} u^{2p}$, because we have a term in u^{2p} divided by n^p. Thus we retain $\frac{u^{2p}}{n^p}$ and neglect $\frac{u^{2(p-\delta}}{n^p}$, although, unless u be large, the former term is of the same or a lower order of magnitude than the latter. That Laplace's method does in a very general manner give an approximation of this kind cannot, I think, be questioned, especially after the verification we have just gone through. But some doubt may perhaps remain, whether such an approximation to the *form* of the function P, if such an expression may be used, is also an approximation to its numerical value, when we consider that in obtaining it we have neglected terms demonstrably larger than those retained.

For two recognized exceptions to the generality of Laplace's investigation, viz. where $\phi\epsilon = \frac{1}{\pi}\frac{1}{1+\epsilon^2}$, and the case in which $\mu_1, \mu_2 \ldots$, decrease *in infinitum sine limite*, I shall only refer to p. 10 of Poisson's paper in the *Connaissance des Tems* for 1827. Neither affects the general argument. We now come to Gauss's second method, which is given in the *Theoria Combinationis Observationum*.

GAUSS'S SECOND DEMONSTRATION.

The connexion between the method of Laplace, and that which Gauss followed in the *Theoria Combinationis Observationum*, will be readily understood from the following remarks.

After determining $\mu_1 \ldots \mu_n$ by the condition that P should be a minimum, Laplace remarked that the same result would have been obtained (viz. that $\Sigma\mu^2k^2$ must be a minimum), if the assumed condition had been that the mean error of the result, *i. e.* the mean arithmetical value of $\Sigma\mu\epsilon$ should be a minimum. (I should rather say that he makes a remark equivalent to this, and differing from it only in consequence of a difference of notation, &c.) It is in fact easy to see that the mean value in question is equal to

$$\frac{\int_0^\infty updu}{\int_0^\infty pdu}, \text{ or to } 2\int_0^\infty updu;$$

and as

$$p = \frac{1}{2(\pi\Sigma\mu^2k^2)^{\frac{1}{2}}} e^{-\frac{u^2}{4\Sigma\mu^2k^2}}$$

$$2\int_0^\infty updu = \frac{2(\Sigma\mu^2k^2)^{\frac{1}{2}}}{\sqrt{\pi}},$$

which is of course a minimum when $\Sigma\mu^2k^2$ is so.

Gauss, adopting this way of considering the subject, pointed out that it involved the postulate that the importance of the error $\Sigma\mu\epsilon$, *i.e.* the detriment of which it is the cause, is proportional to its arithmetical magnitude. Now, as he observes, the importance of the error may be just as well supposed to vary as the square of its magnitude: in fact, it does not, strictly speaking, admit of arithmetical evaluation at all. We must *assume* that it is represented by some direct function of its magnitude, such that both vanish together. One assumption is not more arbitrary than another. Let us suppose, therefore, that the importance of the error is represented by $(\Sigma\mu\epsilon)^2$. That is, that $(\Sigma\mu\epsilon)^2$ is the function whose mean value is to be made a minimum. I now proceed to find it.

$$(\Sigma\mu\epsilon)^2 = \Sigma\mu^2\epsilon^2 + 2\,\Sigma\mu_1\mu_2\,\epsilon_1\epsilon_2 \,\ldots\ldots\ldots\, (13).$$

The mean value of ϵ^2 is $\int_{-\infty}^{+\infty} \epsilon^2\,\phi\epsilon d\epsilon = 2k^2$.

Hence, that of $\Sigma\mu^2\epsilon^2$ is $2\,\Sigma\mu^2 k^2$.

The mean value of $\Sigma\mu_1\mu_2\epsilon_1\epsilon_2$ is zero, positive and negative errors of the same magnitude occurring with equal frequency on the long run.

Consequently,

$$\text{mean of } (\Sigma\mu\epsilon)^2 = 2\,\Sigma\mu^2 k^2 \,\ldots\ldots\ldots\, (14);$$

and therefore, as before, $\Sigma\mu^2 k^2$ is to be made a minimum. The rest of the investigation is of course the same as that of Laplace.

Nothing can be simpler or more satisfactory than this demonstration. It is free from all analytical difficulty, and applicable whatever be the number of observations, whereas that of Laplace requires this number to be very large.

Recurring to equation (11), differentiating it for m^2, and then making $m = 0$, we find

$$\int_{-\infty}^{+\infty} pu^2\,du = \int_{-\infty}^{+\infty} \phi\epsilon_1\,d\epsilon_1 \ldots \int_{-\infty}^{+\infty} \phi\epsilon_n\,d\epsilon_n\,(\Sigma\mu\epsilon)^2 = 2\,\Sigma\mu^2 k^2;$$

and as the first member of this equation is evidently the mean value of u^2 or of $(\Sigma\mu\epsilon)^2$, this is a new verification of our analysis.

As an illustration of Gauss's principle, let the fourth power of the error be taken as the measure of its importance:

$$(\Sigma\mu\epsilon)^4 = \Sigma\mu^4\epsilon^4 + 6\Sigma\mu_1^2\mu_2^2\epsilon_1^2\epsilon_2^2 + \text{terms involving odd powers of } \epsilon.$$

Therefore,

$$\text{mean of } (\Sigma\mu\epsilon)^4 = 2\Sigma\mu^4\kappa^4 + 24\,\Sigma\mu_1^2\mu_2^2k_1^2k_2^2 \quad \text{.........} \quad (15)$$

and $\mu_1 \dots \mu_n$ must be so determined that this may be a minimum.

I have already said that the results given by what Laplace called the most advantageous system of factors are not strictly speaking the most probable of all possible results.

As the distinction involved in this remark seems to me to be essential to a right apprehension of the subject, I will endeavour to illustrate it more fully.

Recurring to the equations of condition, as given in p. 18, we see that the values Laplace assigns to the factors $\mu_1\mu_2$ &c. are independent of V_1V_2 &c. They depend merely on the coefficients ab &c., which are quantities known *à priori, i.e.* before observation has assigned certain more or less accurate values to the magnitudes V_1V_2 &c. All we then can say is, that if we employ Laplace's system of factors, and also any other, in a large number of cases (the coefficients ab &c., being the same in all) we shall be right within certain limits in a larger proportion of cases when the former system of factors is made use of than when we employ the latter. And this conclusion is wholly irrespective of the values of V_1V_2 &c., and consequently of those which we are led in each particular case to assign to xy &c. The comparison is one of *methods*, and not at all one of *results*. But when V_1V_2 &c. are known, another way of considering any particular case presents itself. We can then compare the probability of different *results*. For, let us consider a large number of sets of equations of condition (in each of which not only are ab &c. equal, as in the former case, but also V_1V_2 &c.). The true values of the elements xy &c. may be different in each. But in affirming that $\xi\eta$ &c., are the most *probable* values of xy &c., we affirm that the true values of xy &c. are more frequently equal to $\xi\eta$ &c. than to any other quantities whatever. Here we have no concern with the method by which the values $\xi\eta$ &c. were obtained. The comparison is merely one of *results*.

As for one particular law of error (that considered in p. 15), the results of the method of least squares are the most *probable* possible; and as the function by which this law of error is expressed occurs in Laplace's demonstration of that method, it has been thought that his approximation involved an undue assumption, and that in fact his proof was invalid unless that particular law of error was supposed to obtain.

It is easily seen that the method of least squares can give the most probable results only for that law of error (if we except another which involves a discontinuous function). Mr Ivory attempted to show that Laplace's conclusions might be applied to prove that the results of the method were, in effect, the most probable possible, and thence drew the inference which I have already mentioned. After some consideration, I have decided on not entering on an analysis of his reasoning, which it would be difficult to make intelligible, without adding too much to the length of this communication. It is set forth with a good deal of confidence; Laplace's conclusions are pronounced invalid on the authority of an indirect argument, and without any examination of the process by which he was led to them. I may just mention that in the whole of Mr Ivory's reasoning, the probability that $\Sigma\mu\epsilon$ is precisely equal to any assigned magnitude, is, to all appearance at least, considered a finite quantity, though it is perfectly certain that it must be infinitesimal.

It would seem as if he had taken Laplace's expression of the probability in question, viz.

$$\frac{c^{-\frac{kl^2}{4k''a^2S_{\prime}m^{(1)2}}}}{2a\sqrt{\pi}\sqrt{\frac{k''}{k}Sm^{(1)2}}},$$

without being aware that in Laplace's notation l and a are infinite, and that consequently the expression is infinitesimal. (Vide *Tilloch's Magazine*, LXV. p. 81.)

MR IVORY'S DEMONSTRATIONS.

They are three in number. Two appeared in the sixty-fifth, and a third in the sixty-seventh volumes of *Tilloch's Magazine*.

The aim of all three is the same, namely, to demonstrate the rule of least squares without recourse to the theory of probabilities, which appeared to him to be foreign to the question. The grounds of this opinion he has not clearly developed: perhaps the best refutation of it will be found in the unsatisfactory character of the demonstrations which he proposed to substitute for the methods of Laplace and Poisson. In common with many others, Mr Ivory appears to have looked with some distrust on the results obtained by means of this theory: a not unnatural consequence of the extravagant pretensions sometimes advanced on its behalf.

The first of his demonstrations rests upon what I cannot help considering a vague analogy. In the equation of condition

$$e = ax - V,$$

he remarks that the influence of the error e on the value of x increases as a decreases, and *versâ vice:* that consequently the case is precisely similar to that of a lever which is to produce a given effect, as of course the length of the arm must vary inversely as the weight which it supports.

Consequently, he argues, the condition to be fulfilled, in order that the equations of condition may be combined in the most advantageous manner, is the same as what would be the condition of equilibrium, were $a\ a'\ a''$ &c. weights on a lever, acting at arms $e\ e'\ e''$ &c. This condition is of course

$$\Sigma ae = 0, \text{ whence } \Sigma(ax - V)\ a = 0,$$

the result given by the method of least squares.

But, granting that the influence of an error e, ought to be greater when a is less, and *versâ vice*, how are we entitled to assume that the case is *precisely similar* to that of equilibrium on a lever? Apart from this assumption, there seems to be no reason for inferring that because this influence increases as a decreases, it must therefore vary inversely as a. By what function of a the influence of e ought to be represented, is the very essence of the question; to determine, by introducing the extraneous idea of equilibrium on a lever, that $\frac{1}{a}$ is the function required, seems to be little else than a *petitio principii*, concealed by a metaphor*.

* I have omitted to notice some remarks which Mr Ivory appends to this demonstration, as they do not appear to affect the view taken in the text.

The second demonstration may be thus briefly stated.

The values of different sets of observations might be compared if we knew the average error in each set, or if we knew the average value of the squares of the errors in each. In either case that would be the best set of observations in which the quantity taken as the *measure of precision* was the smallest.

Similarly, by assigning different values to the unknown quantities x, y, &c. involved in a system of equations of condition, we can make it *appear* that the mean of the squares of the errors has a greater or less value. Therefore as of sets of observations, that is the best in which this quantity is least; so of different sets of results deduced from one set of observations, the same is also true; and therefore the sum of the squares of the apparent errors is to be made a minimum.

There seems to be involved in this reasoning a confusion of two distinct ideas; the precision of a set of observations is undoubtedly measured by the average of the errors *actually committed*, and if we knew this average, we should be able to compare the values of different sets of observations. But it is not measured by the average of the *calculated* errors, namely, those which are determined from the equations of condition when particular values have been assigned to x, y, &c.

The problem to be solved may be stated thus. Given that the single observations of which the set is composed are liable to a certain average of error, to combine them so that the resulting values of the unknown quantities may be liable to the smallest average of error.

This problem Laplace and Gauss have both solved. Their solutions differ, because they estimated the average error in different manners.

But how are we justified in assuming that to be the best mode of combining the observations which merely gives the *appearance* of precision *not* to the final results, but only to the individual observations, and which, with reference to them, gives no estimation of the probability that this appearance of accuracy is not altogether illusory?

The third of Mr Ivory's demonstrations is not, I think, more satisfactory than the other two.

The kind of observations to which the method of least

squares is applicable, are such, Mr Ivory observes, that there exists no bias tending regularly to produce error in one direction, and that the error in one case is supposed to have no influence whatever on the error in any other case.

From this principle he attempts to show that the method of least squares is the only one which is consistent with the independence of the errors.

When, however, we speak of the errors as being independent of one another, only this can be meant, that the circumstances under which one observation takes place do not affect the others. In *rerum naturâ* the errors are independent of one another. Nevertheless, with reference to our knowledge they are not so, that is to say, if we know one error we know all, at least in the case in which the equations of condition involve only one unknown quantity, which is that considered by Mr Ivory. For the knowledge of one error would imply the knowledge of the true value of the unknown quantity, and thence that of all the other errors.

Mr Ivory states the following equations of condition:

$$e = ax - m$$

$$e' = a'x - m'$$

$$\&c. = \&c.$$

He thence deduces the following value of x:

$$x = \frac{\Sigma ae}{\Sigma a^2} + \frac{\Sigma am}{\Sigma a^2}, \text{ and those of } ee' \text{ are}$$

$$e = -m + \frac{a\Sigma am}{\Sigma a^2} + \frac{a\Sigma ae}{\Sigma a^2}$$

$$e' = -m' + \frac{a'\Sigma am}{\Sigma a^2} + \frac{a'\Sigma ae}{\Sigma a^2}. \&c. = \&c.$$

He remarks that these errors are not independent of one another, as all depend on the single quantity Σae, which may be eliminated between any two of the last-written equations: but that there is one case in which they are independent of one another, namely, when we assume $\Sigma ae = 0$, which of course leads to the method of least squares, and that in this case, as we shall have

$$e = -m + \frac{a\Sigma am}{\Sigma a^2} \&c. = \&c.$$

each error is determined by "the quantities of its own experiment." But this reasoning is perfectly inconclusive. In the case supposed, $e\,e'$ &c. are as much connected together as in any other, as may be shown by eliminating Σam between the equations

$$e = -m + \frac{a\Sigma am}{\Sigma a^2}, \quad e' = -m' + \frac{a'\Sigma am}{\Sigma a^2} \text{ \&c.} = \text{\&c.};$$

and besides, apart from any mathematical reasoning, it is clear that as if we know one error we know all, so also if we assign any value to one, we have in effect assigned values to all, whether we use the method of least squares or any other.

Moreover, e is not determined by the quantities of its own experiment alone, since Σam involves the results of all the experiments; there is no difference between this and the general case, except that Σae has ceased to appear in the equations. But suppose we multiplied the equations of condition by any function of a, we might deduce the following values of x and e:

$$x = \frac{\Sigma\phi a \,.\, e}{\Sigma a \,.\, \phi a} + \frac{\Sigma\phi a \,.\, m}{\Sigma a \,.\, \phi a}$$

$$e = -m + \frac{a\Sigma\phi a \,.\, m}{\Sigma a \,.\, \phi a} + \frac{a\Sigma\phi a \,.\, e}{\Sigma a \,.\, \phi a}$$

$$e' = -m' + \frac{a'\Sigma\phi a \,.\, m}{\Sigma a \,.\, \phi a} + \frac{a'\Sigma\phi a \,.\, e}{\Sigma a \,.\, \phi a}.$$

Mr Ivory's reasoning would apply word for word as before, and would show that the best mode of combining the equations of condition was to employ the factors ϕa, $\phi a'$, &c. whatever be the form of ϕ. As it thus would serve to establish, at least apparently, an infinity of contradictory results, the inference is that in no case has it any validity.

I have now completed, though in an imperfect manner, the design indicated at the outset of this paper, namely, to give an account of the different modes in which the subject has been treated, and to simplify the analytical investigations. If I have succeeded in doing this, the present communication may tend to make a very curious subject more accessible than it has hitherto been.

SOME REMARKS ON THE THEORY OF MATTER.*

IN the present state of Science, there are few subjects of greater interest than the enquiry whether all the phenomena of the universe are to be explained by the agency of mechanical force, and if not whether the new principles of causation, such as chemical affinity, and vital action, are to be conceived of as wholly independent of mechanical force, or in some way not hitherto explained cognate and connected with it. One reason among many which makes this enquiry interesting is the circumstance that the application of mathematics to natural philosophy has, up to the present time, either been confined to phenomena, which were supposed to be explicable without assuming any other principle of causation than ordinary "push and pull" forces, or as in Fourier's theory of heat and Ohm's theory of the galvanic circuit, has been based on proximate empirical principles.

2. The intention of the remarks which I have the honour to offer to the Society is to suggest reasons for believing that while on the one hand it is impossible not merely from the short-comings of our analysis but from the nature of the case to reduce, as it appears that Laplace wished to do, all the phenomena of the universe to one great dynamical problem, we cannot recognise the existence of any principle of causation wholly disconnected with ordinary mechanical force, or of which the nature could be explained without a reference to local motion: in other words, that the idea of "qualitative action" in the sense which the phrase naturally suggests must be rejected. It will be seen from the explanations I am about to attempt that the objection which Leibnitz has opposed to the atomic, and in effect to any mechanical philosophy, namely, that on such prin-

* *Transactions of the Cambridge Philosophical Society,* Vol. VIII. p. 600. [Read May 22, 1848.]

ciples a finite intelligence might be conceived to exist by which all the phenomena of the universe would be fully comprehended, does not (whatever may be thought of its validity) appear to apply to the views which I have been led to entertain. For these views essentially depend on the conception of what may be called a hierarchy of causes, to which we have no reason for assigning any finite limit. Of this series of principles of causation, ordinary mechanical force is the first term.

3. With respect to the first point, namely, the impossibility of explaining all phenomena mechanically, it may be remarked, that we are met, in the attempt to discuss it, by the difficulty which always attends the establishment of a negative proposition. It is clear that as in the present state of our knowledge we are far from being able to enumerate and classify the phenomena which are or which might be produced by the combined agency of conceivable mechanical forces, we are not in a position to decide *à priori* that any given phenomenon might not be thus produced. *Non constat*, but that the impossibility we find in the attempt to explain the causes of its existence may have no higher origin than the imperfect command which we have as yet obtained of the principles of mechanical causation. We meet, it may be said, with a multitude of ordinary dynamical problems which have as yet received no adequate solution—why then should we have recourse to new kinds of causes, while we have not as yet exhausted the resources, if the expression may thus be used, of those which we already recognise? To this enquiry no conclusive answer can be given, but the following considerations will I think naturally suggest themselves.

4. In the first place, no even moderately successful attempt has, I think, yet been made to explain any chemical phenomenon on mechanical principles. It is quite true that we are unable, to take a particular instance, fully to comprehend the mechanical constitution of the luminiferous ether; the determinations which have as yet been attempted of the law of attraction between its molecules cannot, I apprehend, be accepted as any thing more than hypothetical or provisional results, and there are other points involved in yet greater obscurity. Nevertheless the undulatory theory of light has, as we all know, given consistent and satisfactory explanations of a great variety of phenomena.

Thus it appears, and the same remark might be educed from other though similar considerations, that we are by no means absolutely estopped by the imperfection of our mechanical philosophy, from explaining phenomena really due to mechanical forces, even when these phenomena are connected with subjects not as yet fully comprehended: why then cannot some progress be made in the mechanical explanation of chemical phenomena, or of those, to mention no other class, which we are in the habit of referring to vital action? In these cases, we see or seem to see that the action of mechanical laws is modified or suspended; and though it is not demonstrably impossible that this is not really the case, and that no other causes are at work beside the "push and pull" forces of ordinary mechanics, yet we are at least much tempted to believe, that the difficulties we meet with do not arise from what may be called the disguised action of mechanical forces but from the presence of an agency of a distinct nature. And to this view we find that most of those incline who have made themselves familiar with the science of chemistry or with that which has been called biology; and further that, (with reference to the latter science) the insufficiency not only of a mechanical but even of a chemical physiology has been generally admitted.

Secondly, it is to be observed that even if it be considered doubtful whether a mechanical philosophy be not after all sufficient for the explanation of all phenomena, it is at least certain that it has not been proved to be so: and that by rejecting other conceivable modes of action than those which are recognised by it, we unnecessarily and arbitrarily limit the problem which the universe presents to us; falling thereby into an error similar to that of the atomists, who starting from the assumption that the ἀρχαί, or first principles of all things, are atoms and a vacuum proceeded to construct an imaginary world, in accordance with this arbitrary hypothesis. At the same time it must be granted that a purely mechanical* system such as that of Boscovich is more self-consistent and contains, so to

* The word *mechanical* is of course not used in antithesis to *dynamical*, in the sense in which the latter is commonly employed by the philosophical writers of Germany. The antithesis in question is foreign to the scope of the present essay, and I have accordingly elsewhere used the word dynamical in its ordinary acceptation.

speak, less that is discontinuous, than any which should recognise other principles, for instance chemical affinity, distinct from force without enquiring into the relation which subsists between them.

5. It may however be asserted that this enquiry is altogether superfluous—that the power of exerting attractive or repulsive force is one property of matter, that chemical affinity (and so in other cases) is another—that the two are not merely distinct, but absolutely independent and heterogeneous. But to this view the arguments which seem to have led to the adoption of a purely mechanical system, appear to prevent our assenting. I shall therefore attempt to state what I conceive these arguments to have been.

6. It is a fundamental principle of the secondary mechanical sciences, for instance of the theory of light, that the secondary qualities of bodies are to be explained by means of the primary. Every substance, to use for a moment the language of Leibnitz, is essentially active; in other words it is to be conceived of as the formal cause of the sensible qualities which are referred to it. If we ask why gold is yellow and silver white, the answer at once presents itself that the difference of colour corresponds and is due to a difference between the essential constitution of the two substances. Now the essential constitution here spoken of, and consequently the differences which individuate it in different cases, may conceivably be something altogether incognisable to the human intellect. The notion that it is so was expressed scholastically by saying that substantial forms are not cognoscible. But if, setting aside this opinion, we affirm that the essential constitution of each substance is a matter of which the mind can take cognisance, we are led at once to the distinction between primary and secondary qualities. The first are ascribed to each substance as its essential attributes, in virtue of which it is that which it is—the second result from the primary*, (by which as we have said the essential or formal constitution of the substance in question is determined,) and have reference to the mind by which they are perceived, while the primary are ascribed to it independently of any reference to a percipient

* Or that which in its formation it was to be, τὸ τί ἦν εἶναι.

mind: and a distinction, analogous or identical with that between primary and secondary qualities, has accordingly been expressed by the antithesis between that which is *a parte hominis* and that which is *a parte universi*. That the distinction between primary and secondary qualities is necessary *on the hypothesis on which we are proceeding*, appears at once from the consideration that if we affirm that all the qualities of bodies of which we can form any conception are equally subjective and phenomenal, nothing will remain of which the mind can take cognisance, and by means of which our conception of the nature of any one substance can be discriminated from that of any other*. Let it be granted therefore that the distinction of primary and secondary qualities is a necessary element of physical science. It follows from this that the secondary qualities in a manner disappear when we look at the universe from the scientific point of view. Instead of colours we have vibrations of the luminiferous ether—instead of sounds vibrations of the ambient air, and so on. Now from hence it follows that all the phenomena which we see produced, of whatever nature they may be, are all in reality dependent on the primary qualities of matter. Furthermore, these primary qualities themselves all involve the idea of motion or of a tendency to motion. A body changes its form in virtue of the local motion (absolute or relative) of some of its parts; and when I press a stone between my hands, I find that I can produce no sensible change of form, while contrariwise the stone reacts against my hands, tending to make them move in opposite directions. I then say that the stone is hard as a mode of expressing this, viz. that when an attempt is made to produce relative local motion of its parts, it resists it in virtue of its reactive tendency to produce motion in that which acts upon it. Again, a body whose parts are readily susceptible of relative local motion is said to be soft or fluid, and when a sensible change of form is accompanied by a tendency to such motion as shall restore the original form, it is said to be elastic, and so on. We thus arrive at a point of view at which all

* The doctrine of the cognoscibility of substantial forms, which is intimately connected with this distinction, is, as Leibnitz in effect remarks, as it were the common character of those who with more or less success attempted in the seventeenth century, the restoration of science. Vid. Leibnitz, *Epist. ad Thomas*, I.

secondary qualities having disappeared, and all primary ones* having been resolved into motion and tendency to motion, the sciences which relate to phenomena appear to be resolved into the general doctrine of motion. But if this be true the universe can it is said present to us nothing but one great dynamical problem. Motion, and force the cause of motion, belong essentially to the domain of mechanics: and if chemical affinity be a cause of local motion, that is, if in virtue of its action† a particle of matter finds itself at a given time in a position different from that which it would else have occupied, chemical affinity is not really distinct from mechanical force (which looked at from the dynamical point of view includes everything which is a cause of motion); whereas if it be not a cause of motion the enquiry at once presents itself of what is it? In illustration of this view we may refer to any chemical experiment. If an acid is dropped into a glass containing any vegetable blue, the colour is changed to red. But to say this is to say that the liquid when the acid is introduced into it begins to act on the luminiferous vibrations which exist near it in a different manner from that in which it had previously acted. The whole change, whether we call it a chemical phenomenon or not, consists in the introduction of new forms of motion in virtue of the action of mechanical force.

7. From considerations of this kind it appears to follow that a complete explanation of all phenomena would introduce no principles beyond those with which the science of mechanics is conversant. And in truth if the conclusion drawn had been that all phenomena might, if our knowledge of nature were sufficiently extensive, be reduced to cinematical considerations (using the word cinematics in the large sense in which it is equivalent to the doctrine of motion), I do not see how on our fundamental hypothesis we could refuse to assent to it. But the conclusion drawn by the maintainers of the all-sufficiency of a mechanical philosophy is something different from this—and as I conceive the error they appear to have committed is to be sought for in this discrepancy. But before entering into the discussion of this point, I will make a few remarks on certain points in the history of what may be called the theory of matter.

* That is, all that are commonly enumerated as primary qualities.

† As, for instance, in the phenomenon of crystallization.

8. If we suppose the maxim that secondary qualities are to be explained by means of the primary to have been accepted (either in that or in some equivalent form), or if not formally accepted, at least unconsciously assumed, at a time when the idea of mechanical force was as yet very imperfectly apprehended—the natural result of this state of things is the formation of an atomic theory. For in order to individuate the constitution of any given body, we could only have had recourse to the configuration or motion of its parts. Gold, to return to our previous example, was said to be yellow in virtue of such and such a configuration of its parts; since except configuration there appeared to be no disposable circumstance*, if I may so speak, whereby gold was in its intimate constitution to be distinguished from silver or from anything else. But this configuration must be independent of the body's visible and external form, since changes of the latter do not affect the body's sensible qualities. Hence it must be a configuration of small parts, and we are thus at once led to the primitive form of the atomic theory. In this the atoms possess the primary qualities of larger bodies — they are of various forms and act if the expression may be used by their forms, not by being centres of attractive forces. Such was the atomistic system of the school of Democritus†—a system which we know found no little favour among the scientific reformers of the seventeenth century‡. As an instance of the influence it exerted, I need only mention the great work of Cudworth, in which it is presented apart from the atheistical doctrines with which it had often been connected. Cudworth goes so far as to affirm that Democritus and his followers had corrupted and degraded the atomistic system which was originally altogether free from any irreligious tendency, and which he sought to restore to its first estate.

But as the imperfections of the atomic system became mani-

* Specific differences of motion seem for more than one reason not to have been used in giving an account of the differences of bodies.

† See for a more favourable and, I think, a juster view of the philosophy of Democritus than that which we commonly meet with in the writings of modern historians of philosophy, Zeller's *Philosophie Der Griechen*, I. § 10.

‡ The physical theories of Des Cartes, though not properly atomistic, since he proceeded on the hypothesis of a plenum, yet in many respects are akin to those of which we are speaking.

fest, and on the other hand mechanical conceptions came to be more developed, a new form of this system arose. The atoms, retaining their forms and those which are commonly called their primary qualities, were now supposed to act as centres of attractive force, in other words, each atom was to the rest a cause of motion. But as the ordinary "primary qualities" of bodies may as we have seen be analysed into conceptions which involve nothing beside motion and force, this new form of the doctrine may clearly be considered merely as a state of transition to that which is now known by the title of Boscovich's theory *. To Boscovich appears to belong the credit of having perceived that if the atoms were conceived of simply as unextended centres of force the primary qualities of bodies might sufficiently be accounted for without supposing them to result from the primary qualities of their constituent atoms—a mode of explanation of which, though there has been something like a return to it in some recent speculations, it may be observed that it explains nothing. Boscovich's theory seems to have been so completely in accordance with the direction in which mathematical physics have of late been moving, that it was adopted as it were unconsciously—almost all modern investigations on subjects connected with molecular action are in effect based on his views, though his name is, comparatively speaking, but seldom mentioned. And this theory, (whether or not the hypothesis of the existence of *discrete* centres of action be or be not essential to it, a question connected with that which in former times caused so much perplexity, namely, the nature of continuity, and which it is not necessary to my present purpose to consider), is in truth the highest developement which the mathematical theory of matter has as yet received—it is that on which the pretensions of mathematical physicists to vindicate for their own methods the right, so to speak, if not the power, to explain all phenomena mainly depend. Adopting for the sake of definite conception the received form of this theory, that

* It is, I believe, known that Boscovich's fundamental idea was deduced by a not unnatural filiation from the monadism of Leibnitz. Yet the scope and limits which he proposed to himself differ essentially from those of the German philosopher, inasmuch as they are essentially physical. Moreover, the latter would have objected on the principle of sufficient reason to the want of any thing to individuate the atoms of Boscovich; and, at least in the latter years of his life, to the "Ferne Wirkung," on which the whole theory depends.

namely in which the centres of force are discrete and at insensible distances from each other, I now shall attempt to show what ulterior developements it admits of, and how by means of these the error noticed at the close of the last Section, namely, the confounding the admission that all phenomena are to be explained *cinematically* with the assertion that they can all be explained *mechanically* may be met, and, as it seems to me, sufficiently refuted.

9. I begin by observing that though we speak and shall continue to do so of the *action* of matter on matter, yet that no part of the views I am about to state depends on the hypothesis we adopt touching the nature of causation. They would remain unchanged whether we accept a theory of pre-established harmony, or one of physical influence, or whether we abstain from all theories on the subject. This being understood, we may, I think, lay down the axiom that whatever property we ascribe to matter, we may also ascribe to it, the property of producing in other portions of matter the former property. Of this axiom the present state of Boscovich's theory affords a familiar illustration. Every portion of matter is locally moveable, therefore we may ascribe to any portion of matter the power of producing motion in any other, hereby giving rise to the whole doctrine of attractive and repulsive forces. At this point we have hitherto stopped, but for no satisfactory reason. We may proceed farther, and we are therefore bound, in constructing the most general possible hypothesis, to do so: we may ascribe to each portion of matter the power of engendering in any other that which we call force, in other words the power of producing the power of actuating the potential mobility of matter. It is not *a priori* at all more easy to conceive that A should have the power of setting B in motion, or of changing the velocity it already has, than that C should have the power of enabling A to act on B, or of changing the mode of action which A already possesses. And let it be observed, that the new power thus ascribed to C is as distinct from force, as force is from velocity. The two are related as cause and effect, but formally are wholly independent. Now unless this hypothetically possible mode of action can be shown to have no existence *in rerum naturâ*, it is clear that the inference from the conclusion that no phenomenon can be

imagined not resoluble *en dernière analyse,* into local motion to the assertion that mechanical force is the only agency to be recognised in the material universe is altogether illusory. For matter may act on matter in a manner wholly distinct from force, and yet this kind of action shall, ultimately and indirectly, manifest itself in modifications of local motion. Furthermore, if for an instant we call this kind of action $(\text{force})^2$, we shall at once be led to recognise a hypothetically possible mode of action of matter on matter which in accordance with analogy we shall call $(\text{force})^3$, which consists in the power of modifying $(\text{force})^2$. And so on, *sine limite.*

10. If we compare the language in which the relation between mechanical force and chemical affinity is commonly spoken of, we shall I think perceive its analogy with that which I have used in describing the mode of action which we have called $(\text{force})^2$. Its chemical affinity is spoken of as something which suspends or modifies the action of force, as something distinct from it, but which yet interferes with its effects. Or again, if in physiological writings we observe the manner in which vital action* is described we recognise, or seem at least to do so, the possibility of referring its effects to that mode of action which we have called $(\text{force})^3$. I do not however wish to lay much stress on these similarities, because I think the kind of reasoning we have pursued shows more satisfactorily than they can do, that if chemical affinity and vital action are not resoluble into force, they must be referred to some of the modes of action we have pointed out.

It would be useless to remark on the many points of speculation which here present themselves. The expansion of bodies by heat may however be particularly mentioned, because notwithstanding what has been learnt with relation to the theory of heat, nothing like a mechanical explanation of this phenomenon has as yet been discovered. It seems to depend not on the introduction of new mechanical forces, but on a modification of those which already exist; such modification, in cases of ordinary conduction, being propagated from one part of the body to that which is next it.—It is easy to conceive that by an altera-

* I am, of course, not to be understood as suggesting a materialistic explanation of phenomena of thought or volition.

tion in the function which expresses the mutual action of the molecules, the body may pass into a new state of equilibrium in which the average distance between adjacent molecules may be increased or diminished. If such an explanation could be established, we should have a case of the action of $(\text{force})^2$.

11. In conclusion, it may be well to remark that mathematical analysis is conceivably as applicable to these new modes of action of matter on matter as to ordinary questions in dynamics. It is, however, easily seen that as in these we deal chiefly with differential equations of the second order, and in merely cinematical questions with equations of the first only, so contrariwise when we introduce higher powers of force (so to call them) we shall correspondingly have to do with equations of higher orders. I venture to predict with a degree of confidence, which doubtless I shall not communicate to many, that if we ever succeed in establishing a mathematical theory of chemistry, it will be as much conversant with equations of the third or of a higher order, as physical astronomy is with equations of the second.

REMARKS
ON THE FUNDAMENTAL PRINCIPLE OF THE THEORY OF PROBABILITIES*.

I WISH to make an addition to the remarks on the foundation of the theory of probabilities which were offered some years since to the notice of the Society†. My intention in doing so is to consider, in what way the proposition, which I conceive to be the fundamental principle of the theory, may be the most clearly and conveniently expressed. This principle may for the moment be thus stated: "On a long run of similar trials, every possible event tends ultimately to recur in a definite ratio of frequency." Our conviction of the truth of this proposition is, I think, intuitive,—the word being used, as in all similar cases, with reference to the intuitions of a mind, which has fully and clearly apprehended the subject before it, and to which therefore to have arrived at the truth and to perceive that it has done so are inseparable elements of the same act of thought. If we endeavour to translate the proposition just stated into ordinary philosophical language, we may in the first place remark that the phrase "similar trials," expresses the notion of a group or genus of phenomena to which the different results are subordinated as distinct species. If the trial is the throwing of a die, this may be regarded as the generic character; the occurrence of ace, deuce, &c. constituting different species. Thus much is clear; but it is less obvious how the idea expressed by a "long run of trials" in "definite series of experiments," and the like, is to be expressed, so as to make the analogy between the fundamental principle of the theory of probabilities and those of other sciences more obvious than it has hitherto been. The idea in question

* *Transactions of the Cambridge Philosophical Society*, Vol. IX. p. 605. [Read Nov. 13, 1854.]

† *Transactions of the Cambridge Philosophical Society*, Vol. VIII. p. 1. [p. 1 of this volume.]

is not readily expressed in any way, because in its own nature it is negative and indefinite. The phrases I have just quoted imply merely the absence of the limitations inseparable from individual cases, or from any finite number of such cases, whether contemplated as actually existent or as about to be developed within definite limits of space and time.

When individual cases are considered, we have no conviction that the ratios of frequency of occurrence depend on the circumstances common to all the trials. On the contrary, we recognise in the determining circumstances of their occurrence an extraneous element, an element, that is, extraneous to the idea of the genus and its species. Contingency and limitation come in (so to speak) together; and both alike disappear when we consider the genus in its entirety, or (which is the same thing), in what may be called an ideal and practically impossible realization of all which it potentially contains. If this be granted, it seems to follow that the fundamental principle of the theory of probabilities may be regarded as included in the following statement;—"The conception of a genus implies that of numerical relations among the species subordinated to it."

2. But in what relation, it may be asked, do these conceptions stand to outward realities? How can they be made the foundation of a real science, that is, of a science relating to things as they really exist? We are by such questions led back to what was long the great controversy of philosophy;—I mean the contest between the realists and the nominalists. The former in asserting the reality of universals did not maintain that what we think of when we use a general term is an actually existing thing. Like every one else they admitted, that in one sense nothing can exist but the individual, nevertheless they held that universals are not mere figments of the mind, but that they have a reality of their own which is the foundation of the truth of general propositions. To assert therefore that the theory of probabilities has for its foundation a statement touching genera and their species, and is at the same time a real science, is to take a realistic view of its nature. And this I believe is what, on consideration, we cannot avoid doing.

If it be said that the grouping phenomena together is merely a mental act wholly disconnected from outward reality and altogether arbitrary, it may be replied that no mental act can be so. Why and how facts and ideas correspond is no doubt one of the great questions of philosophy; but the answer to it is surely to be developed from the consideration, that man in relation to the universe is not *spectator ab extra*, but in some sort a part of that which he contemplates, and that the *rebus avolsa ratio*, which is in truth the fundamental postulate of nominalism, is therefore inconcessible. The thoughts we think are, it is true, ours, but so far as they are not mere error and confusion, so far as they have anything of truth and soundness, they are something and much more. The *veritas essendi* (to recur to the language of the schoolmen) is the fountain from whence the *veritas cognoscendi* is derived. The meaning which these phrases were intended to convey is expressed in more modern language by Leibnitz in the passage which I have cited in the note*. In every science the fact and the idea correspond because the former is the realization of the latter, but as this realization is of necessity partial and incomplete—or rather because in the same fact are simultaneously realized a variety of separate ideas, separate, that is, as we conceive them—this correspondence is but imperfect and approximate. It is only when in thought we remove the action of disturbing causes to an indefinite distance, that we can conceive the absolute verification of any *à priori* law. Only on the horizon of our mental prospect earth and sky, the fact and the idea, are seen to meet, though in reality the atmosphere is everywhere present. Everywhere it surrounds and interpenetrates the γῆ μέλαινα on which we stand;—making it put forth and sustain all the numberless forms of organization and of life. The indefinitely prolonged series of trials, which enters into the ordinary statement of the fundamental principle of the theory of probabilities, is analogous to the infinite and infinitely smooth horizontal plane, which would enable us to verify the first law of motion.

3. The simple negative notion of the absence of disturbing forces is perpetually confounded with that of a tendency inherent

* C'est Dieu qui est la dernière raison des choses, et la connaissance de Dieu n'est pas moins le principe des sciences, que son essence et sa volonté sont les principes des êtres. [Erdmann, p. 106.] A little further on he adds: C'est sanctifier la philosophie, que de faire couler ses ruisseaux de la fontaine des attributs de Dieu.

in a series of successively developed results to restore the balance of frequency of occurrence, when this has been by accidental circumstances temporarily deranged. It is commonly thought that this notion, which, as we know, is the foundation of many unsuccessful attempts to circumvent fortune, is sufficiently refuted by saying, that what is past can exert no influence on what is yet to come. But in reality the past influences the future in a thousand different ways; and it is only in idea that we can secure the possibility of an indefinite series of trials, of which those which we regard as the permanent circumstances are not progressively, however slowly, undergoing alteration. The dice box for example wears smooth, and the edges of the die are rounded; and though, in this example, we cannot say what result is facilitated by the change, yet this is not always the case. Such progressive alterations may tend so to alter the ratio of frequency of occurrence, as to restore the balance which the result of past trials has disturbed. There is thus nothing absurd in the notion of a restorative and balancing tendency, though the grounds on which it is commonly assumed indicate much confusion of thought. It would for instance be perfectly reasonable to inquire, whether in the succession of seasons hot years are not oftener followed by cold and cold by hot, than *vice versâ*. Such questions indicate a branch of the theory of methods of observation to which hitherto but little attention has been paid.

REMARKS ON AN ALLEGED PROOF OF THE METHOD OF LEAST SQUARES, CONTAINED IN A LATE NUMBER OF THE *EDINBURGH REVIEW*. IN A LETTER ADDRESSED TO PROFESSOR J. D. FORBES*.

MY DEAR SIR,

THE review of Quetelet's *Lettres à S. A. R. le Duc régnant de Saxe Cobourg et Gotha*, which appeared in the July Number of the *Edinburgh Review*, contains a new demonstration of the method of least squares which ought not, I think, to pass unnoticed. If it is correct, it is so much simpler than those which have hitherto been received, that it ought to supersede them; and if not, the sooner its incorrectness is pointed out the better.

Some years since, in a paper published in the *Cambridge Transactions* for 1844, I made an analysis of all the demonstrations, or professed demonstrations of the method of least squares, with which I was then acquainted, and I therefore read this new one with more attention than you perhaps have given to it.

The reviewer gives some account of the history of the subject, and remarks that the demonstration of the least squares was first attempted by Gauss, but that his proof is no proof at all, because it assumes that in the case of a single element the arithmetical mean of the observed values is in all cases the most probable value, "a thing to be demonstrated, not assumed." Gauss afterwards gave another demonstration, which is perfectly rigorous; but of this the reviewer takes no notice, though it is mentioned in at least one of the works on the theory of probabilities which he has recommended to the attention of students. However, in the proof which the reviewer refers to, which is contained in the tract entitled *Theoria*

* *Philosophical Magazine*, November, 1850.

Motûs Elliptici, Gauss undoubtedly does assume that the arithmetical mean is the most probable value in the case of direct observations of a single element. From this assumption, he shows that the probability, that the magnitude of an error lies between x and $x + dx$, must be

$$\frac{h}{\sqrt{\pi}} e^{-h^2x^2} dx,$$

h being an indeterminate constant. It follows from this, that the results of the method of least squares are always the most probable values that can be assigned to the unknown elements. Without referring to the *Theoria Motûs*, you can see the details of Gauss's reasoning in a paper by Bessel, of which a translation appeared in Taylor's *Scientific Memoirs*. The reviewer is right in saying that Gauss was not entitled to assume that the arithmetical mean is the most probable value. But when he speaks of this as a thing to be proved, and not assumed, we are led to suppose that he believes that subsequent writers have actually proved it. In truth this appears, not only from his statements, but also from the illustrations of which he has made use. Thus he states that if shots are fired at a wafer which is afterwards removed, and we are asked to determine from the position of the shot-marks the most probable position of the wafer, "the theory of probabilities affords a ready and precise rule, applicable not only to this but to far more intricate cases;" and he goes on to say that it may be shown that the most probable position of the wafer is the centre of gravity of the marks. Now this result is only then true when the law of probability of error, which is implied in Gauss's assumption, really obtains; so that, according to the reviewer, the demonstration of the principle of least squares must amount to showing that this law obtains universally; or, which is the same thing, that the arithmetical mean is always the most probable value in the case of direct observations of a single element. If this can be proved, it is doubtless a very curious conclusion; but it is at any rate certain that Laplace has not proved it, of whom however the reviewer asserts that he has given a rigorous demonstration of the principle of least squares. From one end of Laplace's great work to the other, there is nothing to justify the assertion that the centre of gravity of the shot-marks is the most probable position that can be assigned for the wafer; that is, that the concurrent existence of the deviations or errors which must have taken place if the wafer really occupied this position is more probable than that of those which are similarly implied in any other hypothesis as to its place. If you find anybody sceptical as to this, pray ask them to point out the passage, either in the introductory essay, or in the work itself, or in the supplements.

What, then, did Laplace demonstrate? something so unlike this, that one is disposed to wonder how he can have been thus misunderstood. The method of least squares is simply a method for the combination of linear equations, of which the unknown quantities are the elements to be determined; the constant term of each being a direct result of observation, and therefore affected by an unknown error, while the coefficients are supposed absolutely known.

If there are more equations than requisite, that is, more than elements to be determined, what is the best way of combining them? In the first place, they must clearly be combined by some system of constant multipliers, else the resulting equations, not being linear, would generally be insoluble. This condition, however, though absolutely necessary in practice, is in no way derived from the theory of probabilities. It is a merely practical limitation. The question thus narrowed is simply to determine the system of factors to be employed for obtaining the value of any particular element. The factors must of course be such that, in the final equation, the coefficient of this element may be unity, and those of the others severally equal to zero.

These conditions being fulfilled, we get a value for the element in question which is affected by an unknown error, namely the sum of the errors of observation multiplied respectively by the corresponding factors. The mean arithmetical value of this sum may in theory at least be determined, if we know the law of probability of error for each observation; and Laplace calls that system of factors the most advantageous which makes this mean value a minimum. If, however, the law of probability of error is unknown, the mean value of the error cannot be determined. Nevertheless, if the number of observations is very large, this mean value approximates to a certain limit, the form of which is independent of the law of probability. The essence of Laplace's demonstration consists in its enabling us to determine this limit. When this is done, it may easily be shown that the most advantageous system of factors, those, namely, which make this limiting mean value of the error a minimum, will give the same value to the element to be determined as the system of final equations obtained by employing the method of least squares, provided equal positive and negative errors are equally probable. And the same is of course true with respect to the remaining elements. Thus this system of final equations gives to each element a value affected by a smaller average error than any other linear system, if the number of observations is sufficiently large. It nowise follows that these values are the most probable; that is, that the errors which must have been committed if these are the true values, form a combination *à priori* more probable than the errors which in like manner have been committed if any other set of values are the true ones. The most advantageous set of factors for determining any element depends only on the coefficients of the equations to be discussed, and not on their constant terms, which are the direct result of observation. Thus these factors are determinable, *à priori*, before the observations are made. But it is only after the observations have been made that the most probable values of the elements can be found, and then only if we know the law of probability of error. Laplace has pointed out the difference between the two investigations.

This difference, however, the reviewer does not seem to have apprehended. He plainly supposes that Laplace proves the results of the method of least squares to be the most probable results, which can only be the case, as Gauss had in effect shown, if a special law of error obtains. He therefore undertakes to prove, that for all kinds of observations this is actually the only possible law.

But for the supposed authority of Laplace, he would probably have perceived that nothing can be more unlikely than that the errors committed in all classes of observations should follow the same law; and that at any rate this proposition, if true, could only be proved inductively, and not by an *à priori* demonstration. For it is beyond question distinctly conceivable, that different laws may exist in different classes of observation; and that which is distinctly conceivable is *à priori* possible. So that we cannot prove it to be impossible, though we may be able to show empirically that it is not true.

You will probably agree with me in thinking that a wrong notion of Laplace's reasoning lies at the root of the reviewer's new demonstration. But we now come to the demonstration itself. The assumption that the law of error is in all cases the same, is, we are told, justified by our ignorance of the causes on which errors of observation depend. The law "must necessarily be general, and apply alike to all cases, since the causes of error are supposed alike unknown in all." Two remarks are suggested by this statement: in the first place, that our ignorance of the causes of error is not so great but that we have exceedingly good reason to believe that they operate differently in different classes of observations; and in the second, that mere ignorance is no ground for any inference whatever. *Ex nihilo nihil.* It cannot be that because we are ignorant of the matter we know something about it. Or are we to believe that the assumption is legitimate, inasmuch as it in a manner corresponds to and represents our ignorance? But then what reason have we for believing that it can lead us to conclusions which correspond to and represent outward realities? And yet the reviewer at the conclusion of his proof asserts, that, on the long run, and *exceptis excipiendis,* the results of observation "will be found to group themselves......according to one invariable law." Thus the assumption, though "it is nothing more than the expression of our state of complete ignorance of the causes of error and their mode of action," leads us by a few steps of reasoning to the knowledge of a positive fact, and makes us acquainted with a general law, which is as independent of our knowledge or our ignorance as the law of gravitation.

Let us, however, suppose it to be true that the law of error is always the same, and that equal positive and negative errors are equally probable. To determine the special form of the law, the reviewer employs a particular case—he supposes a stone to be dropt with the intention that it shall fall on a given mark. Deviation from this mark is error; and the probability of an error r may be expressed by the function $f(r^2)$ or $f(x^2+y^2)$, the origin of co-ordinates being placed at the mark. It is of course supposed that equal errors in all directions are equally probable. We have now only to determine the form of f. This the reviewer accomplishes in virtue of a new assumption, namely, that the observed deviation is equivalent to two deviations parallel respectively to the co-ordinate axes, "and is therefore a compound event of which they are the simple constituents, therefore its probability will be the product of their separate probabilities. Thus the form of our unknown function comes to be determined from this condition, viz. that the product of such functions of two independent elements is equal to the same function of their sum." Or in other words, we have to solve the functional equation

$$f(x^2)\,f(y^2)=f(x^2+y^2).$$

But it is not true that the probability of a compound event is the product of those of its constituents, unless the simple events into which we resolve it are independent of each other; and there is no shadow of reason for supposing that the occurrence of a deviation in one direction is independent of that of a deviation in another, whether the two directions are at right angles or not. Some notion of an analogy with the composition of forces probably prevented the reviewer from perceiving that, unless it can be shown that a deviation y occurs with the same comparative frequency when x has one value as when it has another, we are not entitled to say that the probability of the concurrence of two deviations x and y is the product of the probabilities of each. Without this subsidiary proof, the rest of the demonstration comes to nothing. The conclusion to which it leads is in itself a *reductio ad absurdum.* Of the above written functional equation the solution is $f(x^2)=e^{mx^2}$, m being a constant, so that the probability of an error of the precise magnitude x is a finite quantity; and I need not point out to you that it follows from hence, that the probability of an error whose magnitude lies between any assigned limits is equal to infinity,—a result of which the interpretation must be left to the reviewer. He may have thought that the exponential factor is the essential part of the expression

$$\frac{h}{\sqrt{\pi}} e^{-h^2x^2}\, dx,$$

and that the others might, for the sake of simplicity, be dropt out. But whatever his views may have been, his conclusion is unintelligible.

The demonstration may, however, be amended so as to avoid this difficulty, and we will suppose that the reviewer meant something different from what he has expressed. Let $f(x^2)\,dx$ be the probability of a deviation parallel to the axis of abscissæ, of which the magnitude lies between x and $x+dx$. Then $f(y^2)\,dy$ is similarly the probability of a deviation parallel to the axis of ordinates, and lying between y and $y+dy$. Thus the probability that the stone drops on the elementary area $dxdy$, of which the corner next the origin has for its co-ordinates x and y, seems to be $f(x^2)\,f(y^2)\,dxdy$; and as all deviations of equal magnitude are equally probable, this probability must remain unchanged as long as the sum of the squares of x and y remains the same; so that we have for determining the unknown function the equation

$$f(x^2)\,f(y^2) = f(0)\,f(x^2+y^2),$$

of which the solution is

$$f(x^2) = Ae^{mx^2};$$

and as the deviation must of necessity have some magnitude included between positive and negative infinity, we must have

$$A\int_{-\infty}^{+\infty} e^{mx^2}\,dx = 1.$$

Hence m must be negative; if we call it $-h^2$, it is easy to show that A is equal to $\frac{h}{\sqrt{\pi}}$; so that finally

$$f(x^2) = \frac{h}{\sqrt{\pi}} e^{-h^2x^2},$$

which is what may be called Gauss's function.

But to this demonstration, though it leads to an intelligible conclusion, the original objection still applies: the probability that the stone drops on the elementary area $dxdy$ is not, generally speaking, equal to $f(x^2) f(y^2)\, dxdy$; so that the equation for determining the form of the function, namely,

$$f(x^2) f(y^2) = f(0) f(x^2 + y^2),$$

is not legitimately established.

To illustrate this, let $\varpi(xy)\, dxdy$ be the probability that the stone falls on the elementary area in question; then the condition that the probability of a deviation of given magnitude is constant will be expressed by

$$\varpi(xy) = \varpi(\sqrt{x^2 + y^2} \,.\, 0) \ldots\ldots\ldots\ldots\ldots\ldots\ldots\ldots (A).$$

Moreover, we shall plainly have

$$f(x^2) = \int_{-\infty}^{+\infty} \varpi(xy)\, dy$$

and

$$f(y^2) = \int_{-\infty}^{+\infty} \varpi(xy)\, dx;$$

and in order that the demonstration may be valid, we must have

$$f(x^2) f(y^2) = \varpi(xy),$$

or

$$\int_{-\infty}^{+\infty} \varpi(xy)\, dy \int_{-\infty}^{+\infty} \varpi(xy)\, dx = \varpi(xy) \ldots\ldots\ldots\ldots\ldots (B).$$

If this be true, then, and then only, equation (A) may be replaced by

$$f(x^2) f(y^2) = f(0) f(x^2 + y^2).$$

But in order that (B) may be true, $\varpi(xy)$ must evidently be the product of two factors; one of them a function of y only, and the other of x, and the integral of each factor taken between infinite limits must be equal to unity. Combining this conclusion with (A), we find that

$$\varpi(xy) = \frac{h^2}{\pi} e^{-h^2(x^2+y^2)},$$

and consequently

$$f(x^2) = \frac{h}{\sqrt{\pi}} e^{-h^2x^2}.$$

Consequently the equation for determining the form of f results from a tacit predetermination of that function.

The assumption expressed by

$$f(x^2)f(y^2) = \varpi(xy)$$

is therefore either a simple mistake or a *petitio principii:* the former, if it is deduced from the general principle that the probability of a compound event is equal to the product of those of its elements; the latter, if it is made to depend on the particular form assigned to $f(x^2)$.

After all, too, if the demonstration were right instead of wrong, it would not prove what is wanted. For if the law of probability of a deviation parallel to a fixed axis is expressed by the function

$$\frac{h}{\sqrt{\pi}} e^{-h^2x^2}\,dx,$$

which is what the amended demonstration tends to show, the probability that the stone falls on the area $dxdy$ is plainly

$$\frac{h^2}{\pi} e^{-h^2(x^2+y^2)}\,dxdy.$$

Transforming this to polar co-ordinates, and integrating from 0 to 2π for the angle vector, we get $2h^2e^{-h^2r^2}\,rdr$ for the probability that the deviation from the mark lies between r and $r+dr$; a result which may be verified by integrating for r from zero to infinity, the integral between these limits being equal to unity. Thus if the deviations measured parallel to fixed axes follow the law which the reviewer supposes to be universally true, the deviations from the centre or origin follow quite another; and hence it appears that his illustration is altogether wrong. For if $2h^2e^{-h^2r^2}\,rdr$ is the probability of an error lying between r and $r+dr$, the centre of gravity of the shot-marks is not the most probable position of the wafer. So that his hypothesis is self-contradictory.

The original source of his error was probably the analogy between Gauss's law, and the limiting function in Laplace's investigation.

I am, my dear Sir,

Most truly yours,

R. L. Ellis.

Brighton, *Sept.* 19.

NOTE TO A FORMER PAPER ON AN ALLEGED PROOF OF THE METHOD OF LEAST SQUARES*.

To the Editors of the Philosophical Magazine and Journal.

GENTLEMEN,

ALLOW me to correct an error in my letter to Professor Forbes, published in your last Number. The Edinburgh reviewer, on whose proof of the method of least squares I was commenting, says that the most probable position of the wafer is the centre of gravity of the shot-marks; of course on the supposition that in this, as in all other cases, the probability of a deviation or error r is equal or proportional to a certain constant base raised to the power $-r^2$.

Now, admitting this supposition to be true, the centre of gravity is not the most probable position of the wafer. But, on the contrary, if the function mentioned at the close of my former communication, viz. $2h^2e^{-h^2r^2}rdr$, expresses the probability of an error r, then the centre of gravity *is* the most probable position. I thus not only omitted to notice that the reviewer's conclusion would not follow from his own hypothesis, but by this omission was led to introduce an error of my own.

It is unnecessary to trouble you with a proof of what I have now said, as the matter does not affect the general question.

I am, Gentlemen,

Your obedient Servant,

R. L. ELLIS.

BRIGHTON, *Nov.* 7.

* *Philosophical Magazine*, December, 1850.

ON SOME PROPERTIES OF THE PARABOLA*.

THERE are many very interesting properties of the Conic Sections which are not to be found in the usual works on the subject, but are scattered through various memoirs in scientific Journals. Those relating to the properties of polygons inscribed in and circumscribed round conic sections, have been investigated by a great many writers both in France and England. Pascal was the first who engaged in these researches, and was led by the curious properties which he discovered to call one of these polygons the "hexagramme mystique." After him Maclaurin gave a proof of a theorem which is not only beautiful in itself, but also very fertile in its consequences. In more recent times Brianchon has demonstrated the remarkable theorems, that in all hexagons either inscribed in or circumscribed round a conic section, the three diagonals joining opposite angles will intersect in one point. Subsequently, Davies in this country, and Dandelin in Belgium, proved in different ways the same propositions along with others. The latter adopted a very peculiar method, deducing these and many other properties of sections of the cone by considering the cone as a particular case of the "hyperboloide gauche." Generally speaking, the Geometrical method is more easily applied than the Analytical to these cases, and accordingly all the proofs given have depended on geometry, with the exception of one published by Mr Lubbock in the Number of the Philosophical Magazine for August 1838. He has there demonstrated, by analysis, Brianchon's Theorem for a circumscribing hexagon in the particular case where the conic section is a parabola; but his method is tedious, and not remarkable for symmetry and elegance, so that another proof is still desirable. The following one is founded on the

* *Cambridge Mathematical Journal*, No. V. Vol. I. p 204, February, 1839.

form of the equation to the tangent of the parabola which is given in Art. 2 of our first Number*.

Let the parabola be referred to its vertex, then the equation to its tangent by that article is

$$y = \frac{x}{\alpha} + m\alpha,$$

where α is the tangent of the angle which the tangent makes with the axis of y. If α' be the corresponding quantity for another tangent, its equation will be

$$y = \frac{x}{\alpha'} + m\alpha'.$$

Combining these equations, we shall find for the co-ordinates of the point of intersection of the two tangents

$$x = m\alpha\alpha', \quad y = m(\alpha + \alpha').$$

We shall distinguish the tangents which form the different sides of the hexagon by suffixing numbers to the α which determines their position, and we shall likewise distinguish the co-ordinates of the summits of the hexagon by suffix letters.

The equations to the three diagonals are these:

(1) $$y(\alpha_4\alpha_5 - \alpha_1\alpha_2) - x(\alpha_4 + \alpha_5 - \alpha_1 - \alpha_2) = m\{(\alpha_1 + \alpha_2)\alpha_4\alpha_5 - (\alpha_4 + \alpha_5)\alpha_1\alpha_2\}.$$

(2) $$y(\alpha_5\alpha_6 - \alpha_2\alpha_3) - x(\alpha_5 + \alpha_6 - \alpha_2 - \alpha_3) = m\{\alpha_2 + \alpha_3)\alpha_5\alpha_6 - (\alpha_5 + \alpha_6)\alpha_2\alpha_3\}.$$

(3) $$y(\alpha_6\alpha_1 - \alpha_3\alpha_4) - x(\alpha_6 + \alpha_1 - \alpha_3 - \alpha_4) = m\{(\alpha_3 + \alpha_4)\alpha_1\alpha_6 - (\alpha_1 + \alpha_6)\alpha_3\alpha_4\}.$$

Expressions which, as they ought to be, are symmetrical with respect to the α's.

Multiply (1) by α_6, (2) by $-\alpha_4$, (3) by α_2, and add. Then y will disappear, and we shall find

$$x = m\,\frac{\alpha_3\alpha_6(\alpha_4\alpha_5 - \alpha_1\alpha_2) - \alpha_4\alpha_1(\alpha_5\alpha_6 - \alpha_2\alpha_3) + \alpha_5\alpha_2(\alpha_6\alpha_1 - \alpha_3\alpha_4)}{\alpha_1\alpha_2 - \alpha_2\alpha_3 + \alpha_3\alpha_4 - \alpha_4\alpha_5 + \alpha_5\alpha_6 - \alpha_6\alpha_1}.$$

Again, multiply (1) by α_3, (2) by $-\alpha_1$, (3) by α_5, and add: as before, y will disappear, and we shall find the same value for x. Consequently two straight lines whose equations are

* *Cambridge Mathematical Journal,* Art. 2, No. I. Vol. I. p. 9.

$$(1)\ \alpha_6 - (2)\ \alpha_4 = 0,$$
$$\text{and } (1)\ \alpha_3 - (2)\ \alpha_1 = 0,$$

and which have a point in common, cut (3) in points whose abscissæ are equal, and which therefore coincide. Hence either two straight lines enclose a space, or (3) passes through the intersection of (1) and (2). Thus the existence of the point common to the three diagonals has been proved, and its abscissa found. To determine its ordinate, add (1), (2), (3), when x disappears, and we have

$$y = m\,\{\alpha_1\alpha_2(\alpha_4+\alpha_5) - \alpha_2\alpha_3(\alpha_5+\alpha_6) + \alpha_3\alpha_4(\alpha_6+\alpha_1) - \alpha_4\alpha_5(\alpha_1+\alpha_2) + \alpha_5\alpha_6(\alpha_2+\alpha_3) + \alpha_6\alpha_1(\alpha_3+\alpha_4)\}$$

divided by

$$\alpha_1\alpha_2 - \alpha_2\alpha_3 + \alpha_3\alpha_4 - \alpha_4\alpha_5 + \alpha_5\alpha_6 - \alpha_6\alpha_1.$$

If we call the co-ordinates of the point where the third and sixth sides of the hexagon meet $x_{\text{III}}\ y_{\text{III}}$, and so of the other two points, these expressions for x and y become

$$x = \frac{x_{\text{III}}(x_4 - x_1) - x_{\text{IV}}(x_5 - x_2) + x_{\text{V}}(x_6 - x_3)}{x_1 - x_2 + x_3 - x_4 + x_5 - x_6}$$

$$y = \frac{x_1y_4 - x_2y_5 + x_3y_6 - x_4y_1 + x_5y_2 - x_6y_3}{x_1 - x_2 + x_3 - x_4 + x_5 - x_6}.$$

These expressions, as of course we should expect, are symmetrical.

In the last Number of this Journal a demonstration was given of a property of a parabola: That the circle which passes through the intersections of three tangents also passes through the focus. Although six demonstrations of this theorem have already appeared, yet the following is so simple that its insertion here may not be inappropriate.

Referring the parabola to the focus as origin, we can put the equation to the tangent under the form

$$y - \frac{x}{m} = a\left(m + \frac{1}{m}\right),$$

where a is one-fourth of the parameter, and m the trigonometrical tangent of the angle which the tangent makes with the axis of y. Hence if x_1, y_1 be the co-ordinates of the point of intersection of

$$y - \frac{x}{m} = a\left(m + \frac{1}{m}\right),$$

$$\text{with } y - \frac{x}{m'} = a\left(m' + \frac{1}{m'}\right),$$

$$\text{we have } x_1 = a\,(mm' - 1),$$

$$y_1 = a\,(m + m'),$$

$$\text{or putting } m = \frac{\sin\alpha}{\cos\alpha},\quad m' = \frac{\sin\alpha'}{\cos\alpha'},$$

$$x_1 = a\,\frac{\cos(\alpha + \alpha')}{\cos\alpha\cos\alpha'},$$

$$y_1 = a\,\frac{\sin(\alpha + \alpha')}{\cos\alpha\cos\alpha'}.$$

To simplify these expressions turn the axes through an angle $= -(\alpha + \alpha' + \alpha'')$, and if x'', y'' be the new values of the co-ordinates, we find, after some simple reductions,

$$x'' = \frac{a\cos\alpha''}{\cos\alpha\cos\alpha'},\quad y'' = -\frac{a\sin\alpha''}{\cos\alpha\cos\alpha'}.$$

Squaring these and adding,

$$x''^2 + y''^2 = \frac{a^2}{\cos^2\alpha\cos^2\alpha'} = \frac{a}{\cos\alpha\cos\alpha'\cos\alpha''}\cdot\frac{a\cos\alpha''}{\cos\alpha\cos\alpha'},$$

$$\text{or } x''^2 + y''^2 = \frac{ax''}{\cos\alpha\cos\alpha'\cos\alpha''}.$$

Now this being symmetrical between α, α', α'', will hold equally true of the three points of intersection, and it is the equation to a circle passing through the origin which is the focus, whose diameter coincides with the axis of x, and whose radius is

$$\frac{a}{2\cos\alpha\cos\alpha'\cos\alpha''}.$$

The chief advantage of this method besides its simplicity is, that it gives us very readily the radius of the circle, and the position of the diameter which passes through the focus.

It is easily seen that the distances from the focus of the three points of intersection of the tangents are respectively

$$r'' = \frac{a}{\cos\alpha\cos\alpha'},\quad r' = \frac{a}{\cos\alpha\cos\alpha''},\quad r = \frac{a}{\cos\alpha'\cos\alpha''}.$$

The area of the triangle formed by the intersection of the tangents, can be expressed by an elegant symmetrical function of

$\tan\alpha$, $\tan\alpha'$, $\tan\alpha''$, that is, of m, m', m''. Since the lines joining the origin with the vertices of the triangle make angles α, α', α'' with the diameter of the circle or the axis of x, the angles they make with each other are $\alpha'-\alpha$, $\alpha''-\alpha'$, $\alpha''-\alpha$, and the area of the triangle will be

$$\tfrac{1}{2}rr'\sin(\alpha'-\alpha)+\tfrac{1}{2}r'r''\sin(\alpha''-\alpha')-\tfrac{1}{2}rr''\sin(\alpha''-\alpha).$$

Substituting for r, r', and r'' their values, this becomes

$$\frac{a^2}{2}\left\{\frac{\sin(\alpha'-\alpha)}{\cos^2\alpha''\cos\alpha\cos\alpha'}+\frac{\sin(\alpha''-\alpha')}{\cos^2\alpha\cos\alpha'\cos\alpha''}-\frac{\sin(\alpha''-\alpha)}{\cos^2\alpha'\cos\alpha\cos\alpha''}\right\}.$$

Expanding the sines and making obvious reductions, we get

$$\frac{a^2}{2}\left\{\frac{\tan\alpha'-\tan\alpha}{\cos^2\alpha''}+\frac{\tan\alpha''-\tan\alpha'}{\cos^2\alpha}+\frac{\tan\alpha-\tan\alpha''}{\cos^2\alpha'}\right\};$$

or grouping differently, and putting $\sec^2\alpha$ for $\dfrac{1}{\cos^2\alpha}$, and so on,

$$\frac{a^2}{2}\{\tan\alpha(\sec^2\alpha'-\sec^2\alpha'')+\tan\alpha'(\sec^2\alpha''-\sec^2\alpha)$$
$$+\tan\alpha''(\sec^2\alpha-\sec^2\alpha')\}.$$

Lastly, putting $1+\tan^2\alpha=1+m^2$ for $\sec^2\alpha$, and so on, we find the area of the triangle to be

$$\frac{a^2}{2}\{m(m'^2-m''^2)+m'(m''^2-m^2)+m''(m^2-m'^2)\},$$

which is quite symmetrical with respect to m, m', m''.

It will be easily seen, that the sides of the triangle are respectively

$$a\frac{m''-m'}{\cos\alpha},\quad a\frac{m-m''}{\cos\alpha'},\quad a\frac{m'-m}{\cos\alpha''}.$$

If these be called p, p', p'', and if ρ be the radius of the circle, by reduction, we obtain

$$p=2\rho\sin(\alpha''-\alpha'),\quad p'=2\rho\sin(\alpha-\alpha''),\quad p''=2\rho\sin(\alpha'-\alpha).$$

If the values of the sines derived from these equations be substituted in the first expression for the area, it becomes

$$\frac{a}{2}\left(\frac{p}{\cos\alpha}+\frac{p'}{\cos\alpha'}+\frac{p''}{\cos\alpha''}\right).$$

ON THE EXISTENCE OF A RELATION AMONG THE COEFFICIENTS OF THE EQUATION OF THE SQUARES OF THE DIFFERENCES OF THE ROOTS OF AN EQUATION*.

THE equation of the squares of the differences of the roots gives the means of ascertaining whether any assigned equation has all its roots real; for if they be so, all the roots of the equation of differences must be real and positive, and consequently, by Descartes' rule of signs, all its coefficients must be alternately positive and negative. Accordingly, Waring applied it to this purpose, and in the *Philosophical Transactions* for 1763 gave the conditions of the reality of all the roots in equations of the fourth and fifth degree.

There will be as many conditions as there are coefficients—that is, as there are units in the degree of the equation of the squares of the differences; and therefore, for an equation of the nth order the equation of the squares of the differences will of course be of the $\frac{n\,.\,n-1}{2}$ th order. Thus, in the third order there would be three conditions, in the fourth, six, and so on.

Lagrange remarked, however, that the number of conditions in these two cases reduced itself to two and three respectively; and he suggested, that a similar simplification might be possible in the ten conditions of the fifth order.

Sturm's theorem, which, however different in form, is still in substance intimately connected with the theory of the equation of the squares of the differences, enables us to ascertain the true number of independent conditions.

By this theorem, we deduce from a given equation $f(x) = 0$ a series of n functions. These, with the original $f(x)$, make

* *Cambridge Mathematical Journal*, No. VI. Vol. I. p. 256, May, 1839.

$n+1$ functions of x. We substitute in them the limits a and b, and the number of changes of sign lost between these limits is the number of real roots of n, which are to be found in this interval. Consequently we have only to write plus and minus infinity in Sturm's functions, to get the whole number of real roots belonging to the equation.

The signs of each function, when $\pm\frac{1}{0}$ is put for x, will of course be that of the first term, supposing each function to be arranged in a series of decreasing powers of x. And if the first term of each be positive, the series of signs at the superior limit will be all permanences, and at the inferior all alternations; that is, all the roots of the equation will be real.

Hence the reality of all the roots depends on the signs of $n+1$ terms. But of these, the sign of $f(x)$ is determined at the limits $\pm\frac{1}{0}$; so is that of $\frac{df(x)}{dx}$, which is Sturm's first function. Consequently there remain but $n-1$ terms on the sign of which the reality of roots depends. Instead, therefore, of $\frac{n \,.\, n-1}{2}$ conditions, there are in reality but $n-1$.

Thus, in the equation of the third degree we find two conditions, in that of the fourth, three, and so on, agreeing with what Lagrange found in these cases, and suspected in that of the fifth degree.

It is not very difficult to see why some of the coefficients of the equations of differences must be so connected with the rest, as not to give any independent condition.

In order to get an idea of this connection, let us imagine $n-1$ independent conditions, that is, $n-1$ functions of the coefficients of $f(x)=0$, which have a definite sign when all the roots are real. These functions are coefficients of the equation of the squares of the differences. Let them all become equal to zero, then we have $n-1$ relations among the n roots, which will give every one of these roots, except one, in terms of that root. Now the relations $b=a$, $c=a$, ... $k=a$, will fulfil our $n-1$ equations, because these evidently make all the coefficients of the equation of the squares of the differences vanish. Hence these are the relations implied in the $n-1$ equations we have assumed; and these would, it has just been said, make all the coefficients $= 0$. Consequently, $n-1$ independent relations are all we can have, and it follows, that if $n-1$ independent coefficients of the equation of differences become $= 0$, all will be so.

We may take the matter somewhat differently, still using the case of equal roots in $f(x) = 0$ to show the relations among the coefficients of the equation of the squares of the differences.

If $f(x) = 0$ has m equal roots, there will be $\frac{m(m-1)}{2}$ roots of the equation of the squares of the differences, or $\Delta = 0$, equal to zero. Let $f(x) = 0$ get another root equal to these m roots by any change in its coefficients, then there will be $\frac{(m+1)m}{2}$ roots in $\Delta = 0$ equal to zero; the difference is m. Thus, a single fresh relation among the coefficients of $f(x) = 0$ makes m coefficients of $\Delta = 0$ vanish; for obviously the last coefficients of this equation disappear whenever it gets roots equal to zero.

We may easily see, too, that the constant function (the last in Sturm's process) is the same as the term independent of u in $\Delta = 0$.

The equation $\Delta = 0$ may, theoretically, be got by eliminating x between the two equations

$$f(x) = 0 \text{ and } f'x + \frac{u}{2}f''x + \frac{u^2}{2\,.\,3}f'''x + \dots u^{n-1} = 0,$$

(*Lagrange*, p. 7); Δ will be the term independent of x: put, then, $u = 0$, Δ reduces itself to its last term, and the process becomes simply that of finding the common measure of $f(x)$ and $f'(x)$, which, abstracting the changes of sign, is exactly Sturm's process; hence his term independent of x, will be "aux signes près" the constant term in $\Delta = 0$.

The development of this idea would undoubtedly lead to the general theory of Sturm's method, and would make it more than a happy artifice, by showing its intimate connection with the equation of the squares of the differences. As is generally the case, the different ways in which the subject may be viewed, ultimately coalesce.

ON THE ACHROMATISM OF EYE-PIECES OF TELESCOPES AND MICROSCOPES*.

MR AIRY, in a paper published in the Second Volume of the *Cambridge Transactions*, has investigated the conditions under which a system of lenses is achromatic, *per se*—that is, when only one kind of glass is made use of. The enquiry, on account of its immediate application to the eye-pieces of telescopes and microscopes, is one of considerable importance; and as the way in which it is conducted in the paper just mentioned appears to be more complicated than necessary, and does not lead to the most general solution of the problem, perhaps the following attempt may be not wholly without interest.

The difficulty of the question consists in finding the angle which a ray of light makes with the axis of the system of lenses after having passed through it. When this is done, we have only to take the chromatic variation, and equate it to zero, to get the general equation of achromatism for the system.

In what follows, lines are considered positive when measured in a direction opposite to that of the incident ray.

Let y_n be the tangent of the angle which the ray makes with the axis before its incidence on the nth lens of the system; let z_n be the distance from the axis of the point where it impinges on that lens; and take a_{n-1} to signify the distance of the nth from the $n-1^{\text{th}}$ lens. Then $\frac{z_n}{y_n}$ and $\frac{z_n}{y_{n+1}}$ are the distances of the conjugate foci from the nth lens, and by the ordinary formula

$$\frac{y_{n+1}}{z_n} - \frac{y_n}{z_n} = \rho_n,$$

ρ being the reciprocal of the focal length; therefore

$$y_{n+1} - y_n = \rho_n z_n.$$

* *Cambridge Mathematical Journal*, No. VI. Vol. I. p. 269, May, 1839.

Again, we have the simply geometrical relation

$$\frac{z_{n+1}}{y_{n+1}} - \frac{z_n}{y_{n+1}} = a_n,$$

$$\text{or} \quad z_{n+1} - z_n = a_n y_{n+1}.$$

The advantage gained by introducing y_n is, that we thus have precisely similar equations for the optical and geometrical conditions of the problem. Nothing is now easier than by successive substitutions to determine the value of y_n in terms of y and z, and y_n is the tangent of the "visual angle," which we are seeking. As an instance, let it be proposed to determine the conditions of achromatism in a system of three lenses. Mr Airy has done this only for rays originally parallel to the axis: the method here proposed applies with equal facility to the general case.

The equations required are these,

$$y_2 - y_1 = \rho_1 z_1$$

$$z_2 - z_1 = a_1 y_2$$

$$y_3 - y_2 = \rho_2 z_2$$

$$z_3 - z_2 = a_2 y_3$$

$$y_4 - y_3 = \rho_3 z_3$$

Hence

$$y_2 = y_1 + \rho_1 z_1$$

$$z_2 = (1 + a_1\rho_1) z_1 + a_1 y_1$$

$$y_3 = [1 + a_1\rho_2] y_1 + [\rho_1 + \rho_2 + a_1\rho_1\rho_2] z_1$$

$$z_3 = (1 + a_1\rho_1 + a_2\rho_1 + a_2\rho_2 + a_1a_2\rho_1\rho_2) z_1 + (a_1 + a_2 + a_1a_2\rho_2) y_1$$

$$y_4 = [1 + a_1\rho_2 + a_1\rho_3 + a_2\rho_3 + a_1a_2\rho_2\rho_3] y_1 + [\rho_1 + \rho_2 + \rho_3 + a_1\rho_1\rho_2 + a_1\rho_1\rho_3 + a_2\rho_1\rho_3 + a_2\rho_2\rho_3 + a_1a_2\rho_1\rho_2\rho_3] z_1.$$

Taking the chromatic variations of the two terms in the usual way, and equating each to zero, we find

$$a_1\rho_2 + a_1\rho_3 + a_2\rho_3 + 2a_1a_2\rho_2\rho_3 = 0$$

$$\rho_1 + \rho_2 + \rho_3 + 2a_1\rho_1\rho_2 + 2a_1\rho_1\rho_3 + 2a_2\rho_1\rho_3 + 2a_2\rho_2\rho_3 + 3a_1a_2\rho_1\rho_2\rho_3 = 0.$$

The first of these equations becomes unnecessary in the particular case considered by Mr Airy, viz. that in which $y_1 = 0$; the second is identical with that given by him at p. 245, when attention is paid to the signs. Taken together they determine the relative positions of three given lenses, which shall form a combination achromatic for rays of any degree of obliquity.

In the particular case in which the focal distances of all the lenses are equal, and the intervals a_1, a_2, &c. are also equal, the general equations degenerate into a system of simultaneous equations in finite differences. They are then

$$y_{n+1} - y_n = \rho z_n, \quad z_{n+1} - z_n = a y_{n+1}.$$

Eliminating z_n, we get

$$y_{n+2} - (\rho a + 2)\, y_{n+1} + y_n = 0.$$

The general solution of this will be

$$y_n = cA^n + c_1 A^{-n},$$

A being a root of the recurring quadratic equation

$$x^2 - (\rho a + 2)\, x + 1 = 0,$$

c and c_1 are to be found by the conditions

$$y_1 = cA + c_1 A^{-1}, \quad y_1 + \rho z_1 = cA^2 + c_1 A^{-2}.$$

The general solution of the system of quasi-equations employed in the enquiry must involve some functional operation which degenerates into the radical contained in A.

It would be perhaps worth considering how far we might be able to present this operation in a distinct form, defined and distinguished by a particular symbol; but the subject is not one which can be discussed at present. At any rate, we see that the research of the general expression for y_n is one of considerable difficulty.

The greater part of the investigation given by Mr Airy in the conclusion of his paper, with respect to the achromatism of microscopes, becomes unnecessary by employing the general expression given above for y_4. His object is to determine the distance of an object-glass of given focal length from a diaphragm whose distance from the field-glass of a given eye-piece of three lenses is given.

Let a_0 be the distance of the diaphragm from the field-glass; therefore we have $z_1 = a_0 y_1$, and putting this value for z_1, we get an expression for y_4 of the form $y_4 = y_1 R$. The chromatic variation of this is to be zero, and consequently that of its logarithm;

$$\therefore\ 0 = \frac{\Delta y_1}{y_1} + \frac{\Delta R}{R}.$$

Now Mr Airy has shown that, (adopting the notation of this paper)

$$\frac{\Delta y_1}{y_1} = -x\rho_0 \frac{\delta\mu}{\mu - 1},$$

x being the distance of the object-glass from the diaphragm, and ρ_0 its vergency, or the reciprocal of its focal length. Putting for R its value, we get at once $x\rho_0 =$

$$[a_0[\rho_1 + \rho_2 + \rho_3] + a_1[\rho_2 + \rho_3] + a_2\rho_3 + 2a_0a_1[\rho_1\rho_2 + \rho_1\rho_3]$$
$$+ 2a_0a_2[\rho_1\rho_3 + \rho_2\rho_3] + 2a_1a_2\rho_2\rho_3 + 3a_0a_1a_2\rho_1\rho_2\rho_3]$$

divided by

$$[1 + a[\rho_1 + \rho_2 + \rho_3] + a_1[\rho_2 + \rho_3] + a_2\rho_3 + a_0a_1[\rho_1\rho_2 + \rho_1\rho_3]$$
$$+ a_1a_2[\rho_1\rho_3 + \rho_2\rho_3] + a_1a_2\rho_2\rho_3 + a_0a_1a_2\rho_1\rho_2\rho_3]$$

which is identical with his result.

ON THE CONDITION OF EQUILIBRIUM OF A SYSTEM OF MUTUALLY ATTRACTIVE FLUID PARTICLES*.

THE generally received theory of the Equilibrium of Fluids, (due in its present form to Euler,) assigns one condition as necessary and sufficient in every case. Mr Ivory conceives, that when a fluid is acted on by forces arising from the mutual attraction of its particles, a second condition is requisite for equilibrium, and has developed the considerations which have led him to this result, in several papers published in the *Phil. Trans.*, and also in the *Phil. Mag.* The authority of Mr Ivory on any point of mathematical physics is very great: his decision on one to which he has long directed his attention, would be almost final, were it not opposed to the views of Euler, Laplace, and Poisson. The object of this paper is, to examine how far Mr Ivory, in a paper published in the *Phil. Mag.* Vol. XIII. p. 321, has demonstrated the necessity of the subsidiary condition in question. The writer feels it unnecessary to express the diffidence with which he attempts to consider so difficult a subject; he regrets also his inability to discuss Mr Ivory's views more at large than the present limits would permit.

In the paper just mentioned, Mr Ivory states the principal steps of the investigation by which Clairaut was led to the condition of equilibrium of a fluid acted on by forces directed to fixed centres; and proceeds to consider the modifications re-

* *Cambridge Mathematical Journal*, No. VII. Vol. II. p. 18, November, 1839.

quired to adapt the method to the case of a fluid whose particles are mutually attractive. Clairaut first supposes a mass of fluid in equilibrium, and conceives an infinitesimal stratum added to it, which shall produce equable pressure over the whole surface;—the equilibrium of the original mass A will not be disturbed, and the increased mass $A + \delta A$ will be in equilibrio, when the forces acting on its surface are normal to it. This principle, that forces acting on a free surface must be normal to it, was laid down by Huygens, and is confessedly true. By a repetition of this process, the original mass can be enlarged to any extent; and the condition that the nucleus must be in equilibrio becomes, Mr Ivory observes, unnecessary, by conceiving it diminished *sine limite.* The mathematical condition of equilibrium is, therefore, the expression of the possibility of adding a stratum which shall produce equable pressure, and at the free surface of which the forces shall be normal to it.

Let us endeavour to put this symbolically. Let the force at the original free surface be F; at the point x, y, z produce the normal, and take a length on it $= \frac{\omega}{F}$, ω being infinitesimal: thus we get a stratum producing an equal pressure ω.

Let $f(x, y, z) = c$ be the equation of the free surface; then F being a function of (x, y, z), all that is requisite for the force at any point of the new free surface to be normal is, that

$$f(x, y, z) = c + \delta c$$

shall be its equation.

Let $\qquad V = \sqrt{(f'x)^2 + (f'y)^2 + (f'z)^2}$; then

$$\delta x = \frac{\omega}{F}\frac{f'x}{V}, \quad \delta y = \frac{\omega}{F}\frac{f'y}{V}, \quad \delta z = \frac{\omega}{F}\frac{f'z}{V} \text{ .... (1)},$$

and $f(x, y, z) = c = f(x', y', z') - [f'x\delta x + f'y\delta y + f'z\delta z]$

(where $x' = x + \delta x$) by Taylor's theorem;

$$\text{therefore } c = f(x', y', z') - \frac{\omega}{F}\frac{1}{V}[(f'x)^2 + (f'y)^2 + (f'z)^2]$$

$$= f(x', y', z') - \omega\frac{V}{F};$$

$$\text{therefore } c = c + \delta c - \omega\frac{V}{F};$$

or $\frac{V}{F}$ = a constant, which we may take for unity; therefore $F = V$. Resolving this force along the axes,

$$X = V \cdot \frac{f'x}{V}; \text{ whence } X = f'x, \text{ and so } Y = f'y, \ Z = f'z.$$

ω is the increment of pressure $= \delta p$; multiplying the three equations (1) by X, Y, Z, and adding, we get

$$X\delta x + Y\delta y + Z\delta z = \delta p \frac{V^2}{F.V};$$

or putting d for δ,

$$dp = Xdx + Ydy + Zdz \text{ .... } (2),$$

the equation of equilibrium of an homogeneous and incompressible fluid, whose density is unity.

An objector to Clairaut's reasoning might urge, that this result, though certainly sufficient, was not shown to be necessary: he might argue, that a way has been shown of building up a fluid mass; but that it has not been proved that every fluid mass is capable of resolution into the smaller masses, by means of which alone Clairaut investigates the conditions of equilibrium. Unless it be made a direct postulate, that every fluid mass in equilibrio will continue in equilibrio, when the part of it contained between the free surface and any level surface is removed, it is difficult to see how this objection can be met, except by showing that the property assigned by Huygens to a free surface, viz. that the force is normal to it, belongs to every surface of equal pressure, and that consequently Clairaut's reasoning is in reality independent of any construction or resolution of a fluid mass into successive strata. When we assert, with Clairaut, that a fluid mass in equilibrium is not disturbed by the addition of a stratum producing equal pressure, we imply that the reaction produced at any point of the surface of A, by the pressures exerted over the rest of the surface, i. e. the effect of the transmitted pressures, is normal to it. For we know that the forces at the surface are so; and unless the inference first stated is correct, there could be no equilibrium. It hence appears, that Clairaut's axiom is equivalent to this—Equable pressure produces a reaction normal to the surface on which it is applied. But if the force at a surface of equal pressure were not normal to it, there could be no equilibrium, because it is only by the transmitted pressures that it can be established.

Clairaut, as his views are represented by Mr Ivory, says nothing of the transmission of pressure; but it is impossible to investigate fluid equilibrium without tacit or expressed reference to some distinctive character of fluidity; and in the principle he makes use of, the idea of the transmission of pressure is essential. It appears, then, that the force at a surface of equal pressure is normal to it; and this conclusion is little else than a different way of putting the principles employed by Clairaut. We are now enabled to dispense with any process of constructing a fluid mass.

On referring to the mathematical reasoning employed above, we shall easily see that, substituting two infinitesimally near surfaces of equal pressure for the consecutive free surfaces of Clairaut, the result we arrive at is simply the symbolical expression of the principle just laid down, viz. that the force at a surface of equal pressure is normal to it. A very little attention will show, that (2) is true in every case of fluid equilibrium, and that it is completely equivalent to the principle which it represents. In translating, so to speak, his fundamental idea from the infinitesimal to the fluxionary conception, that namely of successive generation, Clairaut has tacitly introduced a new condition, namely, that a surface of equal pressure will necessarily be a free surface of equilibrium, the superincumbent part being removed.

Mr Ivory remarks—"The investigation of Clairaut is clear and definite. It evidently assumes that there is no cause tending to disturb the equilibrium of A, except the action of the forces at the surface of A upon the matter of δA. On this account his method fails when there is a mutual attraction between the mass A and the stratum δA. If the mass A attract the matter of the stratum δA, and cause it to press, it follows necessarily that the matter of δA will react, and by its attraction will urge the particles of A to move from their places. In this case, therefore, the equilibrium of A is disturbed by a force which Clairaut has not attended to; and unless the effect of this new force is counteracted, the body of fluid $A + \delta A$ will not be in equilibrium. The principle of the method suggests a remedy for this omission, for it is easy to prove that the equilibrium of A will not be disturbed by the attraction of the stratum δA, if the resultant of that attraction on every particle in the surface of A be directed perpendicularly to it."

This reasoning satisfactorily shows, that if a fluid mass of attractive matter be increased by a stratum producing equal pressure over the free surface, the equilibrium will be destroyed unless a certain condition is fulfilled, of which the symbolical expression is

$$c = \int [Pdx + Qdy + Rdz],$$

P, Q, R being the attractions, parallel to the axes of co-ordinates, of an element of that part of a fluid mass which is external to a given level surface. But the necessity of this condition cannot be proved, unless it is shown to be impossible in any way to increase the mass A, without destroying the equilibrium, supposing it not fulfilled. All that has been shown is, that the mass cannot be increased by a stratum producing equable pressure over the free surface. Now, generally speaking, the mass so increased will not fulfil the condition of having the forces at the new free surface normal to it, those acting at the original free surface being of course so. We cannot, therefore, affirm that we have fallen on a case in which the ordinary condition is fulfilled, without producing equilibrium. If, however, we dispense with the limitation, that the stratum added shall produce equable pressure, we lose the simplicity of Clairaut's method, nor can we make any use of his principle, except by setting aside the construction he employs, which confines him to the particular case in which a surface of equal pressure is potentially a free surface.

This has already been done, and the result is the general equation of equilibrium. It remains to show, that it is in all cases sufficient. It is admitted to be sufficient in the case of a fluid acted on by forces tending to fixed centres. We shall endeavour to reduce the general case to this. Conceive a body acted on by a force directed to a fixed point. It may be so placed, as to remain at rest under the action of the force, that is, the resultant of the force upon it is equal to zero. In this position of the body, the centre of force is some point within it. Let the body, remaining in the same position, diminish *sine limite*, being always similar to itself, the resultant of the force upon it is always equal to zero; and ultimately, when the body becomes a physical point, it coincides in position with the centre of force, and is in the same state with respect to the action of other forces upon it, as if this force did not exist.

This being granted, conceive a homogeneous mass of fluid composed of mutually attractive particles, the free surface of which fulfils the required equation

$$Xdx + Ydy + Zdz = 0.$$

Let the attractive power of each particle be conceived transferred to a fixed centre of force coinciding with it. Then the action of all the other particles on one particle is precisely replaced by that of the fixed centres; and it has been shown, that the resultant of the action of the centre coinciding with a particle on that particle, equals zero. Hence, the supposition we have made does not change, in any way, the forces acting on any particle of the mass. Were the system in its present and former state respectively to move, the motions would be widely different; but in the arbitrary position we have placed it in, the action on it is precisely the same in the two cases. Now, the single equation given above assures its equilibrium, when we regard it as a system acted on by forces directed to fixed centres; and as the hypothesis by which we are enabled to look upon it in this way nowise affects the forces acting on it, it follows, that the system considered as acted on by mutual attraction must be in equilibrium. Consequently, a mass of homogeneous fluid, the particles of which are mutually attractive, will always be in equilibrium when the free surface fulfils the single condition implied in the general equation obtained above. The same reasoning applies to the case of any fluid, elastic or incompressible.

If this demonstration be thought satisfactory, the question raised by Mr Ivory, as to the sufficiency of the general equation, must be looked upon as settled. The suggestions here made with respect to the new condition tacitly introduced in Clairaut's reasoning, will, it is thought, enable us to trace the source of the difference of the view taken by Mr Ivory, and that generally entertained. In one form or other, it seems to recur in every way in which that distinguished mathematician has treated the subject.

MATHEMATICAL NOTE*.

In Vol. I. p. 205†, there were found for the co-ordinates of the point of intersection of two tangents to a parabola, the expressions

$$y = m(\alpha + \alpha'), \quad x = m\alpha\alpha',$$

α, α' being the tangents of the angles which the tangents to the curve make with the axis of y. From these expressions it follows, that if y_1, y_2, &c. x_1, x_2, &c. be the co-ordinates of the angles of any re-entering polygon of $2n$ sides circumscribing a parabola,

$$y_1 - y_2 + y_3, \text{ \&c. } - y_{2n} = 0,$$

$$\text{and } \frac{x_1 x_3 \ldots\ldots x_{2n-1}}{x_2 x_n \ldots\ldots x_{2n}} = 1.$$

Also, the continued product of the abscissæ of the points of intersection of any number of tangents, is equal to the continued product of the abscissæ of the points of contact, provided no three points of intersection lie in the same straight line.

Let x', x'', x''', &c. be the abscissæ of the points of contact, then it is easily seen, from the equation to the parabola, that

$$x' = m\alpha'^2, \quad x'' = m\alpha''^2, \quad x''' = m\alpha'''^2, \text{ \&c.}$$

the continued product of which is

$$x'x''x''' \ldots x^{(n)} = m^n \alpha'^2 \alpha''^2 \alpha'''^2 \ldots \alpha^{(n)2}.$$

And if x_1, x_2, x_3, &c. be the co-ordinates of the points of intersection of the tangents, we have

$$x_1 = m\alpha'\alpha'', \quad x_2 = m\alpha''\alpha''', \quad x_3 = m\alpha''\alpha''', \text{ \&c.}$$

the continued product of which is

$$x_1 x_2 x_3 \ldots x_n = m^n \,.\, \alpha'^2 \alpha''^2 \alpha'''^2 \ldots \alpha^{(n)2},$$

which is equal to the preceding expression. It is necessary to limit the intersections in such a way that no three shall lie in the same line, because otherwise some one of the α's in the second series would appear more than twice.

* *Cambridge Mathematical Journal*, No. VII. Vol. II. p. 48, November, 1839.

† Page 63 of this Volume.

VARIATION OF NODE AND INCLINATION *.

THE following method of finding the variations of the inclination and longitude of the node, is more convenient than that given in Pratt's *Mechanical Philosophy*, p. 336.

Adopting the notation usual in the lunar theory, we have

$$s = k \sin(\theta - \gamma) \quad \ldots\ldots\ldots\ldots\ldots\ldots\ldots (1),$$

$$\text{also } \frac{d^2z}{dt^2} + \frac{\mu z}{r^3} + \frac{dR}{dz} = 0. \quad \ldots\ldots\ldots\ldots\ldots\ldots (2).$$

In the disturbed orbit, (1) and its first derived equation will be true, as if the elements were invariable; which gives the equation

$$\sin(\theta - \gamma)\frac{dk}{dt} - k\cos(\theta - \gamma)\frac{d\gamma}{dt} = 0 \quad \ldots\ldots\ldots\ldots\ldots (3),$$

and differentiating (1) a second time, there is

$$\frac{d^2s}{dt^2} = k\frac{d\theta}{dt} \cdot \frac{d}{d\theta}\left\{\cos(\theta - \gamma)\frac{d\theta}{dt}\right\}$$

$$+ \left\{\cos(\theta - \gamma)\frac{dk}{dt} + k\sin(\theta - \gamma)\frac{d\gamma}{dt}\right\}\frac{d\theta}{dt}.$$

The second and third terms are those due to perturbation. Also, the inclination being very small, the effect of perturbation on $\frac{d^2z}{dt^2}$, or which is the same thing, on $\frac{d^2 . \rho s}{dt^2}$, will be sensible only in the term $\rho\frac{d^2s}{dt^2}$. Hence, equating the perturbation and its effect, we have

$$\rho\frac{d\theta}{dt}\left\{\cos(\theta - \gamma)\frac{dk}{dt} + k\sin(\theta - \gamma)\frac{d\gamma}{dt}\right\} + \frac{dR}{dz} = 0 \ldots\ldots (4),$$

* *Cambridge Mathematical Journal*, No. IX. Vol. II. p. 113, May, 1840.

and eliminating in turn $\frac{dk}{dt}, \frac{d\gamma}{dt}$, by (3), we get

$$\rho \frac{d\theta}{dt} k \frac{d\gamma}{dt} + \frac{dR}{dz} \sin(\theta - \gamma) = 0,$$

$$\rho \frac{d\theta}{dt} \frac{dk}{dt} + \frac{dR}{dz} \cos(\theta - \gamma) = 0.$$

Again, the inclination being small,

$$\frac{dR}{dk} = \frac{dR}{dz} \cdot \frac{dz}{dk} = \rho \frac{dR}{dz} \cdot \frac{ds}{dk} = \rho \frac{dR}{dz} \sin(\theta - \gamma) \text{ nearly},$$

and $\frac{dR}{d\gamma} = \frac{dR}{dz} \cdot \frac{dz}{d\gamma} = \rho \frac{dR}{dz} \cdot \frac{ds}{d\gamma} = -\rho k \frac{dR}{dz} \cos(\theta - \gamma)$;

$$\therefore \rho^2 \frac{d\theta}{dt} k \frac{d\gamma}{dt} = -\frac{dR}{dk},$$

$$\text{and } \rho^2 \frac{d\theta}{dt} k \frac{dk}{dt} = \frac{dR}{d\gamma},$$

$$\rho^2 \frac{d\theta}{dt} = h = \sqrt{\mu a (1 - e^2)} = \frac{\mu \sqrt{1 - e^2}}{na};$$

$$\therefore \frac{d\gamma}{dt} = -\frac{1}{k} \cdot \frac{na}{\mu \sqrt{1 - e^2}} \frac{dR}{dk} \quad \ldots\ldots\ldots\ldots\ldots\ldots \text{(5)},$$

$$\frac{dk}{dt} = \frac{1}{k} \frac{na}{\mu \sqrt{1 - e^2}} \frac{dR}{d\gamma} \quad \ldots\ldots\ldots\ldots\ldots\ldots\ldots \text{(6)},$$

which agree with the known results, k being $= \tan I$, or $\sin I$, *quam proximè*, and γ being what Mr Pratt denotes by Ω. (The squares, &c. of I are neglected throughout.)

INVESTIGATION OF THE ABERRATION IN RIGHT ASCENSION AND DECLINATION*.

THE following investigation of the formulæ for Aberration in Right Ascension and Declination will be found to be more simple than that given in Maddy's *Astronomy*, p. 214.

Let ♈ L be the ecliptic, ♈ E the equator, P its pole, T a point 90° behind the place of the Sun, S the place of the star; then ST will be the plane of aberration.

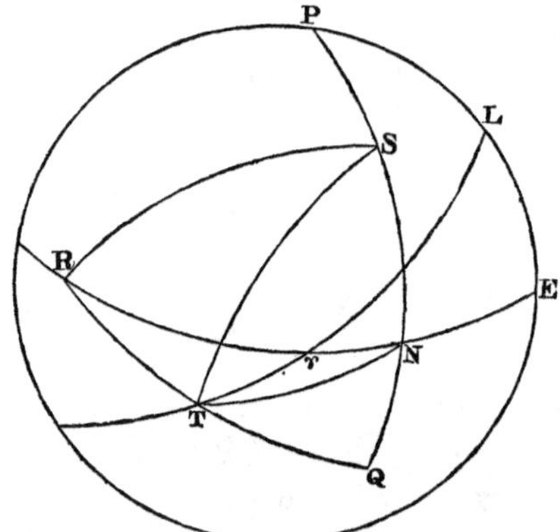

Let ♈ $N = a$, $PN = \delta$, Sun's longitude $= \odot$, and L ♈ $E = \omega$.

1st. For the Aberration in Declination: produce SN to a point Q, such that $SQ = 90^{\circ}$, and join TQ, TN. If A be the coefficient of aberration, and $\Delta\delta$ the aberration in declination,

$$\Delta\delta = - A \sin ST \,.\, \cos TSQ = - A \cos TQ,$$

as TSQ is a quadrantal triangle.

* *Cambridge Mathematical Journal*, No. IX. Vol. II. p. 120, May, 1840.

But $\cos QT = \cos TN . \cos QN + \sin TN . \sin QN . \sin TN♈$,

and $\cos TN = \cos T♈ . \cos ♈ N + \sin T♈ . \sin ♈ N . \cos T♈ N$,

$$= \sin ⊙ \cos a - \cos ⊙ \sin a \cos \omega.$$

Also, $\sin TN . \sin TN♈ = \sin T♈ . \sin T♈ N = \cos ⊙ \sin \omega$.

Substituting these values, and putting $90^{\circ} - \delta$ for QN, we find

$\Delta\delta = - A\{\sin\delta\,(\sin ⊙ \cos a - \cos ⊙ \sin a \cos \omega) + \cos\delta \cos ⊙ \sin \omega\}$.

2nd. For the Aberration in Right Ascension: produce $N♈$ to a point R, such that $NR = 90^{\circ}$, and join RS, RT. Then, if Δa be the aberration in right ascension,

$$\Delta a = -\frac{A}{\cos\delta}\sin ST . \cos TSR = -\frac{A}{\cos\delta}\cos RT.$$

But $\cos RT = \cos T♈ . \cos R♈ + \sin T♈ . \sin R♈ \cos R♈ T$,

$$= \sin ⊙ \sin a + \cos a \sin ⊙ \cos \omega.$$

Consequently,

$$\Delta a = -\frac{A}{\cos\delta}\{\sin ⊙ \sin a + \cos a \sin ⊙ \cos \omega\}.$$

ON THE LINES OF CURVATURE ON AN ELLIPSOID*.

THE following investigation of the Lines of Curvature on an Ellipsoid has the advantages of symmetry and of giving a distinct geometrical conception. The artifice on which it depends may, it is thought, be found useful on other occasions.

The symmetrical equation to the lines of curvature is

$$(b^2-c^2)\,x\,dy\,dz+(c^2-a^2)\,y\,dz\,dx+(a^2-b^2)\,z\,dx\,dy=0\ldots\ldots(1),$$

(see *Mathematical Journal*, Vol. I. p. 142), where xyz are connected by the equation to the surface,

$$\frac{x^2}{a^2}+\frac{y^2}{b^2}+\frac{z^2}{c^2}=1\ldots\ldots\ldots\ldots\ldots\ldots\ldots\ldots(2).$$

$$\text{Put } \frac{x^2}{a^2}=u,\quad \frac{y^2}{b^2}=v,\quad \frac{z^2}{c^2}=w\ldots\ldots\ldots\ldots\ldots\ldots(A).$$

$$\text{Then}\quad x\,dy\,dz=\tfrac{1}{4}a\sqrt{u}\,.\,b\frac{dv}{\sqrt{v}}\,c\frac{dw}{\sqrt{w}}=\frac{abc}{4\sqrt{uvw}}\,u\,dv\,dw.$$

Hence, after the substitution and multiplying by $\dfrac{4\sqrt{uvw}}{abc}$, (1) becomes

$$(b^2-c^2)\,u\,dv\,dw+(c^2-a^2)\,v\,dw\,du+(a^2-b^2)\,w\,du\,dv=0\ldots(3),$$

with the relation

$$u+v+w=1\ldots\ldots\ldots\ldots\ldots\ldots\ldots\ldots(4).$$

Differentiate (3); then, since

$$b^2-c^2+c^2-a^2+a^2-b^2=0,$$

* *Cambridge Mathematical Journal*, No. IX. Vol. II. p. 133, May, 1840.

we get

$$(b^2-c^2)\,ud\,(dv\,dw)+(c^2-a^2)\,vd\,(dw\,du)$$
$$+(a^2-b^2)\,wd\,(du\,dv)=0 \ldots\ldots (5).$$

Now this is satisfied by the assumptions

$$dv\,dw=\frac{1}{f},\quad dw\,du=\frac{1}{g},\quad du\,dv=\frac{1}{h} \ldots\ldots\ldots\ldots (\mathrm{B}),$$

f, g, h, being constants.

But from (4) we deduce

$$du+dv+dw=0 \ldots\ldots\ldots\ldots\ldots\ldots\ldots\ldots (6),$$

and (B) gives

$$du=f\,du\,dv\,dw,\quad dv=g\,du\,dv\,dw,\quad dw=h\,du\,dv\,dw.$$

Hence $$f+g+h=0 \ldots\ldots\ldots\ldots\ldots\ldots\ldots\ldots (7),$$

which establishes a relation among the otherwise arbitrary constants f, g, h.

Now (B) implies the existence of two linear equations in u, v, w. Hence, a *particular solution* of (1) is two linear equations connecting the three variables. But the given equation (4) is linear; hence the solution in question is the one congruent to the problem.

To find the other relation in u, v, w, eliminate the differentials from (3) by means of (B), and there is

$$(b^2-c^2)\frac{u}{f}+(c^2-a^2)\frac{v}{g}+(a^2-b^2)\frac{w}{h}=0 \ldots\ldots\ldots\ldots (8).$$

Equations (4) and (5), with the relation (7), contain the complete solution of the problem. It is obvious that the apparent want of homogeneity of (B) is wholly immaterial.

Keeping in mind the values of u, v, w, given by (A), we see that the geometrical interpretation of (8) is, every line of curvature on an ellipsoid lies on a conical surface of the second order, of which the vertex is the centre of the ellipsoid.

To determine the constants, let the line of curvature pass through a point, for which the values of u, v, w, are u_1, v_1, w_1, we have

$$(b^2-c^2)\frac{u_1}{f}+(c^2-a^2)\frac{v_1}{g}+(a^2-b^2)\frac{w_1}{h}=0,$$

$$f+g+h=0.$$

Hence, after a slight reduction,

$$(b^2-c^2)\,u_1\frac{g}{f}+(c^2-a^2)\,v_1\frac{f}{g}-(a^2-b^2)\,w_1$$
$$+(b^2-c^2)\,u_1+(c^2-a^2)\,v_1=0\,\ldots\ldots\ldots(9),$$

a quadratic in $\frac{g}{f}$, of which the roots are real and of unlike signs. This is obvious, for u_1, v_1, are essentially positive, and a, b, c, being in order of magnitude, the signs of b^2-c^2 and c^2-a^2 are opposite. Similarly, $\frac{g}{h}$ is determined by a quadratic, whose roots are always real and of opposite signs. Thus two lines of curvature pass through every point on the surface of the ellipsoid.

Let us now consider the envelope of the surfaces represented by (8).

Differentiating (7) and (8) for f, g, h, we get

$$(b^2-c^2)\frac{u}{f^2}\,df+(c^2-a^2)\frac{v}{g^2}\,dg+(a^2-b^2)\frac{w}{h^2}\,dh=0\,\ldots\ldots(10),$$

$$df+dg+dh=0\,\ldots\ldots\ldots\ldots\ldots\ldots\ldots\ldots\ldots(11).$$

l being an indeterminate factor, we may put

$$l=(b^2-c^2)\frac{u}{f^2},\quad l=(c^2-a^2)\frac{v}{g^2},\quad l=(a^2-b^2)\frac{w}{h^2};$$

whence, taking the values of f, g, h, to substitute them in (8), we deduce

$$\sqrt{b^2-c^2}\,\sqrt{u}+\sqrt{c^2-a^2}\,\sqrt{v}+\sqrt{a^2-b^2}\,\sqrt{w}=0\,\ldots\ldots(12).$$

As the signs of the radicals are independent, this represents four planes; but c^2-a^2 is negative. Hence the possible part of these planes is their traces on the plane of xz, for which $v=0$. Thus we get the two straight lines

$$\sqrt{b^2-c^2}\,\sqrt{u}+\sqrt{a^2-b^2}\,\sqrt{w}=0\,\ldots\ldots\ldots\ldots\ldots(13),$$

and for the points where they meet the ellipsoid,

$$u+w=1\,\ldots\ldots\ldots\ldots\ldots\ldots\ldots\ldots\ldots\ldots\ldots\ldots(14),$$

whence

$$u_1=\frac{a^2-b^2}{a^2-c^2},\quad v_1=0,\quad w_1=\frac{b^2-c^2}{a^2-c^2}\,\ldots\ldots\ldots\ldots(15).$$

These values belong to the umbilici of the ellipsoid; a result easily anticipated. When they are introduced in (9), it becomes

$$\frac{g^2}{f^2} = 0, \text{ and similarly } \frac{g^2}{h^2} = 0.$$

Hence (8) reduces to

$$v = 0 \ldots\ldots\ldots\ldots\ldots\ldots\ldots\ldots\ldots (16),$$

and represents the principal section of the ellipsoid, which passes through the greatest and least axes. In this case then, as our analysis would lead us to anticipate, the lines of curvature coincide; a result which, although well known, seems not very accurately demonstrated by Leroy. After having shown (p. 309 of the second edition) that the two *directions* of curvature coincide at the umbilical points, he proceeds to integrate, and passes from $\frac{dy}{dx} = 0$ to $y = h$, and thence, determining the constant, to $y = 0$; which last represents the line of curvature sought. But $\frac{dy}{dx}$ has been shown to have the value 0, only for the umbilical points, and we are therefore not at liberty to pass by integration from these to any other points at which this may not hold. Were the process legitimate, it would lead to the strange conclusion, that the lines of curvature through an umbilicus are necessarily plane curves.

As there appears to be still some difficulty with regard to the theory of these singular points, we may enquire whether, in order to determine the lines of curvature through any point whatever, more is requisite than to substitute its co-ordinates in the general equation of the lines of curvature, and thus to get two values for the arbitrary constant; whether the result can ever be indeterminate, except when the lines, as at the extremity of an axis of revolution, are so in reality. In this view we see at once, that the process given by Leroy after Poisson for determining the *directions* of curvature at an umbilicus, is simply the ordinary method for ascertaining the position of the branches of any curve at a multiple point; and that the result arrived at, is not that more than two lines of curvature pass through an umbilicus, but that every point which, with reference to the surface, is umbilical, is, with reference to the lines of curvature, a multiple, or more generally a singular point. These suggestions may, perhaps, show how we must determine the lines of curvature which pass through an umbilicus, a problem distinct from that solved by Leroy of finding the directions of curvature.

Many curious properties may be deduced from the equations we have arrived at. Thus, if we take on two concentric and confocal ellipsoids, a series of pairs of corresponding points, (such as are spoken of in the enunciation of Ivory's theorem,) and if the locus of the points on one of the ellipsoids is a line of curvature, then that of those on the other is so too. Again, the traces on the tangent planes at the extremities of the three axes, made by one of the cones represented by (8), are an ellipse and two hyperbolas respectively. The areas of this ellipse, and of the ellipses conjugate to the two hyperbolas, are so related that their continual product is constant for the same ellipsoid, and for all ellipsoids of the same volume. The method of demonstrating these two theorems is so obvious, that it seems unnecessary to enter more fully on either.

It still remains to be shown how we pass from (8) to the projections of the lines of curvature on the co-ordinate planes. The symmetry of the problem is destroyed by the transition; but as it is in this shape that the results are commonly exhibited, we shall dwell rather more upon it than would otherwise have been necessary.

Putting $C = a^2 - b^2$, $B = c^2 - a^2$, $A = b^2 - c^2$, and eliminating u, v, w, successively between (4) and (8), there result

$$\left.\begin{aligned} \left(\frac{C}{h} - \frac{A}{f}\right) u - \left(\frac{B}{g} - \frac{C}{h}\right) v &= \frac{C}{h} \\ \left(\frac{A}{f} - \frac{B}{g}\right) v - \left(\frac{C}{h} - \frac{A}{f}\right) w &= \frac{A}{f} \\ \left(\frac{B}{g} - \frac{C}{h}\right) w - \left(\frac{A}{f} - \frac{B}{g}\right) u &= \frac{B}{g} \end{aligned}\right\} \ldots\ldots\ldots\ldots (17).$$

Put

$$\frac{C}{h} - \frac{A}{f} = k, \quad \frac{B}{g} - \frac{C}{h} = l, \quad \frac{A}{f} - \frac{B}{g} = m \ldots\ldots\ldots\ldots (18.$$

Then $$kf - lg = C\left(\frac{f+g}{h}\right) - A - B = -(A + B + C) = 0.$$

Thus we get the relations

$$\left.\begin{aligned} kf - lg &= 0 \\ lh - mf &= 0 \\ mg - kh &= 0 \end{aligned}\right\} \dots\dots\dots\dots\dots\dots\dots\dots (19).$$

Hence, $$\frac{A}{mf} - \frac{B}{mg} = 1 = \frac{1}{h}\left(\frac{A}{l} - \frac{B}{k}\right),$$

and consequently

$$\left.\begin{aligned} h &= \frac{A}{l} - \frac{B}{k} \\ g &= \frac{C}{m} - \frac{A}{l} \\ f &= \frac{B}{k} - \frac{C}{m} \end{aligned}\right\} \dots\dots\dots\dots\dots\dots\dots\dots (20).$$

similarly,

By means of (18) and (20) the equations (17) become

$$\left.\begin{aligned} ku - lv &= \frac{Clk}{Ak - Bl} \\ mv - kw &= \frac{Amk}{Bm - Ck} \\ lw - mu &= \frac{Blm}{Cl - Am} \end{aligned}\right\} \dots\dots\dots\dots\dots\dots (21);$$

which, restoring their values to u, v, w, may be written

$$\frac{y^2}{b^2} = \frac{k}{l}\left\{\frac{x^2}{a^2} - \frac{C}{A\frac{k}{l} - B}\right\}, \text{ \&c.} \dots\dots\dots\dots\dots (22).$$

Put $\frac{k}{l}\frac{b^2}{a^2} = m$; $\therefore \frac{k}{l} = m\frac{a^2}{b^2}$, and then we get

$$y^2 = m\left\{x^2 - \frac{a^2}{b^2}\,\frac{a^2 - b^2}{m\frac{a^2}{b^2}(b^2 - c^2) - (c^2 - a^2)}\right\},$$

$$\text{or } y^2 = m\left\{x^2 - \frac{a^2(a^2 - b^2)}{ma^2(b^2 - c^2) - b^2(c^2 - a^2)}\right\} \dots\dots\dots (23),$$

with similar equations for the projections on the other co-ordinate planes. This result is identical with the known one in Leroy, p. 304, or Hymers, p. 201.

It is hoped that the novelty of treating symmetrically a non-integrable equation in three variables, will be admitted as an excuse for the length to which this paper has extended itself.

MATHEMATICAL NOTES*.

1. THE area of a polygon of a given number of sides, circumscribing a given oval figure, will be the least possible when each side is bisected in the point of contact.

This elegant proposition, given in the *Senate-House Problems* for 1836, may be easily demonstrated as follows:—

Let AB, BC, CD, be consecutive sides of the polygon. Produce AB, DC, to meet in E; then BC must, by the condition of the minimum, be in such a position that EBC is a maximum.

Refer the oval to EA, ED, for axes, then the equation to the tangent BC is

$$y'dx - x'dy = y\,dx - x\,dy,$$

y and x being the co-ordinates of the point of contact P.

Put $x' = 0$;

$$\therefore\ y_0 dx = y\,dx - x\,dy,$$

and so
$$-x_0 dy = y\,dx - x\,dy.$$

Also
$$\text{area of } EBC = \tfrac{1}{2} x_0 y_0 \sin E.$$

Hence, $\dfrac{(y\,dx - x\,dy)^2}{dx\,dy}$ is a maximum (the minus sign is immaterial).

Differentiate, considering x as independent; then

$$\frac{y\,dx - x\,dy}{dx\,dy}\,d^2y\left(\frac{y\,dx - x\,dy}{dy} + 2x\right) = 0.$$

The last factor only gives a solution;

$$\therefore\ x = -\tfrac{1}{2}\frac{y\,dx - x\,dy}{dy} = \tfrac{1}{2}x_0,$$

* *Cambridge Mathematical Journal*, No. IX. Vol. II. p. 142, May, 1840.

that is, PM being parallel to EC, $EM = \frac{1}{2}EB$, and $\therefore PB = PC$, or BC is bisected in the point of contact P. The same is true of any other side, and therefore every side is bisected in the point of contact.

2. Let p, p', be two forces into which a given system on a rigid body may be resolved, a, θ, their least distance, and inclination of their directions; $pp'a \sin\theta$ is invariable. (*Senate-House*, 1833.)

Let the line a meet the directions of p and p' in P and P' respectively. At P apply two forces equal and parallel to p', and opposite each other. Thus the system of forces is replaced by the couple $p'a$, and by the force at P, which is the resultant of p and p'. Resolve this, the resultant, along the axis of the couple and in its plane. Then the former component can arise only from the resolved part of p, as p' is wholly in the plane of the couple. Also, as the shortest distance is perpendicular to both lines, it follows that the arm of the couple is perpendicular at P to the plane which contains the two forces p and p'. Hence θ, their mutual inclination, is that of p on the plane of the couple, and therefore $p \sin\theta$ is the part of the general resultant resolved along the axis of the couple. Then, if the general resultant makes an angle ϕ with the axis, we have in the usual notation

$$pp'a \sin\theta = GR\cos\phi = R \, . \, G\cos\phi.$$

Now $G\cos\phi$, as is known, or as may be easily shown, $= G_1$, the minimum maximorum moment of the system;

$$\text{therefore } pp'a \sin\theta = G_1 R,$$

which is constant.

ON THE TAUTOCHRONE IN A RESISTING MEDIUM *.

OUR object is to reduce the problem of the tautochronous curve, when the resistance is equal to kv^2 or to $hv + kv^2$, to the cases in which it is a cycloid, viz. when the resistance is equal to zero or to hv.

In vacuo

$$\sqrt{2g}\,t_1 = \int_0^{s_1} \frac{ds}{\sqrt{x_1 - x}},$$

and the necessary and sufficient condition of tautochronism is $s^2 = Ax$. Hence generally

$$\int_0^{z_1} \frac{dz}{\sqrt{Fz_1 - Fz}}$$

is independent of z_1, provided $z^2 = A \,.\, Fz$. Now, when $R = kv^2$, there is

$$\frac{vdv}{ds} - kv^2 = -g\frac{dx}{ds},$$

$$\text{therefore } e^{-2ks}v^2 = -2g\int e^{-2ks}\frac{dx}{ds}\,ds.$$

$$\text{Let } fs = \int_0 e^{-2ks}\frac{dx}{ds}\,ds,$$

$$\text{therefore } e^{-2ks}v^2 = 2g\,(fs_1 - fs),$$

$$\text{and } \sqrt{2g}\,t_1 = \int_0^{s_1} \frac{e^{-ks}ds}{\sqrt{fs_1 - fs}}.$$

Put $dz = e^{-ks}ds$, and let z and s be equal to zero together,

$$\text{therefore } z = \frac{1}{k}(1 - e^{-ks}), \quad fs = Fz,$$

* *Cambridge Mathematical Journal*, No. X. Vol. II. p. 153, November, 1840.

$$\text{and } \sqrt{2g}\,t_1 = \int_0^{z_1} \frac{dz}{\sqrt{Fz_1 - Fz}}.$$

Therefore for tautochronism

$$z^2 = A \,.\, Fz.$$

$$\text{But } Fz = \int_0 \frac{dx}{ds}(1 - kz)\,dz,$$

$$\text{therefore } \frac{2}{A}z = \alpha z = \frac{dx}{ds}(1 - kz),$$

$$\text{therefore } k\frac{dx}{ds} = \alpha\frac{1 - e^{-ks}}{e^{-ks}},$$

$$\text{or } k\frac{dx}{ds} = \alpha(e^{ks} - 1) \quad \ldots\ldots\ldots\ldots\ldots\ldots\ldots\; (1),$$

the equation of the tautochronous curve.

We shall have, if $t = 0$ when $s = s_1$,

$$z = z_1 \cos\sqrt{\alpha g}\,t.$$

$$\text{Hence } \frac{d^2s}{dt^2} - k\frac{ds^2}{dt^2} = -\alpha g\frac{e^{ks} - 1}{k}$$

must become, when s is expressed in z,

$$\frac{d^2z}{dt^2} = -\alpha g\,z,$$

a result easily verified, for

$$ds = \frac{1}{1 - kz}dz \quad \ldots\ldots\ldots\ldots\ldots\ldots\ldots\ldots\ldots\ldots\ldots\; (\alpha).$$

$$d^2s = \frac{1}{1 - kz}d^2z + \frac{k}{(1 - kz)^2}dz^2 \quad \ldots\ldots\ldots\ldots\ldots\; (\beta).$$

The coefficient of dz in (α) is of course that of d^2z in (β), which shews that if the equation of motion were

$$\frac{d^2s}{dt^2} + h\frac{ds}{dt} - k\frac{ds^2}{dt^2} = -\alpha g\frac{e^{ks} - 1}{k},$$

$$\text{or } R = hv + kv^2,$$

the equation in z would be

$$\frac{d^2z}{dt^2} + h\frac{dz}{dt} = -ag\,z.$$

This is precisely the form of the equation of motion on a cycloid when $R = hv$. And as in that case t_1 deduced from it is independent of s_1, so in this it will be independent of z_1, and consequently of s_1. Hence the curve whose equation is (1), is tautochronous, not only when $R = kv^2$, but also when

$$R = hv + kv^2.$$

For Laplace's abstruse solution, see the first book of the *Mécanique Céleste*, or Mr Whewell's *Dynamics*.

ON THE INTEGRATION OF CERTAIN DIFFERENTIAL EQUATIONS*.

No. I.

It is shown in the theory of the earth's figure, that if the pressure and density at any point be connected by the equation

$$dp = k\rho\, d\rho,$$

where k is a constant, then the ellipticity of the surface may be deduced from the solution of the equation

$$\frac{d^2y}{dx^2} + q^2y = \frac{6y}{x^2}.$$

This equation is not easily integrated. La Place, in the eleventh book of the *Mécanique Céleste* (v. 51), gives a solution of it, but without demonstration; and the lacuna thus left is not supplied in the works on the subject generally made use of in Cambridge.

Mr Gaskin has however effected the integration of

$$\frac{d^2y}{dx^2} + q^2y = \frac{p(p-1)}{x^2}y,$$

when p is integral, in finite terms (vide Hymers' *Diff. Eq.* p. 53), and the proposed equation is a case of this one. But perhaps a more direct analysis is preferable, as it enables us to extend our method to two or three classes of equations of all orders. One of these will be considered in the present paper—another, the solution of which admits of a remarkable symbolical form, will be given in the next number of the *Journal.*

We shall begin with the particular equation which occurs in the theory of the earth's figure, both because from its physical

* *Cambridge Mathematical Journal*, No. X. Vol. II. p. 169. November, 1840.

application it has an interest for some who care but little for pure analysis, and because it will exemplify the general method.

$$\frac{d^2y}{dx^2} + q^2y = 6\frac{y}{x^2} \quad\text{.........................(1).}$$

Let
$$y = \Sigma a_n x^n \quad\text{...........................(2),}$$

$$\therefore\ \{n(n-1) - 6\}\, a_n + q^2 a_{n-2} = 0 \quad\text{...............(3),}$$

$$n(n-1) - 6 = n(n-1) - 3(3-1) = (n-3)(n+2),$$

$$\therefore\ (n-3)(n+2)\, a_n + q^2 a_{n-2} = 0 \quad\text{.............(4).}$$

To get rid of the factor $(n-3)$, assume

$$(n+2)\, a_n = (n-1)\, b_n \quad\text{.....................(5),}$$

$$\therefore\ a_{n-2} = \frac{n-3}{n} b_{n-2},$$

$$\therefore\ n(n-1)\, b_n + q^2 b_{n-2} = 0 \quad\text{.................. (6).}$$

Hence b_n is made to depend on b_1 or b_0 as n is odd or even, and we see at once that

$$\Sigma b_n x^n = b_0 \cos qx + b_1 \sin qx,$$

or changing the constants,

$$\Sigma b_n x^n = C \sin(qx + \alpha) \quad\text{...................... (7).}$$

Also by (5),

$$a_n = b_n - 3\frac{b_n}{n+2} = b_n + \frac{3}{q^2}(n+1)\, b_{n+2} \quad\text{........... (8),}$$

for by (6),

$$(n+1)\, b_{n+2} + \frac{q^2}{n+2} b_n = 0,$$

$$\therefore\ \Sigma a_n x^n = \Sigma b_n x^n + \frac{3}{q^2}\Sigma(n-1)\, b_n x^{n-2}$$

$$= \Sigma b_n x^n + \frac{3}{q^2}\left(\frac{\Sigma b_n x^n}{x}\right)',$$

$$\therefore\ y = C\sin(qx+\alpha) + \frac{3C}{q^2}\left\{\frac{\sin(qx+\alpha)}{x}\right\}' \quad\text{.......... (9),}$$

the complete solution, which may be written thus,

$$y = C\left\{\sin(qx+\alpha)\left(1 - \frac{3}{q^2x^2}\right) + \frac{3}{qx}\cos(qx+\alpha)\right\} \quad\text{......(10).}$$

We now proceed to the more general equation,

$$\frac{d^2y}{dx^2}+q^2y=p(p-1)\frac{y}{x^2} \quad\text{.................. (11).}$$

As before, we shall get

$$\{n(n-1)-p(p-1)\}a_n+q^2a_{n-2}=0 \quad\text{.......... (12).}$$

Now $\quad n(n-1)-p(p-1)=(n-p)(n+p-1)$ ...... (13),

which is the fundamental principle of our analysis,

$$\therefore\ (n-p)(n+p-1)a_n+q^2a_{n-2}=0 \quad\text{......... (14).}$$

Assume $\quad (n+p-1)a_n=(n-p+2)b_n$ .... (15),

$$\therefore\ a_{n-2}=\frac{n-p}{n+p-3}b_{n-2},$$

and $(n-p+2)(n+p-3)b_n+q^2b_{n-2}=0$......... (16).

Again, assume

$$(n+p-3)b_n=(n-p+4)c_n \quad\text{............... (17),}$$

and so on successively. Thus we shall get a series of equations, of which

$$(n-p+\mu)(n+p-\mu-1)l_n+q^2l_{n-2}=0 \quad\text{......... (18),}$$

is the general type, where μ is even.

If p is even, let $p=\mu$, $\therefore p-\mu-1=-1$.

If it is odd, let $p=\mu+1$, $\therefore -p+\mu=-1$:

and in both cases (18) becomes

$$n(n-1)l_n+q^2l_{n-2}=0,$$

and therefore

$$\Sigma l_nx^n=C\sin(qx+\alpha) \quad\text{..................... (19).}$$

Let $\quad \{n-p+(\mu-2)\}(n+p-\mu+1)i_n+q^2i_{n-2}=0,$

$$(n-p+\mu)(n+p-\mu-1)k_n+q^2k_{n-2}=0,$$

be any two consecutive equations; then

$$(n+p-\mu+1)i_n=(n-p+\mu)k_n \quad\text{............(20),}$$

$$n-p+\mu=n+p-\mu+1-2(p-\mu)-1,$$

$$\therefore\ i_n=k_n-\{2(p-\mu)+1\}\frac{k_n}{n+p-\mu+1} \quad\text{....... (21),}$$

and $$\frac{k_n}{n+p-\mu+1} = -\frac{1}{q^2}(n+2-p+\mu)\,k_{n+2},$$

$$\therefore\ \Sigma i_n x^n = \Sigma k_n x^n + \frac{2(p-\mu)+1}{q^2}\,\Sigma\,(n-p+\mu)\,k_n x^{n-2}.$$

Now $$(n-p+\mu)\,k_n x^{n-2} = x^{p-\mu-1}(n-p+\mu)\,k_n x^{n-p+\mu-1}$$

$$= x^{p-\mu-1}\left(\frac{k_n x^n}{x^{p-\mu}}\right)',$$

$$\therefore\ \Sigma i_n x^n = \Sigma k_n x^n + \frac{2(p-\mu)+1}{q^2}\,x^{p-\mu-1}\left(\frac{\Sigma k_n x^n}{x^{p-\mu}}\right)' \quad\text{......(22).}$$

By the application of this formula, y or $\Sigma a_n x^n$ may be deduced by a series of regular operations from $C\sin(qx+a)$.

If p is even, $2(p-\mu)+1$ gives the series 1, 5, 9, &c.

If it is odd, the series is 3, 7, 11, &c.

Particular cases may be solved by (22) with considerable facility. By inspection we have

$$y = C\left\{\sin(qx+\alpha) + \frac{1}{qx}\cos(qx+\alpha)\right\}$$

for the solution of

$$\frac{d^2y}{dx^2} + q^2y = \frac{2y}{x^2}.$$

The solution of $\dfrac{d^2y}{dx^2} + q^2y = \dfrac{20y}{x^2}$, where $p=5$, is easily seen to be

$$y = C\left\{\sin(qx+\alpha) + \frac{3}{q^2}\left(\frac{\sin(qx+\alpha)}{x}\right)'\right\}$$

$$+\frac{7C}{q^2}x^2\left[\frac{1}{x^3}\left\{\sin(qx+\alpha) + \frac{3}{q^2}\left(\frac{\sin(qx+\alpha)}{x}\right)'\right\}\right]'.$$

Lastly, the solution of

$$\frac{d^2y}{dx^2} + q^2y = \frac{12y}{x^2},$$

is $$y = C\left\{\sin(qx+\alpha) + \frac{1}{qx}\cos(qx+\alpha)\right\}$$

$$+\frac{5C}{q^2}\,x\left\{\frac{1}{x^2}\left(\sin(qx+\alpha) + \frac{1}{qx}\cos(qx+\alpha)\right)\right\}'.$$

The second line is equivalent to

$$\frac{5C}{q^2x}\left\{q\cos(qx+\alpha)-\frac{1}{qx^2}\cos(qx+\alpha)-\frac{1}{x}\sin(qx+\alpha)\right.$$
$$\left.-\frac{2}{x}\sin(qx+\alpha)-\frac{2}{qx^2}\cos(qx+\alpha)\right\},$$

and thus

$$y=C\left\{\sin(qx+\alpha)+\frac{6}{qx}\cos(qx+\alpha)-\frac{15}{q^2x^2}\sin(qx+\alpha)\right.$$
$$\left.-\frac{15}{q^3x^3}\cos(qx+\alpha)\right\}.$$

These examples will sufficiently illustrate the general formula.

The same method is applicable to the equation

$$\frac{d^3y}{dx^3}+q^3y=p(p-1)\frac{1}{x^2}\frac{dy}{dx} \dots\dots\dots\dots\dots\dots (23).$$

Here we have

$$n\{(n-1)(n-2)-p(p-1)\}a_n+q^3a_{n-3}=0,$$
$$\therefore\ n(n-p-1)(n+p-2)a_n+q^3a_{n-3}=0.$$

Let $\qquad (n+p-2)a_n=(n-p-1+3)b_n,$

$$\therefore\ n(n-p-1+3)(n+p-2-3)b_n+q^3b_{n-3}=0;$$

and generally

$$n(n-p-1+\nu)(n+p-2-\nu)l_n+q^3l_{n-3}=0 \dots\dots (24),$$

where ν is divisible by 3.

If p is so too, let $p=\nu$,

$$\therefore\ n-p-1+\nu=n-1 \text{ and } n+p-2-\nu=n-2.$$

If $p-1$ is divisible by 3, let $p-1=\nu$,

$$\therefore\ -p-1+\nu=-2 \text{ and } p-2-\nu=-1;$$

and in both cases (24) becomes

$$n(n-1)(n-2)l_n+q^3l_{n-3}=0,$$

and therefore Σl_nx^n fulfils the equation

$$\frac{d^3z}{dx^3}+q^3z=0 \dots\dots\dots\dots\dots\dots\dots\dots (25).$$

Again, if

$$n(n-p-1+\nu-3)(n+p-2-\nu+3)\,l_n+q^3 l_{n-3}=0,$$

$$n(n-p-1+\nu)(n+p-2-\nu)\,k_n+q^3 k_{n-3}=0,$$

be any consecutive equations, we have

$$(n+p-2-\nu+3)\,i_n=(n-p-1+\nu)\,k_n,$$

$$\therefore\ i_n=k_n-2(p+1-\nu)\frac{k_n}{n+p+1-\nu}.$$

Also, $$\frac{k_n}{n+p+1-\nu}=-\frac{1}{q^3}(n+3)(n+3)+p-1+\nu)\,k_{n+3},$$

$$\therefore\ \Sigma i_n x^n=\Sigma k_n x^n+\frac{2(p+1-\nu)}{q^3}\Sigma n(n-p-1+\nu)\,k_n x^{n-3},$$

$$\text{or } \Sigma i_n x^n=\Sigma k_n x^n+\frac{2(p+1-\nu)}{q^3}x^{p-\nu-1}\left\{\frac{(\Sigma k_n x^n)'}{x^{p-\nu}}\right\}'\ldots\ldots(26),$$

as may easily be seen *à priori*, or verified by differentiation. The formula (26) is used in the same way as (22), to which it is analogous.

We will give one instance of its application,

$$\frac{d^3y}{dx^3}+q^3y=\frac{6}{x^2}\frac{dy}{dx}.$$

The solution of

$$\frac{d^3z}{dx^3}+q^3z=0, \text{ is}$$

$$z=C_1e^{-qx}+C_2e^{\frac{q}{2}x}\sin\left(\frac{\sqrt{3}}{2}qx+\alpha\right),$$

and by (26),

$$y=z+\frac{2.(3+1-3)}{q^3}x^{-1}\left(\frac{dz}{x^0dx}\right)',$$

$$\text{or } y=z+\frac{2}{q^3x}\frac{d^2z}{d^2x}, \text{ which gives}$$

$$y=C_1e^{-qx}\left(1+\frac{2}{qx}\right)$$

$$+C_2e^{\frac{q}{2}x}\left\{\sin\left(\frac{\sqrt{3}}{2}qx+\alpha\right)\left(1-\frac{1}{qx}\right)+\frac{\sqrt{3}}{qx}\cos\left(\frac{\sqrt{3}}{2}qx+\alpha\right)\right\},$$

for the complete solution of the proposed equation.

It is obvious that analogous equations exist in all orders, and that when p is of certain forms,

$$\frac{d^m y}{dx^m} + q^m y = p(p-1)\frac{1}{x^2}\frac{d^{m-2}}{dx^{m-2}}y$$

may be integrated in finite terms.

It will be sufficient, after what has been said for the cases of $m = 2$ and $= 3$, to state the results of the general investigation; they may be very readily deduced by the same method as that we have already used.

The process succeeds when either of the factors p or $p-1$ is divisible by m, and the general formula of which (22) and (26) are cases, is

$$\Sigma i_n x^n = \Sigma k_n x^n + \frac{2(p-rm)+m-1}{q^m} x^{p-rm-1}\left(x^{-(p-rm)}\frac{d^{m-2}}{dx^{m-2}}\Sigma k_n x^n\right)' \quad \text{...... (27).}$$

Particular cases may however be easily solved without reference to this formula; thus, if we had

$$\frac{d^4 y}{dx^4} - q^4 y = \frac{12}{x^2}\frac{d^2 y}{dx^2},$$

we should proceed as follows:

$$n(n-1)\{(n-2)(n-3) - 4.3\} a_n - q^4 a_{n-4} = 0,$$

$$n(n-1)(n-6)(n+1) a_n - q^4 a_{n-4} = 0,$$

$$(n+1) a_n = (n-2) b_n,$$

$$\therefore\ n(n-1)(n-2)(n-3) b_n - q^4 b_{n-4} = 0,$$

$$\therefore\ (\Sigma b_n x^n)^{iv} - q^4 \Sigma b_n x^n = 0,$$

$$\therefore\ \Sigma b_n x^n = C_1 e^{qx} + C_2 e^{-qx} + C_3 \sin(qx + \alpha),$$

and $a_n = b_n - \dfrac{3}{n+1} b_n = b_n - \dfrac{3}{q^4}(n+4)(n+3)(n+2) b_{n-4},$

$$\therefore\ y = \Sigma b_n x^n - \frac{3}{q^4}\Sigma n(n-1)(n-2) b_n x^{n-4},$$

$$= \Sigma b_n x^n - \frac{3}{q^4 x}\frac{d^3}{dx^3}\Sigma b_n x^n,$$

$$\therefore\ y = C_1 e^{qx}\left(1 - \frac{3}{qx}\right) + C_2 e^{-qx}\left(1 + \frac{3}{qx}\right)$$

$$+ C_3\left\{\sin(qx+\alpha) + \frac{3}{qx}\cos(qx+\alpha)\right\}.$$

The principle of our analysis, it has already been remarked, is contained in the equation

$$n(n-1) - p(p-1) = (n-p)(n+p-1);$$

and this consideration suggests an extension of it.

For it is obvious that the coefficients of a_n in

$$\frac{d^m y}{dx^m}, \text{ and in } p(p-1)\frac{d^s}{dx^s}\left(\frac{1}{x^2}\frac{d^{m-s-2}}{dx^{m-s-2}}y\right)$$

differ only in this, that where the first has the factors

$$(n-m+s+2)(n-m+s+1),$$

the second has $p(p-1)$.

Thus the same transformation applies; and if either p, or $p-1$ is divisible by m, the solution of

$$\frac{d^m y}{dx^m} + q^m y = p(p-1)\frac{d^s}{dx^s}\left(\frac{1}{x^2}\frac{d^{m-s-2}}{dx^{m-s-2}}y\right) \text{.........} (28),$$

may be made to depend on that of

$$\frac{d^m y}{dx^m} + q^m y = 0,$$

and thus effected in finite terms.

The formula of reduction in this case is a little more complicated than those already given, and we will not dwell longer upon it, our object being rather to point out the integrability of certain classes of equations than actually to integrate them.

The equation

$$n(n-1) - p(p-1) = (n-p)(n+p-1)$$

is a particular case of

$$n(n-\mu) - p(p-\mu) = (n-p)(n+p-\mu),$$

and the latter will give us various formulæ of reduction according to the value of μ. Thus

$$\frac{d^2 y}{dx^2} - \frac{1}{x}\frac{dy}{dx} + q^2 y = p(p-2)\frac{y}{x^2},$$

may be reduced to

$$\frac{d^2y}{dx^2} - \frac{1}{x}\frac{dy}{dx} + q^2y = 0,$$

for the coefficient of a_n in the former is

$$n(n-2) - p(p-2) = (n-p)(n+p-2),$$

which, provided p is even, may be reduced to $n(n-2)$. But in this, and in analogous cases, the auxiliary equation is, apparently, insoluble.

The applicability of our transformation would, it is evident, not be affected, if the equation were, instead of (25),

$$\frac{d^my}{dx^m} + q^{m-t}\frac{d^ty}{dx} = 0, \text{ \&c. } \ldots\ldots\ldots\ldots\ldots\ldots \text{ (29);}$$

and, provided p or $p-1$ were divisible by $m-t$, (29) might be reduced to

$$\frac{d^my}{dx^m} + q^{m-t}\frac{d^ty}{dx^t} = 0.$$

But this case requires more care than those already considered, as if certain factors which apparently disappear are neglected, our solution is incomplete, or erroneous.

An instance will make this clear,

$$\frac{d^2y}{dx^2} + q\frac{dy}{dx} = 2\frac{y}{x^2}.$$

Here $\qquad (n-2)(n+1)a_n + (n-1)^q a_{n-1} = 0 \ldots\ldots\ldots \text{(A)}.$

Let $\qquad (n+1)a_n = (n-1)b_n \ldots\ldots\ldots\ldots\ldots\ldots (\alpha),$

$$\therefore\ (n-2)(n-1)nb_n + (n-2)(n-1)^q b_{n-1} = 0 \ldots\ldots \text{(B)}.$$

The factor $n-2$ may be safely neglected. But $n-1$ is essential, because the solution of the auxiliary equation

$$\frac{d^2z}{dx^2} + q\frac{dz}{dx} = 0$$

gives $\qquad (n-1)(nb_n + qb_{n-1}) = 0,$

and would be incomplete if we omitted the first factor.

From (α) we get

$$a_n - b_n - 2\frac{b_n}{n+1},$$

and, as except when $n=1$, there is

$$nb_n + qb_{n-1} = 0, \quad \frac{b_n}{n+1} = -\frac{1}{q} b_{n+1}, \text{ unless } n = 0,$$

$$\therefore\ a_n = b_n + \frac{2}{q} b_{n+1} \dots\dots\dots\dots\dots (\alpha').$$

Now the solution of the auxiliary equation is

$$z = c_1 + c_2 e^{-qx};$$

and from (α') we deduce

$$y = z + \frac{2}{qx} z;$$

therefore
$$y = \left(1 + \frac{2}{qx}\right)(c_1 + c_2 e^{-qx})$$

is apparently the solution of the proposed equation. But it will be found not to satisfy it, unless $c_1 = 0$, and then

$$y = c_2 \left(1 + \frac{2}{qx}\right) e^{-qx}$$

is only a particular solution. The reason is, that in laying down (α') as generally true, we imply that

$$nb_n + qb_{n-1} = 0$$

is true for $n=1$; whereas the equation which contains the solution of $\frac{d^2z}{dx^2} + q\frac{dz}{dx} = 0$, viz.

$$(n-1)(nb_n + qb_{n-1}) = 0,$$

shows that b_1 is not *necessarily* connected with b_0; and that if we assume such connection, we get only a particular solution. Hence our formula of reduction implies the connection of b_1 and b_0; while their independence is implied in the general solution of the auxiliary equation, to which this formula is applied; and these contradictory suppositions lead to an erroneous result. To put $c_1 = 0$, is to connect b_0 and b_1, or, which is the same thing, to neglect the factor $n-1$; and the value of y thus got is therefore a solution, but not the complete solution of the proposed equation.

To complete it, we must, bearing in mind the independence of b_0, recur to (α), which is always true,

$$\therefore\ a_0 = -b_0;$$

and from (α'), which is true for $n = -1$, we get

$$a_{-1} = b_{-1} + \frac{2}{q} b_0.$$

Now b_{-1} is obviously $= 0$,

$$\therefore \; a_{-1} = -\frac{2}{q} a_0;$$

and these two quantities are independent of a_1, a_2, &c.,

$$\therefore \; y = a_0 \left(1 - \frac{2}{qx}\right)$$

is a particular solution, and

$$y = c_1 \left(1 - \frac{2}{qx}\right) + c_2 \left(1 + \frac{2}{qx}\right) e^{-qx}$$

is the complete solution of the proposed equation.

The method of proceeding suggested by this example is to obtain a solution, neglecting all factors analogous to $(n - 1)$, and then to complete it by reference to the assumptions of transformation, such as (α), which have been made use of.

The equations which we have solved are not a very numerous nor perhaps an important class. But one of them, at least, is susceptible of a physical application of great interest; and so few equations of the higher orders are integrable in finite terms, that the discussion of those which are, has always some degree of value.

ON THE INTEGRATION OF CERTAIN DIFFERENTIAL EQUATIONS*.

No. II.

In the last number of the Journal†, a method was given for the investigation of a class of differential equations, by means of successive reductions.

The present communication contains solutions of some analogous equations effected by a similar process. The results will however be exhibited in a very different form.

We begin by taking a particular case of the equations in question,

$$\frac{d^m y}{dx^m} + ky = \frac{pm}{x}\frac{d^{m-1}y}{dx^{m-1}} \ldots\ldots\ldots\ldots\ldots\ldots\ldots(1),$$

where p is an integer.

Let
$$y = \Sigma a_n x^n,$$

$$\therefore\ n(n-1)\ldots(n-m+2)(n-m+1-pm)\,a_n + ka_{n-m} = 0\ldots(2).$$

Assume

$$a_n = \{n-m+1-(p-1)m\}\{n-m+1-(p-2)m\} \\ \ldots(n-m+1)f(k)\,b_n\ldots\ldots(3),$$

$f(k)$ being some function of k, to be determined hereafter. Then

$$a_{n-m} = (n-m+1-pm)\{n-m+1-(p-1)m\} \\ \ldots(n-m+1-m)f(k)\,b_{n-m}\ldots\ldots(4).$$

* *Cambridge Mathematical Journal*, No. XI. Vol. II. p. 193, February, 1841.

† Page 96 of this volume.

If we substitute these values in (2), every factor of (4) and every factor, except the last, of (3), will disappear, and the resulting equation will be

$$n(n-1)\dots(n-m+2)(n-m+1)\,b_n + kb_{n-m} = 0 \dots\dots (5).$$

This is what (2) would be, were $p=0$. Hence $y = \Sigma . b_n x^n$ fulfils the equation $\frac{d^m y}{dx^m} + ky = 0$, to which (1) would, in that case, be reduced. Let $y = X$ be the ordinary form of the solution of the last-written equation. Then we must obviously have

$$\Sigma . b_n x^n = \phi(k) . X,$$

where $\phi(k)$ may be any function of k. A very little attention to the mode of integrating linear equations with constant coefficients will show, that in X, x always occurs in conjunction with $k^{\frac{1}{m}}$.

If we put

$$X = \Sigma . A_n x^n,$$

we must consequently have

$$A_n = Nk^{\frac{n}{m}},$$

where N is a function of n, except for values of $n < m$, when it is an arbitrary constant; therefore

$$b_n = N\phi(k)\,k^{\frac{n}{m}}.$$

Recurring to (3), inverting the factors and multiplying and dividing by m^p, we shall easily deduce the following equation,

$$a_n = m^p \left(\frac{n-m+1}{m}\right)\left(\frac{n-m+1}{m} - 1\right)$$
$$\dots \left\{\frac{n-m+1}{m} - (p-1)\right\} N . f(k)\,\phi(k)\,k^{\frac{n}{m}}.$$

The form of these p decreasing factors naturally suggests the idea of making $f(k) = k^{-p}$; and if we then put $\phi(k) = k^{-\frac{m-1}{m}}$, we get

$$a_n = m^p \left(\frac{n-m+1}{m}\right)\left(\frac{n-m+1}{m} - 1\right)$$
$$\dots \left\{\frac{n-m+1}{m} - (p-1)\right\} Nk^{\frac{n-m+1}{m} - p};$$

$$\therefore \ a_n = m^p \frac{d^p}{dk^p} b_n, \text{ and } \therefore \ \Sigma a_n x^m = m^p \frac{d^p}{dk^p} \Sigma b_n x^n.$$

The factor m^p may obviously be neglected, and we shall therefore have, on replacing $\Sigma b_n x^n$ by $\phi(k) X$, i. e. by $\dfrac{X}{k^{\frac{m-1}{m}}}$, the following equation,

$$y = \frac{d^p}{dk^p} \frac{X}{k^{\frac{m-1}{m}}},$$

for the solution of (1), $y = X$ being that of $\dfrac{d^m y}{dx^m} + ky = 0$.

If $m = 2$, $X = C \sin\{\sqrt{(k)}\, x + \alpha\}$, and $\dfrac{m-1}{m} = \frac{1}{2}$. Hence

$$y = \frac{d^p}{dk^p} \frac{C \sin\{\sqrt{(k)}\, x + \alpha\}}{\sqrt{(k)}}$$

is the solution of

$$\frac{d^2 y}{dx^2} + ky = \frac{2p}{x} \frac{dy}{dx}.$$

This result is given in Hymers' *Diff. Equations*, and is, I believe, due to Mr Gaskin.

If the proposed equation were

$$\frac{d^m y}{dx^m} + ky + \frac{pm}{x} \frac{d^{m-1} y}{dx^{m-1}} = 0,$$

we should immediately conclude, from analogy, that its solution must be

$$y = \frac{d^{-p}}{dk^{-p}} \frac{X}{k^{\frac{m-1}{m}}};$$

but it may be as well to establish this conclusion by an independent investigation.

Equation (2) will, in this case, be

$$n(n-1)\ldots(n-m+2)(n-m+1+pm)\, a_n + ka_{n-m} = 0.$$

Assume

$$a_n = f(k) \frac{b_n}{(n-m+1+pm)\{n-m+1+(p-1)m\}\ldots(n-m+1+m)},$$

$$\therefore \ a_{n-m} = f(k) \frac{b_{n-m}}{\{n-m+1+(p-1)m\}\ldots(n-m+1)},$$

$$\therefore \ n(n-1)\ldots(n-m+1)\, b_n + kb_{n-m} = 0.$$

Therefore as before,

$$b_n = N.\phi(k)\,k^{\frac{n}{m}},$$

$$\text{and } a_n = m^{-p} N \frac{f(k)\,\phi(k)\,k^{\frac{n}{m}}}{\left(\frac{n-m+1}{m}+1\right)\dots\left(\frac{n-m+1}{m}+p\right)}.$$

Here we make $f(k) = k^p$, and $\phi(k) = k^{-\frac{m-1}{m}}$,

$$\therefore\ a_n = m^{-p} N \frac{d^{-p}}{dk^{-p}} b_n,$$

no complementary function being added.

Hence, precisely as before, we find that

$$y = \frac{d^{-p}}{dk^{-p}} \frac{X}{k^{\frac{m-1}{m}}}$$

is the solution required.

Let us now consider the more general equation,

$$\frac{d^m y}{dx^m} + ky = pm \frac{d^{m-s-1}}{dx^{m-s-1}}\left(\frac{1}{x}\frac{d^s y}{dx^s}\right) \dots\dots\dots\dots\dots (6).$$

If $y = \Sigma . a_n x^n$, there will be

$$n \dots (n-s+1)(n-s-pm)(n-s-1)\dots$$
$$\dots (n-m+1)\,a_n + ka_{n-m} = 0 \dots\dots (7).$$

This equation is analogous to (2); but $m-1$ is replaced by s. Assume, therefore,

$$a_n = \{n-s-(p-1)\,m\}\{n-s-(p-2)\,m\}\dots(n-s)\,k^{-p}.b_n \dots (8),$$

$$\therefore\ a_{n-m} = (n-s-pm)\{n-s-(p-1)\,m\}\dots(n-s+m)\,k^{-p}b_{n-m},$$

and $n\dots(n-s+1)(n-s)(n-s-1)\dots(n-m+1)\,b_n + kb_{n-m} = 0.$

Hence, as before,

$$\Sigma b_n x^n = \phi(k)\,X,$$

where X denotes the same function of x that it did in the former case.

Consequently $\qquad b_n = N\phi(k)\,k^{\frac{n}{m}}$, and

$$a_n = m^p \left(\frac{n-s}{m}\right)\left(\frac{n-s}{m}-1\right)\dots\left\{\frac{n-s}{m}-(p-1)\right\} N\phi(k)\,k^{\frac{n}{m}-p}.$$

We must, it is evident, make $\phi(k) = k^{-\frac{s}{m}}$, and then

$$a_n = m^p \frac{d^p}{dk^p} N k^{\frac{n-s}{m}} = m^p \frac{d^p}{dk^p} b_n .$$

Hence $y = \dfrac{d^p}{dk^p} \dfrac{X}{k^{\frac{s}{m}}}$ is the solution of the equation

$$\frac{d^m y}{dx^m} + ky = pm \frac{d^{m-s-1}}{dx^{m-s-1}} \left(\frac{1}{x} \frac{d^s y}{dx^s}\right),$$

for as before the factor m^p may be neglected.

A particular case of this result is that in which $s = 0$. If with this value of s we have $m = 2$, the equation to be integrated takes the form

$$\left(\frac{d^2 y}{dx^2} + ky\right) x^2 = 2p \left(x \frac{dy}{dx} - y\right),$$

and the solution is

$$y = C \frac{d^p}{dk^p} \sin \{\sqrt{(k)}\, x + \alpha\}.$$

Equation (7) is the most general one in which the coefficient of a_n differs in one factor only from what it is in the case of

$$\frac{d^m y}{dx^m} + ky = 0.$$

But our method is applicable in other cases.

Let us resume the equation discussed in the last number of the Journal,

$$\frac{d^m y}{dx^m} + ky = p(p-1) \frac{1}{x^2} \frac{d^{m-2}}{dx^{m-2}} y \;\ldots\ldots\ldots\ldots\ldots (8).$$

By the usual method of making $y = \Sigma a_n x^n$, we get

$$n(n-1)\ldots(n-m+3)\{(n-m+2)(n-m+1) - p\,.\,(p-1)\}\, a_n + k a_{n-m} = 0,$$

or $$n(n-1)\ldots(n-m+3)\{(n-m+2-p)(n-m+1+p)\}\, a_n + k a_{n-m} = 0 \ldots\ldots (9).$$

It may be remembered that we found it necessary that p or $p-1$ should be divisible by m. Suppose then that p is so divisible, and that the quotient is q.

Let

$$a_n = fk \frac{\{(n-m+2-(p-m)\}\{n-m+2-(p-2m)\}\dots(n-m+2)}{(n-m+1+p)\{n-m+1+(p-m)\}\dots(n-m+1+m)} b_n \qquad \dots\dots(10),$$

$$\therefore\ a_{n-m} = fk \frac{(n-m+2-p)\{n-m+2-(p-m)\}\dots\dots}{\{n-m+1+(p-m)\}\dots\dots(n-m+1)} b_{n-m},$$

(there are q factors in both numerator and denominator); equation (9) becomes

$$n(n-1)\dots(n-m+3)(n-m+2)(n-m+1)\, b_n + kb_{n-m} = 0;$$

and, as in the two preceding cases, we shall have

$$\Sigma b_n x^n = \phi(k)\, X,$$

$$\text{and } b_n = N.\, k^{\frac{n}{m}} \phi(k).$$

(10) may be written thus, as $p = qm$,

$$a_n = f(k) \frac{\left(\frac{n-m+2}{m}\right)\dots\dots\left(\frac{n-m+2}{m} - q + 1\right)}{\left(\frac{n-m+1}{m} + 1\right)\dots\dots\left(\frac{n-m+1}{m} + q\right)} b_n.$$

Now $\dfrac{d^q}{dk^q} k^{\frac{n-m+2}{m}} = \left(\dfrac{n-m+2}{m}\right)\dots\left(\dfrac{n-m+2}{m} - q + 1\right) k^{\frac{n-m+2}{m} - q}$;

$$\text{therefore } \frac{d^{-q}}{dk^{-q}}\left(k^{q-\frac{1}{m}} \frac{d^q}{dk^q}.\, k^{\frac{n-m+2}{m}}\right)$$

$$= \frac{\left(\frac{n-m+2}{m}\right)\dots\dots\left(\frac{n-m+2}{m} - q + 1\right)}{\left(\frac{n-m+1}{m} + 1\right)\dots\dots\left(\frac{n-m+1}{m} + q\right)}.\, k^{\frac{n-m+1}{m} - q}.$$

But if we make $\phi(k) = k^{\frac{-m+2}{m}}$, then

$$b_n = N.\, k^{\frac{n-m+2}{m}}.$$

Let us also put $f(k) = k^{q-\frac{1}{m}}$; therefore

$$a_n = \frac{\left(\frac{n-m+2}{m}\right)\dots\ \&c.}{\left(\frac{n-m+1}{m} + 1\right)\dots\ \&c.} N k^{\frac{n-m+1}{m} + q}$$

$$= N \frac{d^{-q}}{dk^{-q}}\left(k^{q-\frac{1}{m}} \frac{d^q}{dk^q} k^{\frac{n-m+2}{m}}\right),$$

$$\text{or } a_n = \frac{d^{-q}}{dk^{-q}}\left(k^{q-\frac{1}{m}}\frac{d^q}{dk^q}b_n\right).$$

Hence
$$y = \frac{d^{-q}}{dk^{-q}}\left(k^{q-\frac{1}{m}}\frac{d^q}{dk^q}\frac{X}{k^{\frac{m-2}{m}}}\right) \text{..................} (11)$$

is the solution required.

It admits also of another form, which it may be worth while to remark.

$$\frac{d^{-q}}{dk^{-q}}k^{\frac{n-m+1}{m}} = \frac{1}{\left(\frac{n-m+1}{m}+1\right)\ldots\ldots\left(\frac{n-m+1}{m}+q\right)}k^{\frac{n-m+1}{m}+q},$$

$$\text{therefore } \frac{d^q}{dk^q}\left(k^{-q+\frac{1}{m}}\frac{d^{-q}}{dk^{-q}}k^{\frac{n-m+1}{m}}\right)$$

$$= \frac{\left(\frac{n-m+2}{m}\right)\ldots\ldots\left(\frac{n-m+2}{m}-q+1\right)}{\left(\frac{n-m+1}{m}+1\right)\ldots\ldots\left(\frac{n-m+1}{m}+q\right)}k^{\frac{n-m+2}{m}-q},$$

Here we must make $\phi(k) = k^{\frac{-m+1}{m}}$ and $f(k) = k^{-q+\frac{1}{m}}$, and then

$$b_n = Nk^{\frac{n-m+1}{m}} \text{ and } a_n = N(\ldots\ldots)\,k^{\frac{n-m+2}{m}-q},$$

$$\text{therefore } a_n = \frac{d^q}{dk^q}\left(k^{-q+\frac{1}{m}}\frac{d^{-q}}{dk^{-q}}b_n\right)$$

$$\text{and } y = \frac{d^q}{dk^q}\left(k^{-q+\frac{1}{m}}\frac{d^{-q}}{dk^{-q}}\frac{X}{k^{\frac{m-1}{m}}}\right) \text{..................} (12).$$

The value of y, deduced from the development of (12), can of course differ only in a factor of some function of k from that which is given by (11), and it will easily appear, on comparing the values of a_n in the two cases, that this factor is $k^{-2q+\frac{1}{m}}$.

Let us now consider the case in which p is not divisible by m, while $p-1$ is so. And let $p-1 = qm$. The two factors of (9) on which our reduction operates, viz.

$$(n-m+2-p)(n-m+1+p),$$

may be written thus,

$$\{n-m+2+(p-1)\}\{n-m+1-(p-1)\},$$

$$\text{or } (n-m+2+qm)(n-m+1-qm).$$

The change which this will introduce in the process, is not difficult to perceive. We must assume

$$a_n = f(k) \frac{\frac{n-m+1}{m} \ldots\ldots \frac{n-m+1}{m} - q + 1}{\frac{n-m+2}{m} + 1 \ldots\ldots \frac{n-m+2}{m} + q} b_n$$

as before, the transformed equation will be

$$n \ldots\ldots (n-m+2)(n-m+1)\, b_n + k b_{n-m} = 0.$$

Now

$$\frac{d^q}{dk^q} k^{\frac{n-m+1}{m}} = \left(\frac{n-m+1}{m}\right) \ldots\ldots \left(\frac{n-m+1}{m} - q + 1\right) k^{\frac{n-m+1}{m} - q},$$

$$\text{therefore } \frac{d^{-q}}{dk^{-q}} \left(k^{q+\frac{1}{m}} \frac{d^q}{dk^q} k^{\frac{n-m+1}{m}}\right)$$

$$= \frac{\left(\frac{n-m+1}{m}\right) \ldots\ldots \left(\frac{n-m+1}{m} - q + 1\right)}{\left(\frac{n-m+2}{m} + 1\right) \ldots\ldots \left(\frac{n-m+2}{m} + q\right)} k^{\frac{n-m+2}{m} + q}.$$

If then we make $f(k) = k^{q+\frac{1}{m}}$ and $\phi(k) = k^{\frac{-m+1}{m}}$, we get

$$b_n = N k^{\frac{n-m+1}{m}}; \quad a_n = N \frac{\left(\frac{n-m+1}{m}\right) \ldots \&c.}{\left(\frac{n-m+2}{m} + 1\right) \ldots \&c.} k^{\frac{n-m+2}{m} + q},$$

$$\text{or } a_n = N \frac{d^{-q}}{dk^{-q}} \left(k^{q+\frac{1}{m}} \frac{d^q}{dk^q} k^{\frac{n-m+1}{m}}\right).$$

Hence, finally,

$$y = \frac{d^{-q}}{dk^{-q}} \left(k^{q+\frac{1}{m}} \frac{d^q}{dk^q} \frac{X}{k^{\frac{m-1}{m}}}\right) \ldots\ldots\ldots\ldots\ldots\ldots\ldots (13).$$

As an illustration, let us take the case of the equation which occurs in the theory of the figure of the earth,

$$\frac{d^2y}{dx^2} + ky = \frac{6y}{x^2}.$$

Here $m = 2$, $p = 3$, $p - 1 = 2$: hence $q = 1$, and as p is not a multiple of m, the formula (13) is to be used.

It is in this case, as $X = C\sin\{\sqrt{(k)}\,x + \alpha\}$,

$$y = \frac{d^{-1}}{dk^{-1}}\left[k^{\frac{3}{2}}\frac{d}{dk}\,\frac{C\sin\{\sqrt{(k)}\,x + \alpha\}}{k^{\frac{1}{2}}}\right],$$

$$\frac{d}{dk}\,\frac{\sin\{\sqrt{(k)}\,x + \alpha\}}{k^{\frac{1}{2}}} = \tfrac{1}{2}\frac{x}{k}\cos\{\sqrt{(k)}\,x + \alpha\} - \tfrac{1}{2}\frac{\sin\{\sqrt{(k)}\,x + \alpha\}}{k^{\frac{3}{2}}},$$

therefore

$$y = \tfrac{1}{2}C\frac{d^{-1}}{dk^{-1}}\left[xk^{\frac{1}{2}}\cos\{\sqrt{(k)}\,x + \alpha\} - \sin\{\sqrt{(k)}\,x + \alpha\}\right],$$

or integrating the first term by parts,

$$y = kC\sin\{\sqrt{(k)}\,x + \alpha\} - \tfrac{3}{2}C\frac{d^{-1}}{dk^{-1}}\sin\{\sqrt{(k)}\,x + \alpha\}.$$

But

$$\frac{d^{-1}}{dk^{-1}}\sin\{\sqrt{(k)}\,x + \alpha\} = \frac{2k^{\frac{1}{2}}}{x}\cos\{\sqrt{(k)}\,x + \alpha\} + \frac{2}{x^2}\sin\{\sqrt{(k)}\,x + \alpha\},$$

therefore, if $kC = C_1$,

$$y = C_1\left[\sin\{\sqrt{(k)}\,x + \alpha\} + \frac{3}{xk^{\frac{1}{2}}}\cos\{\sqrt{(k)}\,x + \alpha\} - \frac{3}{kx^2}\sin\{\sqrt{(k)}\,x + \alpha\}\right],$$

which is the required solution.

Equation (13) corresponds to (11): but there is another form of the solution in the case of $p - 1 = qm$, which we shall just mention, and which is the counterpart of (12). It is

$$y = \frac{d^q}{dk^q}\left(k^{-q-\frac{1}{m}}\frac{d^{-q}}{dk^{-q}}\frac{X}{k^{\frac{m-2}{m}}}\right)\ldots\ldots\ldots\ldots\ldots\ldots(14).$$

It would be a needless repetition to go through the steps which lead to this result.

All the operations indicated in these symbolical solutions are practicable. This will appear by considering the nature of the function X, which, in its most general form, consists of the sum of terms, of which the type is

$$Ce^{[\alpha+\beta\sqrt{(-1)}]k^{\frac{1}{m}}x}.$$

If we make $k = \kappa^m$, this will become

$$Ce^{[\alpha+\beta\sqrt{(-1)}]\kappa x},$$

which may be integrated any number of times for κ, and consequently, if it is multiplied by any rational and integral function of κ, it may still be integrated by parts as often as we please. Now $\frac{dk}{k^{\frac{s}{m}}}$ will become $m\kappa^{m-s-1}d\kappa$; and as m and s are integral, the method of parts applies, provided s is not greater than $m-1$, which it is in none of our formulæ.

Fourier's expression, by means of definite integrals for the ith differential coefficient of any function, would enable us to extend our solutions to the cases in which p is fractional. But merely analytical transformations of the results at which we have arrived are not of much interest, and the methods of effecting them are direct and obvious.

Equation (1) admits of another symbolical solution besides the one already given.

It is easily seen, that if

$$\Sigma a_n x^n = x^{mp}\frac{d}{dx}\frac{1}{x^{m-1}}\cdot\frac{d}{dx}\frac{1}{x^{m-1}}\cdots\Sigma b_n x^n,\ (p \text{ factors}),$$

$$a_n = (n-pm+1)\ldots(n-m+1)\,b_n,$$

which is what (3) is, when $f(k)=1$. $\left(\frac{d}{dx}\right.$ applies to all that follows it.$\left.\right)$

Hence we shall clearly have

$$y = x^{pm}\frac{d}{dx}\frac{1}{x^{m-1}}\frac{d}{dx}\frac{1}{x^{m-1}}\cdots\cdots\frac{d}{dx}\frac{1}{x^{m-1}}X$$

for the solution of (1).

Similarly, the solution of (6) is

$$y = x^{(p-1)m+(s-1)}\cdot\frac{d}{dx}\frac{1}{x^{m-1}}\cdots\cdots\frac{d}{dx}\frac{1}{x^{s}}X.$$

Many applications and modifications of the method we have employed will readily present themselves, but the subject is not of sufficient importance to deserve a fuller discussion. It is not difficult to multiply artifices, by means of which particular equations may be solved, but the results will, generally speaking, be of little value.

ANALYTICAL DEMONSTRATIONS OF DR MATTHEW STEWART'S THEOREMS*.

IN 1746 Dr Matthew Stewart, the father of Dugald Stewart, published his "General Theorems." He was at that time a candidate for the chair of mathematics at Edinburgh, then vacant by the death of Maclaurin; and his success is attributed to the celebrity which these remarkable propositions immediately acquired. They were enunciated by Dr Stewart without demonstrations, and remained undemonstrated till 1805. Mr Glenie, in the *Edinburgh Transactions* for that year, has given a geometrical method by which the General Theorems and other similar results may be established.

But as yet they have not, I believe, been proved, except by Geometry; and in an article in the 17th volume of the *Edinburgh Review*, ascribed to Playfair, they are strongly recommended to the attention of analysts. It is hoped, therefore, that the following attempt will have some degree of interest.

We shall begin by establishing a general proposition, from which all the theorems in question, and many others, may be deduced.

LEMMA. If $f\phi$ is a rational and integral function of $\sin\phi$ and $\cos\phi$, then a value may always be assigned to n, such that

$$f(\phi)+f\left(\phi+\frac{2\pi}{n}\right)+\dots f\left(\phi+\frac{n-1}{n}2\pi\right)$$

shall be independent of ϕ.

The preceding expression is equivalent to

$$(1+D+\dots D^{n-1})f\phi, \quad \left(\text{where } D\phi=\phi+\frac{2\pi}{n}\right),$$

* *Cambridge Mathematical Journal*, No. XII. Vol. II. p. 271, May, 1841.

and therefore to

$$\frac{D^n f\phi - f\phi}{D-1} = \Delta^{-1}\{f(\phi+2\pi) - f\phi\} = \Delta^{-1}0.$$

Now $\quad \Delta^{-1}0 = \Sigma a_m \sin mn\phi + \Sigma b_m \cos mn\phi, \ (m \text{ integral}).$

Hence $\quad f(\phi) + \ldots + f\left(\phi + \frac{n-1}{n}2\pi\right) = \Sigma a_m \sin mn\phi + \&\text{c}.$

Let the index of the highest power of $\sin\phi$ or $\cos\phi$ in $f\phi$ be p; then it is easily seen that when $f\phi$ is developed, as it may always be, in a series of sines and cosines of the multiple arcs, $p\phi$ will be the largest arc that can enter into the development. But if n is greater than p, $mn\phi$ will be greater than $p\phi$, except when m is zero. Hence the development

$$\Sigma a_m \sin mn\phi + \&\text{c}.$$

cannot coincide with that obtained by summing the separate developments of $f\phi$, $f\left(\phi + \frac{2\pi}{n}\right)$, &c., unless a_m and b_m are $=0$ in every case, except when $m=0$. Hence as $\sin mn\phi = 0$ when $m=0$, the expression will be reduced to b_0, and we shall have, when $n > p$,

$$f(\phi) + \ldots f\left(\phi + \frac{n-1}{n}2\pi\right) = b_0 \ldots \text{a constant.}$$

Q. E. D.

The constant b_0 will of course be the sum of the constant parts of the developments of $f\phi$, &c.; and as these are all equal and are n in number, it will be n times the constant term in $f\phi$. Now by Fourier's theorem this is equal to $\frac{1}{2\pi}\int_0^{2\pi} f\phi \, . \, d\phi$, as indeed is obvious. Hence

$$f\phi + \ldots f\left(\phi + \frac{n-1}{n}2\pi\right) = \frac{n}{2\pi}\int_0^{2\pi} f\phi d\phi, \text{ when } n > p,$$

which is our fundamental formula.

The first of Stewart's propositions is the following:

From any point in the circumference of a circle draw perpendiculars p, p_1, &c. to the sides of a regular n-sided polygon circumscribed about it; then, if r is the radius,

$$2\Sigma p^3 = 5nr^3 \ldots\ldots (1) \ldots\ldots n > 3.$$

DEM. Let the assumed point subtend at the centre an angle ϕ from the adjacent point of contact. Then

$$p = r(1 - \cos\phi),\quad p_1 = r\left\{1 - \cos\left(\phi + \frac{2\pi}{n}\right)\right\},\ \&c.\ \&c.$$

$$\therefore\ \Sigma p^3 = r^3\,\Sigma(1 - \cos\phi)^3;$$

and by the general formula,

$$\Sigma(1 - \cos\phi)^3 = \frac{n}{2\pi}\int_0^{2\pi}(1 - \cos\phi)^3\,d\phi \ldots\ldots\ n \text{ being} > 3.$$

$$\text{Now } \int_0^{2\pi}(1 - \cos\phi)^3\,d\phi = 2^4\int_0^{\pi}\sin^6\theta\,d\theta \ldots\ (\theta = \tfrac{1}{2}\phi),$$

$$\text{and } \int_0^{\pi}\sin^6\theta\,d\theta = \frac{5\,.\,3\,.\,1}{6\,.\,4\,.\,2}\,.\,\pi = \frac{5}{2^5}\,2\pi;$$

$$\therefore\ \Sigma(1 - \cos\phi)^3 = \frac{5n}{2},$$

and therefore $\quad 2\Sigma p^3 = 5nr^3.$ Q. E. D.

This is a particular case of the second proposition in which the assumed point is not confined to the circumference of the circle, but may have any position whatever. Let l be its distance from the centre; then

$$2\Sigma p^3 = 2nr^3 + 3nl^2r \ldots n > 3 \ldots\ldots\ldots\ldots\ (2).$$

DEM. In this case $p = r - l\cos\phi$, &c. = &c.

$$\therefore\ \Sigma p^3 = nr^3 - 3r^2l\,\Sigma\cos\phi + 3rl^2\,\Sigma\cos^2\phi - l^3\,\Sigma\cos^3\phi.$$

$$\text{But } \int_0^{2\pi}\cos\phi\,d\phi = 0,\quad \int_0^{2\pi}\cos^2\phi\,d\phi = \pi,\quad \int_0^{2\pi}\cos^3\phi\,d\phi = 0;$$

$$\therefore\ \Sigma\cos\phi = 0,\quad \Sigma\cos^2\phi = \frac{n}{2},\quad \Sigma\cos^3\phi = 0,$$

and $2\Sigma p^3 = 2nr^3 + 3nl^2r.$ Q. E. D.

In the third proposition, a regular n-sided polygon is *inscribed* in the circle, and lines c, c_1, &c. are drawn from its corners to a point assumed in the circumference; then

$$\Sigma c^4 = 6nr^4 \ldots\ldots\ldots\ldots\ldots\ldots\ldots\ (3).$$

DEM. The assumed point and adjacent corner subtending an angle ϕ at the centre, we have

$$c^2 = 2r^2 (1 - \cos\phi);$$

$$\therefore\ \Sigma c^4 = 4r^4 \Sigma (1 - \cos\phi)^2,$$

$$\int_0^{2\pi} (1 - \cos\phi)^2 \, d\phi = 2^3 \cdot \frac{3 \cdot 1}{4 \cdot 2} \pi = 2\pi \frac{3}{2};$$

$$\therefore\ \Sigma (1 - \cos\phi)^2 = \frac{3n}{2};$$

$$\therefore\ \Sigma c^4 = 6nr^4. \qquad \text{Q. E. D.}$$

The fourth proposition includes the third. The assumed point may now have any position we please. Let l be its distance from the centre. Here we have

$$c^2 = r^2 + l^2 - 2rl \cos\phi,$$

and $\Sigma c^4 = n (r^4 + l^4 + 2r^2 l^2) - 4rl (r^2 + l^2) \Sigma \cos\phi + 4r^2 l^2 \Sigma \cos^2\phi.$

By the values above given for $\Sigma \cos\phi$ and $\Sigma \cos^2\phi$, this becomes

$$\Sigma c^4 = nr^4 + 4nr^2 l^2 + nl^4 \ldots\ldots\ldots\ldots\ldots\ldots (4),$$

which is the proposition in question.

In the fifth proposition we return to the circumscribed polygon, and our object is to determine the sum of the fourth powers of the perpendiculars. As before,

$$\Sigma p^4 = r^4 \Sigma (1 - \cos\phi)^4,$$

and $\displaystyle\int_0^{2\pi} (1 - \cos\phi)^4 \, d\phi = 2^5 \frac{7 \cdot 5 \cdot 3 \cdot 1}{8 \cdot 6 \cdot 4 \cdot 2} \pi = 2\pi \cdot \frac{35}{8};$

therefore $\quad 3\, \Sigma p^4 = 35n \cdot r^4 \ldots\ldots\ldots\ldots\ldots\ldots\ldots\ldots (5).$

Q. E. D.

In the general case, when l is the distance of the assumed point from the centre,

$$\Sigma p^4 = nr^4 - 4r^3 l \Sigma \cos\phi + 6r^2 l^2 \Sigma \cos^2\phi - 4rl^3 \Sigma \cos^3\phi + l^4 \Sigma \cos^4\phi,$$

and $\displaystyle \Sigma \cos^4\phi = \frac{n}{2} \cdot \frac{3 \cdot 1}{4 \cdot 2} = \frac{3n}{8};$

$$\therefore\ \Sigma p^4 = nr^4 + 6r^2l^2 \cdot \frac{n}{2} + \frac{3}{8}nl^4,$$

$$\text{or } 8\,\Sigma p^4 = 8nr^4 + 24nr^2l^2 + 3nl^4 \ldots\ldots\ldots\ldots\ldots\ldots (6).$$

This is the sixth proposition.

The seventh is, for the mth power of the perpendiculars, what the first and fifth are for the third and fourth powers respectively. It is this:

$$\Sigma p^m = n\,\frac{2m-1\,.\,2m-3\ldots 1}{m\,.\,m-1\ldots 1}\,r^m \ldots\ldots (7) \ldots\ldots (n > m).$$

DEM. $$\Sigma p^m = r^m \frac{n}{2\pi}\int_0^{2\pi}(1-\cos\phi)^m\,d\phi.$$

Let $\theta = \frac{1}{2}\phi$;

$$\therefore \int_0^{2\pi}(1-\cos\phi)^m\,d\phi = 2^{m+1}\int_0^{\pi}\sin^{2m}\theta\,d\theta$$

$$= 2^{m+1}\,\frac{2m-1\,.\,2m-3\ldots 1}{2m\,.\,2m-2\ldots 2}\,\pi = 2\pi\,.\,\frac{2m-1\ldots 1}{m\,.\,m-1\ldots 1};$$

$$\therefore\ \Sigma p^m = n\,\frac{2m-1\,.\,2m-3\ldots 1}{m\,.\,m-1\ldots 1}\,r^m.$$

Q. E. D.

If the assumed point is at l distance from the centre,

$$\Sigma p^m = nr^m - \frac{m}{1}\,.\,r^{m-1}\,l\,\Sigma\cos\phi + \frac{m\,.\,m-1}{1\,.\,2}\,r^{m-2}\,l^2\,\Sigma\cos^2\phi - \&c.$$

This, it is easily seen, will reduce into the following form,

$$\Sigma p^m = nr^m + n\,\frac{m\,.\,m-1}{1\,.\,2}\,\frac{1}{2}\,r^{m-2}l^2 + n\,\frac{m\ldots m-3}{1\,.\,2\,.\,3\,.\,4}\,\frac{3\,.\,1}{4\,.\,2}\,r^{m-4}\,l^4 + \&c.$$
$$\ldots\ldots (8),$$

which is the eighth proposition......$(n > m)$.

Lastly, let us consider the $2m^{\text{th}}$ powers of the chords in the case of the inscribed polygon: we have already in the third proposition found the value of their sum when $m = 2$.

As before, $$c^2 = 2r^2(1-\cos\phi);$$

$$\therefore\ \Sigma c^{2m} = 2^m r^{2m}\,\Sigma(1-\cos\phi)^m.$$

That is, as we have already seen,

$$\Sigma c^{2m} = n\,\frac{2m-1\,.\,2m-3\ldots 1}{m\,.\,m-1\ldots 1}\,r^{2m} \ldots\ldots\ldots\ldots\ldots (9).$$

We have thus gone through Dr Stewart's properties of the circle, and have arrived at his results by a simple and uniform method.

It is evident that there is no limit to the number of geometrical theorems which may be deduced from the general formula: almost every curve will afford *interpretations*, if the word may be so used, of our analytical conclusions.

Thus in the ellipse: If any n radii vectores be drawn from the centre at equal angles to one another, the sum of the squares of their reciprocals is equal to n times the square of the reciprocal of that radius vector which is equally inclined to the major and minor axes. For we have

$$\frac{1}{r^2} - \frac{1}{b^2}(1 - e^2 \cos^2 \phi);$$

$$\therefore \ \Sigma \frac{1}{r^2} = \frac{n}{b^2} - \frac{e^2}{b^2} \Sigma \cos^2 \phi = n \frac{1}{b^2}(1 - \tfrac{1}{2}e^2),$$

$$\text{and } \tfrac{1}{2} = \cos^2 \frac{\pi}{4};$$

$$\therefore \ \Sigma \frac{1}{r^2} = \frac{n}{b^2}\left(1 - e^2 \cos^2 \frac{\pi}{4}\right),$$

therefore &c. Q. E. D.

It is to be regretted that we have hardly any idea by what considerations Dr Stewart was led to the curious theorems which bear his name. It is said, indeed, that he was engaged on geometrical porisms when he discovered them, and we are told that he would have published them under the title of porisms, but for his unwillingness to interfere with a subject which the researches of his friend, Dr Simson, seemed to have appropriated. Whether they are in reality porismatic, is a question on which it would not be worth while to enter.

The fundamental formula of our analysis is perhaps not new; the geometrical applications which we have made of it appear to be original.

NOTE ON A DEFINITE INTEGRAL.

THE value of $\int_0^{\frac{\pi}{2}} \log \sin\theta\, d\theta$, obviously the same as that of $\int_0^{\frac{\pi}{2}} \log \cos\theta\, d\theta$, was first assigned by Euler, and may be obtained in the following manner.

By Cotes's theorem,

$$z^{2m} - 1 = (z^2 - 1)\left(z^2 - 2z\cos\frac{1}{m}\pi + 1\right)\ldots\left(z^2 - 2z\cos\frac{m-1}{m}\pi + 1\right) \quad \ldots\ldots(1).$$

Let $z = 1$, then

$$m = 2^{2(m-1)} \sin^2\frac{1}{m}\frac{\pi}{2} \ldots\ldots \sin^2\frac{m-1}{m}\frac{\pi}{2}.$$

Take the logarithms of both sides, and divide by m, then

$$\frac{\log m + 2(m-1)\log\frac{1}{2}}{2m} = \left(\log\sin\frac{1}{m}\frac{\pi}{2} + \ldots + \log\sin\frac{m-1}{m}\frac{\pi}{2}\right)\frac{1}{m}.$$

Let m become infinite, and $= \frac{1}{dx}$: the first side becomes equal to $\log\frac{1}{2}$, for $\left(\frac{\log m}{m}\right) = 0$ when $m = \frac{1}{0}$; and the second is transformed into the definite integral $\int_0^1 \log\sin x\frac{\pi}{2}\,.\,dx$; therefore $\int_0^1 \log\sin x\frac{\pi}{2}\,dx = \log\frac{1}{2}$.

Let $\theta = x\frac{\pi}{2}$;

* *Cambridge Mathematical Journal*, No. XII. Vol. II. p. 282, May, 1841.

$$\therefore \int_0^{\frac{\pi}{2}} \log \sin\theta \, d\theta = \frac{\pi}{2} \log \tfrac{1}{2} \text{..................(1).}$$

COR. 1. Integrating by parts, we get

$$\int \log \sin\theta \, d\theta = \theta \log \sin\theta - \int \frac{\theta \, d\theta}{\tan\theta}.$$

The integrated part vanishes at both limits,

$$\therefore \int_0^{\frac{\pi}{2}} \frac{\theta \, d\theta}{\tan\theta} = \frac{\pi}{2} \log 2 \text{ .... (2).}$$

COR. 2. Let $\sin\theta = e^{-\frac{1}{2}x}$; therefore the limits of x are 0 and ∞,

$$\text{and } d\theta = -\tfrac{1}{2} \frac{e^{-\frac{1}{2}x}}{\sqrt{(1-e^{-x})}} dx = -\tfrac{1}{2} \frac{dx}{\sqrt{(e^x - 1)}};$$

$$\therefore \int_0^{\infty} \frac{x \, dx}{\sqrt{(e^x - 1)}} = 2\pi \log 2 \text{..................(3).}$$

COR. 3. In this last integral, if we expand the denominator

$$(\epsilon^x - 1)^{-\frac{1}{2}} = \epsilon^{-\frac{x}{2}} + \frac{1}{2}\epsilon^{-\frac{3x}{2}} + \frac{1 \,.\, 3}{1 \,.\, 2}\frac{1}{2^2}\epsilon^{-\frac{5x}{2}} + \text{\&c.}$$

$$\text{and as } \int_0^{\infty} \epsilon^{-mx} x \, dx = \frac{1}{m^2}, \text{ we find}$$

$$\frac{\pi}{2} \log 2 = 1 + \frac{1}{1} \cdot \frac{1}{2} \frac{1}{3^2} + \frac{1 \,.\, 3}{1 \,.\, 2} \frac{1}{2^2} \frac{1}{5^2} + \frac{1 \,.\, 3 \,.\, 5}{1 \,.\, 2 \,.\, 3} \frac{1}{2^3} \frac{1}{7^2} + \text{\&c.}$$

REMARKS ON THE DISTINCTION BETWEEN ALGEBRAICAL AND FUNCTIONAL EQUATIONS*.

THE distinction which it is usual to make between algebraical and functional equations will not, I think, bear a strict examination. It is generally said that an algebraical equation determines the value of an unknown quantity, while a functional equation determines the form of an unknown function. But, in reality, the unknown quantity in the former case is a function of the coefficients of the equation, and our object in solving it is simply to ascertain the form of this function. Thus it appears, that in both cases the forms of functions are what we seek.

Let us therefore consider the subject in a more general manner, and endeavour to find a more decided point of distinction. The science of symbols is conversant with operations, and not with quantities; and an equation, of whatever species, may be defined to be a congeries of operations, known and unknown, equated to the symbol zero. Every operation implies the existence of a base, or something on which the operation is performed—in the language of Mr Murphy, a subject. But the base of an operation is often the result of a preceding one. Thus, in $\log x^2$, the base of the operation log is x^2, itself the result of the operation expressed by the index on the base x. This in its turn may be considered as the result of an operation performed on the symbol unity. But in every kind of equation there is a point at which the farther analysis of symbols into operations on certain bases becomes irrelevant; and thus we are led in every case to recognize the existence of ultimate bases.

* *Cambridge Mathematical Journal,* No. XIV. Vol. III. p. 92, February, 1842.

To solve an equation of any kind, is to determine the unknown operations by means of the known. If one symbol is said to be a function of another, it is, in reality, the result of an operation performed upon it. Thus the idea of functional dependence pervades the whole science of symbols, and on this idea the following remarks are based.

In order to classify equations, we can make use of two considerations: 1st. The nature of the operations which are combined together; 2nd. The order in which they succeed one another in the congeries of operations which is made equal to zero.

Let us illustrate these remarks by some examples.

If we have an equation of the form

$$x^2 + ax + b = 0 \text{ ....(1)},$$

the bases are a and b; the operations are, first, the unknown one denoted by x, and then certain known ones denoted by the index, the coefficient, &c. All these are what are called algebraical operations.

If again we have an equation of the form

$$\frac{dy}{dx} - x = 0 \text{ ....(2)},$$

the base is x; the operations are, first, the unknown one denoted by y, which is a function of x, then the operation $\frac{d}{dx}$, and lastly, certain algebraical operations. From the presence of the operation $\frac{d}{dx}$, this is called a differential equation. Equations (1) and (2) are discriminated by the nature of the operations combined, on our first principle of classification.

But in one important point these equations agree. In both, the unknown operation is performed immediately on the bases; the known are subsequent to the unknown: but in what are called functional equations this is not so. Thus, in the equation

$$\phi(mx) + x = 0 \text{ ....(3)},$$

the base is x, the unknown operation is ϕ, which is performed, not on x, but on the result of a previous operation. In the preceding example the previous operation is known; but this is not essential. Thus in

$$\phi\phi x + x = 0 \text{..........................} (4)$$

the previous operation denoted by the right-hand ϕ is unknown. The operation $\dfrac{d}{dx}$ may enter into equations where the unknown operation is not performed on the base. Thus we may have an equation of the form

$$\phi \frac{d}{dx} \phi x + x = 0 \text{......................} (5).$$

Equations (3), (4), (5) are functional equations; (3), (4) are ordinary functional equations; (5) is a differential functional equation; (3) is said to be of the first order, (4) of the second.

The introduction of the functional notation appears to be sometimes taken as the essence of functional equations; but if we wrote (1) and (2) thus,

$$\{\phi\,(ab)\}^2 + a\phi\,(ab) + b = 0 \text{ ....} (1)',$$

$$\frac{d}{dx}\phi\,(x) - x = 0 \text{ ....} (2)',$$

they would still be perfectly distinct from (3) or (4) or (5). The name functional equation is not happy; it refers to the notation, and not to the essence of the thing.

A question now arises: To what class shall we refer equations in finite differences? These are generally of the form

$$F\,(x, y_x, y_{x+1} \ldots) = 0 \text{ ....} (6),$$

where y_x is an unknown function, say $\phi\,(x)$ of x; so that (6) may be written thus,

$$F\{x, \phi x, \phi\,(x+1) \ldots\} = 0.$$

Here the unknown operation is ϕ, which in the case of $\phi\,(x+1)$ is performed, not upon the base x, but on $x+1$. Thus it appears, that equations in finite differences are only a case of ordinary functional equations of the first order: and this is the reason why, in researches on functional equations, we perpetually meet with cases in which they may be reduced to equations in finite differences.

The preceding remarks contain, I think, the outline of a natural arrangement of the science of symbols. It is not difficult to overrate the importance of a mere classification; but I hope to be able to show, that the considerations now suggested are not without some degree of utility.

As the distinction between functional and common equations depends on the order of operations, it follows that, when part of the solution of an equation does not vary with the nature of the operation subjected to the resolving process, this part is applicable as much to functional equations as to any other. The special application of this principle to the discussion of a class of differential functional equations will be the object of a subsequent paper.

In the preceding remarks, operations of derivation, such as D, Δ, &c. are supposed to be replaced by functional operations in every case in which this can be effected.

MATHEMATICAL NOTES*.

1. In the Examination Papers for 1834, the following problem is given: "If the chord of a conic section, whose eccentricity is e, subtend at its focus a constant angle 2α, prove that it always touches a conic section having the same focus whose eccentricity is $e \cos \alpha$." A solution of this problem by a peculiar analysis will be found in a preceding article; but the following method may be found not uninteresting.

Let r_1, r_2 be radii vectores to the ends of the chord, $\phi - \alpha$, $\phi + \alpha$, the corresponding angles vectores, p the perpendicular from the focus on the chord;

$$\therefore\ p \times \text{chord} = r_1 r_2 \sin 2\alpha,$$

$$\therefore \frac{1}{p} = \frac{\sqrt{(r_1^2 + r_2^2 - 2r_1 r_2 \cos 2\alpha)}}{r_1 r_2 \sin 2\alpha}$$

$$= \operatorname{cosec} 2\alpha \sqrt{\left(\frac{1}{r_1^2} + \frac{1}{r_2^2} - \frac{2}{r_1 r_2} \cos 2\alpha\right)},$$

$$\frac{1}{r_1} = \frac{1 + e \cos(\phi - \alpha)}{l}, \quad \frac{1}{r_2} = \&\text{c}.$$

For $\cos 2\alpha$ put $1 - 2\sin^2\alpha$; then, by a few obvious steps,

$$\frac{1}{p} = \frac{1}{l \cos \alpha} \sqrt{(1 + 2e \cos \alpha \cos \phi + e^2 \cos^2 \alpha)}.$$

Put $\alpha = 0$; then the chord becomes the tangent, and

$$\frac{1}{p_0} = \frac{1}{l} \sqrt{(1 + 2e \cos \phi + e^2)}.$$

But the general form coincides with this, if we put

$$l \cos \alpha = \lambda \text{ and } e \cos \alpha = \epsilon;$$

for then

$$\frac{1}{p} = \frac{1}{\lambda} \sqrt{(1 + 2\epsilon \cos \phi + \epsilon^2)}.$$

* *Cambridge Mathematical Journal*, No. IX. Vol. III. p. 94, February, 1842.

Hence p is *generally* the perpendicular on a tangent of an ellipse of eccentricity $e \cos \alpha$. Hence the chord touches such an ellipse. The latus rectum is diminished in the same ratio as the eccentricity.

2. The relation between the long inequalities of two mutually disturbing planets, may be easily found without having recourse to the development of the disturbing function.

Let m, m', be the masses of the planets, a, a', the major axes of their orbits, n, n', their mean motions, h, h', twice the areas described in 1″; then we have

$$n = \frac{\mu^{\frac{1}{2}}}{a^{\frac{3}{2}}}, \quad n' = \frac{\mu^{\frac{1}{2}}}{a'^{\frac{3}{2}}},$$

μ being the mass of the Sun, in comparison with which the masses of the planets are neglected, so that it is the same for both. Taking the logarithmic differentials of these equations, and replacing the differentials by differences, we find

$$\frac{\Delta n}{n} = -\frac{3}{2}\frac{\Delta a}{a}, \quad \frac{\Delta n'}{n'} = -\frac{3}{2}\frac{\Delta a'}{a'}.$$

But by the principle of the conservation of areas,

$$mh + m'h' = \text{const.}$$

so that

$$m\Delta h + m'\Delta h' = 0.$$

Now the orbits being supposed circular, we have

$$h = (\mu a)^{\frac{1}{2}}, \quad h' = (\mu a')^{\frac{1}{2}}:$$

hence

$$\frac{\Delta h}{h} = \frac{1}{2}\frac{\Delta a}{a}, \quad \frac{\Delta h'}{h'} = \frac{1}{2}\frac{\Delta a'}{a'}:$$

Therefore we have

$$\frac{\Delta n}{n} : \frac{\Delta n'}{n'} = \frac{\Delta a}{\Delta a'} \cdot \frac{a'}{a} = \frac{\Delta h}{\Delta h'} \cdot \frac{h'}{h} = -\frac{m'}{m}\frac{a'^{\frac{1}{2}}}{a^{\frac{1}{2}}};$$

and $\frac{\Delta n}{n}$ and $\frac{\Delta n'}{n'}$ are the inequalities due to the disturbances, so that their ratio is thus given.

ON THE SOLUTION OF FUNCTIONAL DIFFERENTIAL EQUATIONS*.

It is well known that the solution of a considerable class of differential equations may be effected by means of differentiation. Clairaut's equation is a particular case of this class. We will begin by considering it.

$$y = px + fp \;\ldots\; (1) \qquad \left(p = \frac{dy}{dx}\right),$$

where f denotes any given function.

Differentiating (1), we get

$$(x + f'p)\, q = 0 \;\ldots\ldots\ldots\ldots\ldots\ldots\ldots\ldots\; (2),$$

hence $$q = 0, \text{ or } x + f'p = 0 \;\ldots\ldots\ldots\ldots\ldots\ldots\ldots\ldots\; (3).$$

The first of these equations gives the complete integral. Being twice integrated it becomes $y = ax + b$; and on substitution in (1), we get $b = fa$, therefore $y = ax + fa \ldots (4)$ is the complete integral of (1).

It has always been supposed, in this and similar cases, that f must necessarily be a given function. But this condition is not essential: a differential equation, *e.g.* such as (1), will, when solved, give y as a function of x. Now the function f, which enters into (1) may, instead of being given, as is usually the case, be in some way dependent on the function which y is of x. Thus the form of f is unknown, until that of the latter function has been determined. It is evident that according to the classification proposed in the last number of the Journal, (1) is in all such cases a functional equation. For the unknown operation f is performed on p, which is itself the result of the unknown operation ψ' performed on x (we suppose $y = \psi x$).

* *Cambridge Mathematical Journal,* No. XV. Vol. III. p. 131, May, 1842.

To differential functional equations, ordinary methods of solution do not, generally speaking, apply, because they require a knowledge of the forms of the functions on which they operate. But in the case before us, the differentiation and subsequent substitution, by which (4) was derived from (1), are independent of any knowledge of the nature of f. Consequently (4) is always true.

Let us suppose, for instance, that $f = -m\psi$, m being a constant; then

$$\psi x - x\psi' x + m\psi\psi' x = 0 \text{..................} (5).$$

We are of course obliged to introduce a functional notation: (4) in this case becomes

$$\psi x = ax - m\psi a \text{........................} (6).$$

In order to determine ψa, put $x = a$;

$$\text{then} \quad \psi a = \frac{a^2}{1+m},$$

$$\text{and} \quad \psi x = ax - \frac{m}{1+m} a^2 \text{...............} (7),$$

which is a solution of (5).

In the ordinary cases of Clairaut's equation, the factor $x + f'p = 0$ leads to the singular solution; and so it does when f is an unknown function.

Thus, in the example just considered, as $f' = -m\psi'$, we shall have

$$m\psi'\psi' x = x \text{..........................} (8).$$

Of this a solution is

$$\psi' x = \frac{x}{\sqrt{(m)}}.$$

Hence we get, by integration,

$$\psi x = \frac{x^2}{2\sqrt{(m)}} + C.$$

On substitution it is found that $C = 0$, therefore

$$\psi x = \frac{x^2}{2\sqrt{(m)}} \text{........................} (9)$$

is a new solution of (5), and perfectly distinct from (7).

If $m=1$, (5) and (7) become respectively

$$\psi x - x\psi' x + \psi\psi' x = 0 \text{..................} (5'),$$

$$\psi x = ax - \tfrac{1}{2}a^2 \text{..........................} (7');$$

in this case, (8) admits of a variety of simple solutions. Thus we shall have

$$\psi x = \tfrac{1}{4}c^2 - \tfrac{1}{2}(c-x)^2,$$

$$\psi x = \tfrac{1}{2} + \log x,$$

$$\&c. = \&c.$$

as singular solutions of (5′).

The preceding remarks are sufficient to indicate the existence of a class of functional equations, to which a considerable portion of the theory of singular solutions may be applied. They appear therefore to possess some interest with reference to this theory, independently of the method they suggest for the solution of such equations.

In fact the theory can hardly be considered complete, unless some notice is taken of the equations of which we have been speaking. They have been excluded from it, because the function f, which they involve, is not, as in the ordinary case, a known function. But this, it has been already remarked, is not an essential distinction.

On the other hand, the method by which the singular solution is in the common theory deduced from the complete integral, does not apply to the cases now considered. It appears unnecessary to point out the reason of this difference.

With regard to the class of differential equations, which, like Clairaut's, separate into factors on differentiation, we may refer to Lagrange's *Leçons sur le Calcul des Fonctions*, l. 16me. He there shows that if a differential equation of the first order can be put into the form $M=fN$, where M and N are the values of a and b deduced from

$$F(xyab) = 0,$$

$$\frac{d}{dx}F(xyab) = 0;$$

then, when differentiated, it will resolve itself into two factors, one of which leads to the singular solution, and the other to the complete integral. (The latter is, as may readily be seen,

$$F(xyfb\,.\,b)=0.)$$

The demonstration of this proposition is probably familiar to the majority of my readers, and I shall therefore not dwell upon it. Similar considerations apply to equations of higher orders.

Generalizing the remarks already made, we see that in the equation

$$M=fN,$$

the function f need not be a given one; it may be, in any way we please, dependent on the function which, in virtue of this equation, y is of x. In all such cases the equation in question is functional. Nevertheless, Lagrange's reasoning applies as much in these as in other cases. Let us take one or two examples of what has been said.

The following problem may be proposed.

Any point P of a certain curve is referred to the axis of x in M, and to that of y in N. MP is produced to Q; PQ is taken equal to a, and NQ touches the curve. Find its equation.

Let x, ψx be the co-ordinates of the point where NQ touches the curve.

$$ON=\psi x-x\psi' x. \quad NP=\frac{a}{\psi' x},$$

and as P is a point in the curve,

$$ON=\psi\{NP\}, \text{ or}$$

$$\psi x-x\psi' x=\psi\left(\frac{a}{\psi' x}\right)\ldots\ldots\ldots\ldots\ldots\ldots\ldots(10).$$

This is the equation of the problem. Differentiating it, we get

$$\psi'' x=0,$$

$$x=\frac{a}{(\psi' x)^2}\psi'\left(\frac{a}{\psi' x}\right).$$

The former equation gives the complete integral, but, for a reason I shall hereafter notice, leads to no tangible solution of the problem; the latter corresponds to the singular solution.

In order to solve it, assume

$$\frac{a}{\psi' x} = \chi x;$$

then $$\psi' \frac{a}{\psi' x} = \frac{a}{\chi^2 x} \text{ and } \frac{x}{\chi x} = \frac{\chi x}{\chi^2 x}.$$

Let $x = u_z$, $\chi x = u_{z+1}$, and therefore $\chi^2 x = u_{z+2}$.

Then $\frac{u_z}{u_{z+1}} = \frac{u_{z+1}}{u_{z+2}} = \frac{1}{C}$, where C is arbitrary;

therefore $\chi x = Cx$.

C is a function of z, which does not change when $z+1$ is substituted for z.

We confine ourselves to the only simple case, that in which it is an absolute constant; then

$$\psi' x = \frac{a}{Cx} = \frac{b}{x} \ldots (bC = a)$$

and $\psi x = b \log \frac{x}{c} \ldots c$ being an arbitrary constant.

On substitution, we find $b = ae$; therefore

$$y = ae \log \frac{x}{c} \ldots\ldots\ldots\ldots\ldots\ldots\ldots\ldots (11)$$

is a solution of the problem.

This is the equation of a logarithmic curve, which has therefore the required property. The method employed to resolve the equation in χx, namely,

$$x\chi^2 x = (\chi x)^2,$$

is applicable to every equation of the form

$$F(x \,.\, \chi x \ldots \chi^n x) = 0 \ldots\ldots\ldots\ldots\ldots\ldots (12).$$

Every such equation may be at once reduced to the following equation in finite differences,

$$F(u_z u_{z+1} \ldots u_{z+n}) = 0 \ldots\ldots\ldots\ldots\ldots\ldots (13).$$

This reduction is in reality a particular case of an important transformation due to Mr Babbage, which often enables us to solve functional equations of the higher orders.

In (12) we may write for χx, $\phi f \phi^{-1} x$.

Hence $\chi^2 x = \phi f^2 \phi^{-1} x$ &c. = &c., and (12) becomes

$$F(\phi x \,.\, \phi f x \,\dots\, \phi f^n x) = 0 \;\dots\dots\dots\dots\dots\dots (14),$$

by putting ϕx for x; f being a known function, (14) is a functional equation of the first order.

Such is Mr Babbage's method. Let $fx = 1 + x$; (14) becomes

$$F\{\phi x \,.\, \phi(1+x) \,\dots\, \phi(n+x)\} = 0,$$

and if we denote ϕx by u_x, and replace x by z, we shall obtain (13).

It must be admitted, that it is difficult to prove that the generality of (12) is not restricted by these transformations. They are however often useful, and serve to illustrate what was remarked in the last number, with respect to the affinity of functional equations, and equations in finite differences.

If, instead of (10), we had taken the more general equation

$$\psi x - x\psi' x = \psi\left(\frac{a}{\psi' x}\right) + A \;\dots\dots\dots\dots\dots (15),$$

where A is an arbitrary constant, precisely the same method would have applied. In this case the factor $\psi'' x = 0$ would have led to the result

$$\psi x = \alpha x + \beta,$$

and by substitution $\qquad \beta = a + \beta + A,$

$$\text{therefore} \quad a + A = 0, \quad \text{or } \beta = \infty.$$

Now in the case we have been considering, the former condition is not fulfilled; hence we must have $\beta = \infty$, and the geometrical interpretation of the complete integral is a right line at an infinite distance from the axis of abscissæ.

We not unfrequently meet with similar cases, in which the complete integral becomes nugatory or impossible in the process of introducing the necessary relation between its constants. Under particular conditions, however, this difficulty does not occur, and then we obtain, what in the ordinary methods of discussing functional differential equations, appears to be a *conjugate* solution, unconnected with any other; (15) would be an instance of this, were $a + A = 0$.

I shall next consider a celebrated problem, first proposed by Euler, in the Petersburgh memoirs.

In a certain class of curves, the square of any normal exceeds the square of the ordinate drawn from its foot by a certain quantity a.

Let $y^2 = \psi x$ be the equation of the curve. The subnormal is therefore $\frac{1}{2}\psi' x$, and the equation of the problem consequently is

$$\psi(x + \tfrac{1}{2}\psi' x) = \psi x + \tfrac{1}{4}(\psi' x)^2 - a \ldots\ldots\ldots\ldots (16).$$

Differentiating this, we get

$$\psi'(x + \tfrac{1}{2}\psi' x) = \psi' x;$$

$$\text{or} \quad 1 + \tfrac{1}{2}\psi'' x = 0.$$

The first of these two equations leads to the singular solutions. In order to solve it, let

$$x + \tfrac{1}{2}\psi' x = \chi x,$$

$$\text{then} \quad \chi^2 x - 2\chi x + x = 0.$$

Hence by the transformation already noticed,

$$u_{z+2} - 2u_{z+1} + u_z = 0,$$

$$\text{whence} \quad u_z = Pz + zP_1 z,$$

where Pz and $P_1 z$ are functions of z, which remain unchanged when z increases by unity;

$$\text{therefore} \quad u_{z+1} = Pz + (z+1)P_1 z.$$

Hence we have

$$\left.\begin{aligned} \tfrac{1}{2}\psi' x &= P_1 z \\ x &= Pz + zP_1 z \end{aligned}\right\} \text{ for the required solution.}$$

$$dx = (P'z + P_1 z + zP_1' z)\,dz;$$

$$\text{therefore} \quad ydy = P_1 z\,(P'z + P_1 z + zP_1' z)\,dz;$$

and integrating by parts, we get

$$\left.\begin{aligned} y^2 &= P_1 z\,(2Pz + zP_1 z) + \int (P_1 z)^2\,dz \\ x &= Pz + zP_1 z \end{aligned}\right\} \ldots\ldots\ldots\ldots (17),$$

for a general solution of the proposed problem. (The parameter a is involved in $P_1 z$.)

Let us suppose Pz and P_1z constant;

$$y^2 = \alpha(2b + \alpha z) + \alpha^2 z + C,$$

$$x = b + \alpha z;$$

$$\text{therefore} \quad y^2 = 2\alpha x + C.$$

On substitution we find

$$\alpha^2 = -a.$$

Thus, in order to a real result, we must suppose a negative, *e.g.* let $a = -k^2$; then

$$y^2 = 2kx + C \dots\dots\dots\dots\dots\dots (18),$$

the equation to a parabola, which accordingly is a solution of the problem, and the only simple one it admits of.

When $a=0$, it becomes two straight lines parallel to the axis.

The other factor $1 + \frac{1}{2}\psi''x = 0$ gives, on integration,

$$\psi x + x^2 = \alpha x + \beta \dots\dots\dots\dots\dots\dots (19),$$

the equation to a circle; but on substitution, we find

$$\tfrac{1}{4}\alpha^2 = \tfrac{1}{4}\alpha^2 - a,$$

which leads to no result, unless $a = 0$.

A solution of this problem, by Poisson, is given at p. 591 of the last volume of Lacroix's great work. It is apparently equivalent in point of generality to (17); and the author points out its incompleteness in the case of $a=0$. The preceding views show distinctly the nature and origin of the new solution which then presents itself. Mr Babbage also has considered this problem at the end of his second essay on the Calculus of Functions (vide *Phil. Trans.* 1816, p. 253). But I believe it will be found that his solution is erroneous.

Notwithstanding the length this paper has already reached, I must endeavour to point out, as briefly as possible, my reasons for thinking so.

Mr Babbage confines himself to the case of $a=0$. He begins by demonstrating the existence of a relation, equivalent, excepting a difference of notation, to $\psi'\{x+\frac{1}{2}\psi'x\} = \psi'x$, but in doing this, loses sight of the other factor $1 + \frac{1}{2}\psi''x = 0$.

This relation shows that $\psi' x$ is constant, for a series of points in the curve, and therefore, Mr Babbage reasons, we may consider it as a constant in (16), which thus becomes an equation in finite differences. He integrates it on this supposition, and adds an arbitrary function of $\psi' x$, which has been treated as an absolute constant. The result is therefore

$$\psi x = \tfrac{1}{2} x \psi' x + f \cdot \tfrac{1}{2} \psi' x,$$

$$\text{or}\quad y^2 = xy \frac{dy}{dx} + f\left(y \frac{dy}{dx}\right) \quad\text{.................(20)},$$

which is an ordinary differential equation.

This process appears to have been suggested by an incorrect analogy with the way in which arbitrary functions are introduced into partial differential equations.

A little consideration would have convinced Mr Babbage, that by integrating (16) as an equation in finite differences, he only passed discontinuously from one ordinate of the curve to another, and therefore could not obtain a continuous relation between x and y. The legitimate result of his process is merely,

$$\psi \{x + \tfrac{1}{2} x \psi' x\} - \psi x = \frac{n}{4} (\psi' x)^2,$$

where n is any positive or negative integer. This is quite different from

$$\psi x = \tfrac{1}{2} x \psi' x + f \{\tfrac{1}{2} \psi' x\}.$$

In exemplifying equation (20), Mr Babbage first supposes

$$f\left(y \frac{dy}{dx}\right) = \infty \times y \frac{dy}{dx},$$

and thus obtains the equation of a straight line parallel to Ox, as a solution of the problem, which undoubtedly it is.

In his next example $f\left(y \frac{dy}{dx}\right) = a^2$. By making the constant of integration imaginary, he gets $y^2 = a^2 - x^2$, the equation to a circle. But although this is also a real solution, it has no connection with the relation $\psi' \{x + \tfrac{1}{2} \psi' x\} = \psi' x$, from which it appears to be derived. It is, as we have seen, a particular case of the complete integral. Consequently if the method pursued had been correct, it could not have given this solution.

The preceding pages appear to contain the germ of a general theory of differential functional equations; a subject of great extent, and ultimately, perhaps, of considerable importance. But it cannot be denied, that hitherto the Calculus of Functions has not led to many results of much interest. Its value arises chiefly from the wide views it gives of the science of the combination of symbols.

MATHEMATICAL NOTE*.

Problem from the Papers of 1842.—If $F(x, y, z) = \phi(u, v, w)$, where F is homogeneous of the nth degree in x, y, z, and $u = \frac{dF}{dx}$, $v = \frac{dF}{dy}$, $w = \frac{dF}{dz}$; then

$$x = (n-1)\frac{d\phi}{du}, \quad y = (n-1)\frac{d\phi}{dv}, \quad z = (n-1)\frac{d\phi}{dw}.$$

Since F is homogeneous of n dimensions in x, y, z, we have

$$nF = x\frac{dF}{dx} + y\frac{dF}{dy} + z\frac{dF}{dz} = xu + yv + zw.$$

Hence $\quad ndF = xdu + ydv + zdw + udx + vdy + wdz,$

or $\quad (n-1)\,dF = xdu + ydv + zdw.$

But $(n-1)\,dF = (n-1)\,d\phi = (n-1)\left(\frac{d\phi}{du}du + \frac{d\phi}{dv}dv + \frac{d\phi}{dw}dw\right).$

Therefore equating the coefficients of the differentials,

$$x = (n-1)\frac{d\phi}{du}, \quad y = (n-1)\frac{d\phi}{dv}, \quad z = (n-1)\frac{d\phi}{dw}.$$

* *Cambridge Mathematical Journal*, No. XV. Vol. III. p. 152, May, 1842.

EVALUATION OF CERTAIN DEFINITE INTEGRALS*.

WHEN the value of a definite integral is known, we may, if it involve an arbitrary parameter, integrate it (under certain conditions) with respect to this quantity. The result thus obtained involves an arbitrary constant of integration; in order to eliminate it, we may ascribe two different values to the quantity for which the integration has been effected, and then, of the two corresponding equations thus got, subtract one from the other. In other words, we integrate between limits for the arbitrary parameter, and thus get a new definite integral, involving two arbitrary quantities, namely, the two limiting values ascribed to the single one involved in the original integral. We may integrate again, with respect to either of these, and so on. But this method of proceeding, though it will lead to a variety of particular results, is not well fitted to show the nature of the class of definite integrals to which they all belong, and which may be obtained by repeated integrations for an arbitrary parameter.

If we integrate n times successively, we shall introduce n constants. These may be eliminated *at once*, in the manner I am about to point out. The result thus got, includes for every original definite integral, all that can be deduced from it by n integrations for an arbitrary parameter.

The following theorem will serve to illustrate the general method.

If Fx is a rational and integral function of circular functions of x (sines and cosines), then we may express in finite terms the value of $\int_{-\infty}^{+\infty} \frac{Fx}{x^n}\, dx$, n being a positive integer, and such that $\left.\frac{Fx}{x^n}\right\}_0$ is not infinite.

* *Cambridge Mathematical Journal*, No. XVI. Vol. III. p. 185, Nov. 1842.

This theorem applies to several remarkable definite integrals, some of which occur in the theory of probabilities; there are others which do not seem to have been noticed.

DEM. Fx, as every function of x, may be considered the sum of two functions, one of which remains unchanged when x changes its sign, and the other changes its sign with that of x; its value *aux signes près* remaining unaltered. Hence, whether n is odd or even, we may write

$$\frac{Fx}{x^n} = \frac{fx}{x^n} + \frac{\phi x}{x^n},$$

where $$\frac{f(-x)}{(-x)^n} = \frac{fx}{x^n}, \text{ and } \frac{\phi(-x)}{(-x)^n} = -\frac{\phi x}{x^n}.$$

It is obvious that

$$\int_{-\infty}^{+\infty} \frac{Fx}{x^n}\, dx = 2\int_0^\infty \frac{fx}{x^n}\, dx;$$

and that if n is odd, fx, which is of course a rational and integral function of circular functions (sines and cosines) of x, must be developable in a series of sines exclusively, and if n is even in a series of cosines exclusively.

Thus, we may assume

$$fx = \Sigma A \left.\begin{matrix}\sin\\ \cos\end{matrix}\right\} ax \ldots\ldots\ldots\ldots\ldots\ldots\ldots (1).$$

Now, as $\dfrac{fx}{x^n}$ is not infinite when $x = 0$, the lowest power of x which can enter into fx must be not $< n$, call it m, and develope in powers of x every sine or cosine which appears on the second side of the last written equation. We must have $\Sigma A a^{m-2} = 0$, $\Sigma A a^{m-4} = 0$, &c. ... $\frac{1}{2}m - 1$ equations if m is even, and $\frac{1}{2}(m-1)$ if it is odd.

Let us now consider the definite integral

$$\int_0^\infty e^{-\alpha x} \cos rx dx = \frac{\alpha}{\alpha^2 + r^2}.$$

Integrating it repeatedly for r, we get

$$\int_0^\infty e^{-\alpha x} \frac{\sin rx}{x}\, dx = \tan^{-1}\frac{r}{\alpha},$$

$$\int_0^\infty e^{-\alpha x}\frac{\cos rx}{x^2}\,dx = -r\tan^{-1}\frac{r}{\alpha} + \alpha\log\sqrt{(\alpha^2+r^2)} + C,$$

and generally

$$\int_0^\infty e^{-\alpha x}\frac{\left.\begin{matrix}\sin\\ \cos\end{matrix}\right\} rx}{x^n}\,dx = \pm\frac{r^{n-1}}{[n-1]}\tan^{-1}\frac{r}{\alpha} + \alpha F_1(r\alpha)$$

$$+ Cr^{n-2} + C_1 r^{n-4} + \&\text{c.} \ldots\ldots\ldots\ldots\ldots\ldots \text{(2)},$$

where $F_1(r\alpha)$ does not become infinite for $\alpha = 0$.

Replace r by every quantity represented in (1) by the general symbol a. Multiply each result by the corresponding coefficient A, and add.

Then, in virtue of the conditions,

$$\Sigma A a^{m-2} = 0, \quad \Sigma A a^{m-4} = 0, \ \&\text{c.}$$

we shall have

$$\int_0^\infty e^{-\alpha x}\frac{fx}{x^n}\,dx = \pm\Sigma\frac{\Sigma A a^{n-1}}{[n-1]}\tan^{-1}\frac{a}{\alpha} + \alpha\Sigma A F_1(a\alpha).$$

Put $\alpha = 0$, then $\tan^{-1}\frac{a}{\alpha} = \pm\frac{\pi}{2}$, according to whether a is $>$ or $<$ than zero.

Thus we have

$$\int_0^\infty \frac{fx}{x^n}\,dx = \pm\frac{\pi}{2[n-1]}\,\Sigma \pm A a^{n-1} \ldots\ldots\ldots\ldots \text{(3)}.$$

From hence, the truth of our theorem is obvious.

Of the ambiguous signs outside the symbol of summation, the upper is to be taken when n is of the forms $4p$, or $4p+1$.

When a is positive, we must take the upper of the ambiguous signs under the Σ.

It will be remarked, that in obtaining (3) we have eliminated all the constants at once, instead of getting rid of them one by one by particular conditions at each successive integration, and that the generality of this method enables us to recognize a class of definite integrals, which are all deduced from the known value of $\int_0^\infty e^{-\alpha x}\cos rx\,dx$.

Equation (3) admits of several remarkable applications. Thus let us suppose $n = 3$ and $fx = \sin ax \sin bx \sin cx$: then

$$fx = -\tfrac{1}{4}\{\sin(a+b+c)x - \sin(-a+b+c)x - \sin(a-b+c)x - \sin(a+b-c)x\}.$$

Consequently, we have by (3),

$$\int_0^\infty \frac{\sin ax \sin bx \sin cx}{x^3}\,dx = \frac{\pi}{4}\{s^2 \mp (s-a)^2 \mp (s-b)^2 \mp (s-c)^2\},$$

where, as in trigonometrical formulæ,

$$2s = a + b + c:$$

the upper sign is to be taken when the quantity to which it is affixed is > 0.

Likewise $\displaystyle\int_0^\infty \frac{\sin ax \sin bx \sin cx}{x}\,dx = \frac{\pi}{8}(1 \mp 1 \mp 1 \mp 1)$,

where the signs follow the same rule as in the former case; the different unities involved being the zero powers of s, $s-a$, &c.

Let us now suppose that $fx = \sin^m x \cos zx$; the corresponding integral, viz. $\displaystyle\int_0^\infty \frac{\sin^m x \cos zx}{x^n}\,dx$ occurs in the theory of probabilities. Its value is given at p. 170 of Laplace's *Théorie des Probabilités*, where it is obtained by a method founded on a transition from real to imaginary quantities. The nature of what are called imaginary quantities is certainly better understood than it was some time since; but it seems to have been the opinion of Poisson, as well as of Laplace himself, that results thus obtained require confirmation. In this view I confess I do not acquiesce; but if only in deference to their authority, it may be desirable to show how readily imaginary quantities may be avoided in estimating the value of the integral in question.

$$\sin^m x = \pm \frac{1}{2^{m-1}}\left\{\cos mx - \frac{m}{1}\cos(m-2)x + \&c.\right\} \text{ if } m \text{ is odd,}$$

$$\text{and } = \pm \frac{1}{2^{m-1}}\left\{\sin mx - \frac{m}{1}\sin(m-2)x + \&c.\right\} \text{ if it is even.}$$

Hence, by (3), $\displaystyle\int_0^\infty \frac{\sin^m x \cos zx}{x^n}\,dx =$

$$\frac{\pi}{[n-1]\,2^m}\left\{(m+z)^{n-1} \pm (m-z)^{n-1} - \frac{m}{1}\{(m+z-2)^{n-1} \pm (m-z-2)^{n-1}\} + \&c.\right\}$$

Let us suppose $z > m$; then the lower of each pair of ambiguous signs must be taken, and the expression within the brackets may be written thus

$$(m+z)^{n-1}-\frac{m}{1}(m+z-2)^{n-1}+\&c.\ +\frac{m}{1}(m-z-2)^{n-1}-(m-z)^{n-1}\ldots(q).$$

As, from the nature of the case, m and n are either both odd or both even, if m is even, $n-1$ is odd, and therefore $(m-z)^{n-1} = -(z-m)^{n-1}$; and thus, in every case, (q) equals

$$(1-D^{-1})^m (z+m)^{n-1} \ldots \{D\phi z = \phi(z+2) \quad \text{say,}\}$$

and this is $\qquad \Delta^m D^{-m}(z+m)^{n-1} = \Delta^m (z-m)^{n-1} = 0,$

since m is $> n-1$. Consequently

$$\int_0^\infty \frac{\sin^m x}{x^n} \cos zx\, dx = 0, \text{ when } z \text{ is } > m,$$

a remark not made by Laplace; when $m = n = 1$, its truth is known.

As $(q) = 0$, add it, multiplied by $\frac{\pi}{[n-1]\,2^{m-1}}$, to the value of the integral already found, therefore

$$\int_0^\infty \frac{\sin^m x}{x^n} \cos zx dx = \frac{\pi}{[n-1]\,2^m}\left\{(m+z)^{n-1}-\frac{m}{1}(m+z-2)^{n-1}+\&c.\right\}\ldots(4),$$

where the series stops whenever the next term would introduce *a negative quantity raised to the power* $n-1$. *This is* easily seen to be true, for every such term will have a different sign in (q), and in the definite integral, and thus on addition, all such terms will disappear. Equation (4) is Laplace's form; the discontinuity of the function is now expressed, not by ambiguous signs but by the stopping short of the series at different points.

By a similar method, we find that

$$\int_0^\infty \frac{\sin^m x}{x^{n+1}} \sin zx dx = \frac{\pi}{[n]2^{m-1}}\left\{(m+z)^{n-1}-\frac{m}{1}(m+z-2)^{n-1}+\&c.\right\}\ldots(5),$$

which might have been deduced from (4) by integrating both sides without introducing any complementary quantities. This remark is general; having once established the general form of (4) for any given value of n, we may deduce from it that which corresponds to any other value of n, simply by differentiation or by integration, without bringing in any constants.

I conceive that this remark is general, and if so we may differentiate on both sides with fractional indices. Let the index be $-p$, then as

$$d^p \cos zx = x^p \cos\left(zx + p\,\frac{\pi}{2}\right) dz^p,$$

the first side of (4) will become

$$\int_0^\infty \frac{\sin^m x}{x^n}\, x^{-p} \cos\left(zx - p\,\frac{\pi}{2}\right) dx;$$

and the second will be

$$\frac{\pi}{[n-1+p]\,2^m}\,\{(m+z)^{n-1+p} - \&c.\}$$

Thus, we get

$$\int_0^\infty \frac{\sin^m x}{x^n}\, x^{-p} \cos\left(zx - p\,\frac{\pi}{2}\right) dx =$$

$$\frac{\pi}{[n-1+p]\,2^m}\left\{(m+z)^{n-1+p} - \frac{m}{1}\,(m+z-2)^{n-1+p} + \&c.\right\} \quad \ldots\ldots (6).$$

If we take $n = m$ this is equivalent to Laplace's general formula (p), at p. 168 of the *Théorie.*

The method of this paper leads to some elegant results when applied to the definite integral $\int_0^\infty e^{-ax^2}\,dx$, but it is enough to point out this application, which involves no difficulty whatever.

MATHEMATICAL NOTE*.

Stability of Eccentricities and Inclinations. The equation for proving the stability of the eccentricities and inclinations of the planetary orbits may, as has been shown by Laplace, be deduced from the principle of the conservation of areas, joined to the fact of the invariability of the major axes.

Let m be the mass of a planet, h be twice the area described by the radius vector about the sun in an unit of time projected on the ecliptic, and i the inclination of its orbit to the ecliptic; then, by the principle of the conservation of areas,

$$\Sigma\,(mh) = \text{const.}$$

Every term under the sign of Summation is positive, because all the planets move round the sun in the same direction. But

$$h = \{a\,(1-e^2)\}^{\frac{1}{2}} \cos i = a^{\frac{1}{2}}\,(1-e^2)^{\frac{1}{2}}\,(1+\tan^2 i)^{-\frac{1}{2}}.$$

Now e and i are small at the present time; hence, if we neglect their fourth power and the products of their squares, we have

$$h = a^{\frac{1}{2}}\,\{1 - \tfrac{1}{2}e^2 - \tfrac{1}{2}\tan^2 i\}.$$

Hence $$\Sigma\,\{ma^{\frac{1}{2}}\,(1 - \tfrac{1}{2}e^2 - \tfrac{1}{2}\tan^2 i)\} = \text{const.}$$

But since the major axes have no secular inequalities

$$\Sigma\,(ma^{\frac{1}{2}}) = \text{const.};$$

hence the preceding equation is equivalent to

$$\Sigma\,(ma^{\frac{1}{2}}\,e^2 + ma^{\frac{1}{2}}\tan^2 i) = \text{const.}$$

Now the left-hand side of the equation being small at the present time, the second side is also small, and therefore the first side is always small, and therefore

$$\Sigma\,(ma^{\frac{1}{2}}\,e^2) \text{ and } \Sigma\,(ma^{\frac{1}{2}}\tan^2 i)$$

are both always small.

* *Cambridge Mathematical Journal*, No. XVIII. Vol. III. p. 290, May, 1843.

ON THE EVALUATION OF DEFINITE MULTIPLE INTEGRALS*.

THE following pages contain some general results obtained by means of Fourier's theorem. A few words will be sufficient to explain the manner in which it has been applied.

A definite multiple integral, where the limits are given by the inequality

$$f(xy \ldots\ldots) \overset{=}{>} h_1 \overset{=}{<} h,$$

may be treated as if the limits of the different variables were independent of one another, provided the function under the signs of integration be considered discontinuous, and equal to zero whenever $f(xy \ldots\ldots)$ transgresses the assigned limits h_1 and h. This idea has been made use of by M. Lejeune Dirichlet. It had, however, occurred to me before I was acquainted with his paper on multiple integrals, and the way in which I have applied it is, I believe, new.

Suppose the function to be integrated were of the form $\phi(xy \ldots\ldots)\, \psi\{f(xy \ldots\ldots)\}$: the limits being given by the inequality already mentioned. Then, by Fourier's theorem, writing $f.$ for $f(xy \ldots\ldots)$,

$$\psi f. = \frac{1}{\pi}\int_0^\infty d\alpha \int_{h_1}^{h} \psi u \,.\, \cos\alpha\,(f. - u)\, du,$$

for all values of $f.$ which lie between h_1 and h; moreover for the purposes of integration the formula may be applied, except in particular cases, so as to include these limiting values (see Poisson, *Théorie de la Chaleur*); while for all values of $f.$ which lie without these limits, the second side of the equation is equal to zero. Consequently the integral

* *Cambridge Mathematical Journal*, No. XIX. Vol. IV. p. 1, November, 1843.

$$\int dx \int dy \ldots \phi(xy\ldots)\,\psi\{f(xy\ldots)\}$$

$$= \frac{1}{\pi}\int_{h_1}^{h} \psi u du \int_0^\infty d\alpha \int dx \int dy \ldots \phi(xy\ldots)\cos\alpha\,(f. - u) \ldots (1);$$

the limits on the first side being given in the manner already mentioned. Those of the integrations with respect to x, y, &c. are arbitrary, provided they include all values of the variables which satisfy the given inequality.

Again, if in the given multiple integral the limits were determined by the single relation $f. \leqq h$, joined to the conditions that x, y, &c. were to have no values less than certain assigned limits; *e.g.* if we were to consider only positive values of the variables, the formula (1) would still apply with a slight modification. The inferior limit of integration with respect to u would be arbitrary, provided it included all the values which could be given to $f.$ by admissible values of the variables, while the inferior limits of integration for x, y, &c. would be determined by the particular conditions of the case.

Let us take, as an example of the method, the integral

$$\int dx \int dy \ldots x^{a-1} y^{b-1} z^{c-1} \ldots f(mx + ny + \ldots) \ldots\ldots\ldots (A),$$

m, n, &c. being all positive, and the limits being given by

$$mx + ny + \ldots \leqq h,$$

no negative values of the variables being admitted. In this case (1) becomes

$$\int dx \int dy \ldots x^{a-1} y^{b-1} \ldots f(mx + ny + \ldots)$$

$$= \frac{1}{\pi}\int_{h_1}^{h} fu du \int_0^\infty d\alpha \int_0 dx \int_0 dy \ldots x^{a-1} y^{a-1} \ldots \cos\alpha\,(mx + ny + \ldots - u) \ldots (2),$$

and the integrations with respect to x, y, &c. may be conveniently extended to infinity; (h_1 it should be observed is an arbitrary quantity < 0).

I first seek the value of

$$\int_0^\infty dx \int_0^\infty dy \ldots x^{a-1} y^{b-1} \ldots \cos\alpha\,(mx + ny + \ldots - u) = (B).$$

Integrating first for x, we get

$$(B) = \frac{\Gamma(a)}{m^a \alpha^a}\int_0^\infty dy \ldots y^{b-1} \ldots \cos\left\{a\frac{\pi}{2} + \alpha\,(ny + \ldots - u)\right\}.$$

This follows from the formulæ

$$\left.\begin{aligned}\int_0^\infty x^{a-1}\cos\alpha x dx &= \frac{\Gamma(a)}{\alpha^a}\cos a\frac{\pi}{2}\\ \int_0^\infty x^{a-1}\sin\alpha x dx &= \frac{\Gamma(a)}{\alpha^a}\sin a\frac{\pi}{2}\end{aligned}\right\}\ldots\ldots\ldots\ldots(\gamma).$$

Integrating in a similar manner for y, &c. successively, we get ultimately

$$(\mathrm{B}) = \frac{\Gamma(a)\,\Gamma(b)\ldots}{m^a n^b\ldots\,\alpha^{a+b+\ldots}}\cos\left\{(a+b+\ldots)\frac{\pi}{2}-\alpha u\right\}.$$

Now by (γ), we easily see that

$$\int_0^\infty v^{a+b+\ldots-1}\cos\alpha(v-u)\,dv = \frac{\Gamma(a+b+\ldots)}{\alpha^{a+b+\ldots}}\cos\left\{(a+b+\ldots)\frac{\pi}{2}-\alpha u\right\},$$

and therefore

$$(\mathrm{B}) = \frac{1}{m^a n^b\ldots}\cdot\frac{\Gamma(a)\,\Gamma(b)\ldots}{\Gamma(a+b+\ldots)}\int_0^\infty v^{a+b+\ldots-1}\cos\alpha(v-u)\,dv.$$

Hence $$\int_0^\infty d\alpha\,(\mathrm{B}) = \frac{\pi}{m^a n^b\ldots}\cdot\frac{\Gamma(a)\,\Gamma(b)\ldots}{\Gamma(a+b+\ldots)}u^{a+b+\ldots-1},$$

for all positive values of u, and $=0$ for all negative values, by Fourier's theorem. Consequently the second side of (2) becomes

$$\frac{1}{m^a n^b\ldots}\,\frac{\Gamma(a)\,\Gamma(b)\ldots}{\Gamma(a+b+\ldots)}\int_0^h fu u^{a+b+\ldots-1}\,du + \frac{1}{\pi}\int_{h_1}^0 fu\,.\,0\,.\,du.$$

Thus finally

$$\int dx\int dy\ldots x^{a-1}y^{b-1}\ldots f(mx+ny+\ldots)$$
$$= \frac{1}{m^a n^b\ldots}\,\frac{\Gamma(a)\,\Gamma(b)\ldots}{\Gamma(a+b+\ldots)}\int_0^h fu\,u^{a+b+\ldots-1}\,du\ldots\ldots\ldots\ldots(3);$$

which is equivalent to Liouville's extension of Dirichlet's theorem.

I proceed to evaluate the definite integral

$$\int_0 dx\int_0 dy\ldots e^{-(max+nby+\ldots)}f(mx+ny+\ldots)\ldots\ldots\ldots(\mathrm{E}),$$

$ab\ldots$ and $mn\ldots$ being all positive, and the limits being given by

$$mx+ny+\ldots \lesseqgtr h.$$

By the general formula

$$E = \frac{1}{\pi}\int_{h_1}^{h} fu\,du \int_0^\infty d\alpha \int_0^\infty dx \int_0^\infty dy \ldots e^{-(max+nby+\ldots)} \cos\alpha\,(mx+ny+\ldots-u)$$

$$(h_1 < 0).$$

Let $F = \int_0^\infty dx \int_0^\infty dy \ldots e^{-(max+nby+\ldots)} \cos\alpha\,(mx+ny+\ldots-u)$;

then $F = \cos\alpha u \int_0^\infty dx \int_0^\infty dy \ldots e^{-(max+nby+\ldots)} \cos\alpha\,(mx+ny+\ldots)$

$$+ \sin\alpha u \int_0^\infty dx \int_0^\infty dy \ldots e^{-(max+nby+\ldots)} \sin\alpha\,(mx+ny+\ldots);$$

which we may put equal to $G\cos\alpha u + H\sin\alpha u$.

First to find the value of G

$$= \int_0^\infty dx \int_0^\infty dy \ldots e^{-(max+nby+\ldots)} \cos\alpha\,(mx+ny+\ldots).$$

Develope the cosine; the result is composed of terms, each containing sines or cosines of all the variables.

Also by the formulæ

$$\left.\begin{aligned} \int_0^\infty e^{-max} \cos\alpha\,mx\,dx &= \frac{a}{m\,(a^2+\alpha^2)} \\ \int_0^\infty e^{-max} \sin\alpha\,mx\,dx &= \frac{\alpha}{m\,(a^2+\alpha^2)} \end{aligned}\right\}$$

we see that every factor whether sine or cosine introduces on integration a factor of the form $\frac{1}{m\,(a^2+\alpha^2)}$. Moreover a sine factor introduces α in the numerator, a cosine factor a or b or, &c.

Let P represent the continued product of a, b, c, &c. and D that of $m\,(a^2+\alpha^2)\,n\,(b^2+\alpha^2)$, &c. Then $G = \frac{P}{D}\Sigma C\frac{\alpha^\mu}{ab\ldots}$, where μ is a positive integer less than the whole number of variables in E, and equal to the number of factors in the denominator of $\frac{\alpha^\mu}{ab\ldots}$, and C is some coefficient.

A little consideration shows, that if we develope

$$\cos\alpha\left(\frac{1}{a}+\frac{1}{b}+\ldots\ldots\right)$$

in a series of powers and products of

$$\frac{1}{a}\frac{1}{b}\ldots\ldots\ldots\ldots\ldots\ldots,$$

and neglect all terms involving powers above the first of these quantities, the result will be $=\Sigma C\dfrac{\alpha^\mu}{ab\ldots}$.

Consequently

$$\Sigma C\frac{\alpha^\mu}{ab\ldots}=1-\alpha^2\Sigma\frac{1}{a}\frac{1}{b}+\alpha^4\Sigma\frac{1}{a}\frac{1}{b}\frac{1}{c}\frac{1}{f}-\&c.$$

$\Sigma\dfrac{1}{a}\dfrac{1}{b}$ denoting the combinations two and two of the quantities $\dfrac{1}{a}, \dfrac{1}{b}\ldots$; and so of the rest. For in the development of $(A+B+\ldots)^m$ where m is not greater than the number of quantities A, B, &c. there is always a term involving no power above the first of any of these quantities, and its coefficient is $1.2\ldots m$. This term is obviously the sum of the combinations m and m together of A, B, &c., and that its coefficient is equal to $1.2.3\ldots m$, appears from the polynomial theorem, viz.

$$(A+B+\ldots)^m=\Sigma\frac{1.2\ldots m}{1.2\ldots p.1.2\ldots q\ \&c.}A^hB^v\ldots$$

Thus $$G=\frac{p_t-\alpha^2p_{t-2}+\alpha^4p_{t-4}-\&c.}{mn\ldots(a^2+\alpha^2)(b^2+\alpha^2)\ldots}\ldots\ldots\ldots\ldots\ldots\ldots (4);$$

where p_t, p_{t-2}, &c. are the alternate coefficients of the equation

$$v^t-p_1v^{t-1}+\&c.\quad\pm p_{t-1}v\mp p_t=0,$$

whose roots are a, b, c, &c. (I suppose t to be the number of variables in E.) In precisely the same way we should find

$$H=\frac{\alpha p_{t-1}-\alpha^3p_{t-3}+\&c.}{mn\ldots(a^2+\alpha^2)(b^2+\alpha^2)\ldots}\ldots\ldots\ldots\ldots (5).$$

Next to find the values of

$$\int_0^\infty G\cos\alpha u d\alpha \text{ and} \int_0^\infty H\cos\alpha u d\alpha.$$

Let $$K=\int_0^\infty\frac{\cos\alpha u d\alpha}{(a^2+\alpha^2)(b^2+\alpha^2)\ldots}$$

$$=\Sigma\frac{1}{(b^2-a^2)(c^2-a^2)\ldots}\int_0^\infty\frac{\cos\alpha u d\alpha}{a^2+\alpha^2};$$

$$\therefore\ K = \frac{\pi}{2} \Sigma \frac{1}{a (b^2 - a^2)(c^2 - a^2) \dots} e^{\mp au} \dots\dots\dots\dots\dots (6),$$

the upper sign is to be taken when u is > 0.

By differentiating this for a, we have the value of

$$\int_0^\infty \frac{\alpha \sin \alpha u d\alpha}{(a^2 + \alpha^2)(b^2 + \alpha^2) \dots} = \pm \frac{\pi}{2} \Sigma \frac{1}{(b^2 - a^2)(c^2 - a^2) \dots} e^{\mp au} \dots (7),$$

and so by repeated differentiations we find the values of all the integrals which enter into $\int_0^\infty G \cos \alpha u d\alpha$ and $\int_0^\infty H \sin \alpha u d\alpha$. Thus

$$F = \frac{\pi}{2} \frac{1}{mn \dots} \Sigma \{p_t + a p_{t-1} + \&c. + a^t\} \frac{e^{-au}}{a (b^2 - a^2)(c^2 - a^2) \dots} \dots (8),$$

when u is > 0, and

$$F = \frac{\pi}{2} \frac{1}{mn \dots} \Sigma \{p_t - a p_{t-1} + \&c. \pm a^t\} \frac{e^{+au}}{a (b^2 - a^2)(c^2 - a^2) \dots} \dots (9),$$

when u is < 0.

But $p_t - a p_{t-1} + \&c. \pm a^t = 0$.

Hence $F = 0$, when u is < 0, and therefore in (E) we may make $h_1 = 0$, so that the limits of integration for u are 0 and h.

Again $\quad a^t + p_1 a^{t-1} + \&c. = (a + a)(a + b) \dots$

and $2a (b^2 - a^2)(c^2 - a^2) \dots = \{2a . (a + b) \dots\} \{(b - a)(c - a) \dots\}$.

Therefore

$$\frac{a^t + p_1 a^{t-1} + \&c.}{2a (b^2 - a^2)(c^2 - a^2) \dots} = \frac{1}{(b - a)(c - a) \dots}$$

and

$$F = \frac{\pi}{mn \dots} \Sigma \frac{e^{-au}}{(b - a)(c - a) \dots} \dots\dots\dots\dots\dots (10),$$

and therefore finally

$$\int_0 dx \int_0 dy \dots e^{-(max + nby + \dots)} f(mx + ny + \dots)$$

$$= \frac{1}{mn \dots} \Sigma \frac{\int_0 fu e^{-au} du}{(b - a)(c - a) \dots} \dots\dots\dots\dots (11).$$

If $a = b = c = \&c. = A$, the first side of this equation

$$= \frac{1}{mn \dots} \frac{1}{\Gamma(t)} \int_0^h fu\, e^{-Au} u^{t-1} du \text{ by (3).}$$

As a verification of our analysis we may remark, that in this case

$$\Sigma \frac{e^{-au}}{(b-a)(c-a)\ldots} = e^{-Au} \frac{u^{t-1}}{\Gamma(t)};$$

for we have, what is probably a known result, and which at any rate may be easily proved,

$$\Sigma \frac{F(a)}{(b-a)(c-a)\ldots} = \frac{1}{\Gamma(t)} F^{(t-1)}(A) \text{ when } a = b = \&\text{c.} = A,$$

where $F^{(p)}(A)$ denotes the pth derived function of $F(A)$: and this formula applied to the case where $F(a) = e^{-au}$ gives the above written result.

By differentiating (11) for a, b, c, &c. λ, μ, ν, &c. times respectively, λ, μ, ν, &c. being integral or fractional, and dividing by m^λ, n^μ, p^ν, &c. we should obtain the value of

$$\int_0 dx \int_0 dy \ldots e^{-(max+nby+\ldots)} f(mx+ny+\ldots)\, x^\lambda y^\mu \ldots$$

which would include every case of (3). But the investigation would be complex, and I shall therefore only indicate it.

In a future number of the *Journal* I may perhaps apply the method to some other cases, and particularly with regard to such multiple integrals as

$$\int dx \int dy \ldots \phi(xy\ldots) f[\psi(xy\ldots)] F[\chi(xy\ldots)], \&\text{c.}$$

the limits being given by the series of inequalities,

$$\psi. > h_1 < h,$$

$$\chi. > k_1 < k,$$

$$\&\text{c.} > \&\text{c.} < \&\text{c.}$$

In theory, such an integral is reducible to a multiple integral of as many variables as there are limiting inequalities. But it is not easy to find cases in which this reduction can be actually effected.

MATHEMATICAL NOTES*.

1. If a plane passes through any point of a surface, and makes any function of the intercepts it cuts off from the axes, a maximum or a minimum when it touches the surface, this maximum or minimum value is constant for all points of the surface; and, conversely, if for every point of a surface, a given function of the intercepts of the tangent plane is constant, this function is, with reference to any single point of the surface, a maximum or minimum for the tangent plane.

This appears at once from the following considerations: x, y, z being a point in the surface, x_0, y_0, z_0, the three intercepts, $\phi(x_0, y_0, z_0)$ the given function, if we seek to determine the surface so that ϕ shall be a maximum or minimum, we have the equations

$$\frac{x}{x_0}+\frac{y}{y_0}+\frac{z}{z_0}=1 \quad \text{..................... (1)},$$

$$\frac{d\phi}{dx_0}=\mu\frac{x}{x_0^2} \quad \text{...... (2)}, \qquad \frac{d\phi}{dy_0}=\mu\frac{y}{y_0^2} \quad \text{........ (3)};$$

$$\frac{d\phi}{dz_0}=\mu\frac{z}{z_0^2} \quad \text{....................... (4)},$$

μ being a factor. From these equations we get

$$x_0=f_1(xyz), \qquad y_0=f_2(xyz), \qquad z_0=f_3(xyz),$$

and therefore the differential equation of the surface is

$$\frac{dx}{f_1(xyz)}+\frac{dy}{f_2(xyz)}+\frac{dz}{f_3(xyz)}=0 \quad \text{........... (5)};$$

* *Cambridge Mathematical Journal*, No. XIX. Vol. IV. p. 47, November, 1843.

for by the ordinary equation of the tangent plane we have

$$\frac{1}{x_0} : \frac{1}{y_0} : \frac{1}{z_0} :: \frac{dF}{dx} : \frac{dF}{dy} : \frac{dF}{dz},$$

$F=0$ being the equation to the surface.

Again, if we seek to determine the surface, so that ϕ shall be constant, *i. e.* to find the envelope of all the planes represented by (1), we have (1), (2), (3), (4), as before, and in addition

$$\phi(x_0, y_0, z_0) = c \text{ .... } (6).$$

Thus, as before, $x_0 = f_1$, $y_0 = f_2$, $z_0 = f_3$, and the equation of the surface may be got by integrating (5), and determining the constant so that the result may coincide with (6). And the identity of the equations connecting x_0, y_0, z_0, and x, y, z, in the two cases proves our proposition and its converse. Take as an example the ellipsoid

$$\frac{x^2}{a^2} + \frac{y^2}{b^2} + \frac{z^2}{c^2} = 1. \quad \text{Here } x_0 = \frac{a^2}{x}, \quad y_0 = \frac{b^2}{y}, \quad z_0 = \frac{c^2}{z};$$

$$\therefore \frac{a^2}{x_0^2} + \frac{b^2}{y_0^2} + \frac{c^2}{z_0^2} = 1.$$

Therefore the tangent plane to any point of the ellipsoid makes $\frac{a^2}{x_0^2} + \frac{b^2}{y_0^2} + \frac{c^2}{z_0^2}$, a minimum with reference to any plane passing through that point.

2. To find the value of

$$\frac{fa_1}{(a_2-a_1)(a_3-a_1)\ldots(a_n-a_1)} + \frac{fa_2}{(a_1-a_2)(a_3-a_1)\ldots(a_n-a_2)} + \&c. = A,$$

when $a_1 = a_2 = \&c. = a$.

Let $a_1 = a + z_1$, $a_2 = a + z_2$, &c.

$$\therefore (a_2 - a_1)(a_3 - a_1) \ldots\ldots (a_n - a_1) = (z_2 - z_1)(z_3 - z_1) \ldots\ldots (z_n - z_1),$$

and so of the rest;

$$\therefore A = fa \left\{ \frac{1}{(z_2 - z_1)\ldots\ldots(z_n - z_1)} + \frac{1}{(z_1 - z_2)\ldots\ldots(z_n - z_2)} + \&c. \right\}$$

+ &c.

$$\frac{+f^{(p)}a}{1.2\ldots p}\left\{\frac{z_1^{\,p}}{(z_2-z_1)\ldots\ldots(z_n-z_1)}+\frac{z_2^{\,p}}{(z_1-z_2)\ldots\ldots(z_n-z_2)}+\&\text{c.}\right\}$$

$+$ &c. by Taylor's theorem.

Now, when $z_1=z_2=\&\text{c.}=0$, the coefficient of $f^{(p)}a$ will vanish if $p>n$. And whatever the values of z, the coefficient of $f^{(p)}a$ vanishes if $p<n$: for we know that

$$\Sigma\,\frac{z_1^{\,k}}{(z_2-z_1)\ldots\ldots(z_n-z_1)}=0,$$

k being $<n$, and $=1$ when $k=n$;

$$\therefore\ A=\frac{f^{(n)}a}{1.2\ \ldots\ldots n},$$

which was to be found.

NOTE ON A DEFINITE MULTIPLE INTEGRAL*.

IN the XVIIIth number of this *Journal* Mr Boole pointed out the incorrectness of a theorem given by M. Catalan. The following pages contain a brief demonstration of the result to which he was led. Both he and M. Catalan made use of what is generally known as Liouville's theorem, and thus perhaps rendered their analysis less simple than it would otherwise have been.

Let us transform the integral

$$\int dx_1 \ldots\ldots \int dx_n f(a_1 x_1 + \ldots\ldots + a_n x_n)$$

by the assumption

$$a_1 x_1 + \ldots\ldots a_n x_n = \alpha u_1 \ldots\ldots (\alpha^2 = \Sigma a^2),$$

and by $(n-1)$ other linear relations connecting $x_1 \ldots\ldots x_n$ and $u_1 \ldots\ldots u_n$, and such that

$$\Sigma x^2 = \Sigma u^2.$$

Then, as is well known, $dx_1 \ldots\ldots dx_n$ is to be replaced by $du_1 \ldots\ldots du_n$, and thus

$$\int dx_1 \ldots \int dx_n f(a_1 x_1 + \ldots + a_n x_n) = \int du_1 f \alpha u_1 \int du_2 \ldots \int du_n \ldots (1).$$

Let the integrations on the first side of this equation include all values of the variables which do not transgress the limits

$$\Sigma x^2 = A^2, \quad \Sigma x^2 = B^2,$$

were B is supposed to be greater than A. Then, as $\Sigma x^2 = \Sigma u^2$, the corresponding limits on the second side of the equation are

$$\Sigma u^2 = A^2, \quad \Sigma u^2 = B^2.$$

* *Cambridge Mathematical Journal*, No. XX. Vol. IV. p. 64, February, 1844.

Transform the integral in x by assuming

$$x_1 = r\cos\theta_1,\ \ x_2 = r\sin\theta_1\cos\theta_2 \ldots\ \ x_n = r\sin\theta_1 \ldots \sin\theta_{n-1};$$

and that in u, by similar assumptions,

$$u_1 = r\cos\theta'_1,\ \ u_2 = r\sin\theta'_1\cos\theta'_2 \ldots u_n = r\sin\theta'_1 \ldots \sin\theta'_{n-1}.$$

I may be allowed to mention that this transformation, which appears to have been given for the first time by Mr Boole, in the last number of the *Journal*, had occurred to me before I had seen his paper. His analysis leads at once to the conclusion that $dx_1 \ldots dx_n$ is to be replaced by

$$r^{n-1}\sin^{n-2}\theta_1 \ldots \sin\theta_{n-2}\ \ dr d\theta_1 \ldots d\theta_{n-1}.$$

This may be also proved by successive substitutions in the manner pointed out in the case of three variables by Mr A. Smith, in the first volume of the *Journal*.

Thus (1) becomes, since $\Sigma x^2 = \Sigma u^2 = r^2$,

$$\int_A^B r^{n-1} dr \int \sin^{n-2}\theta_1 d\theta_1 \ldots \int d\theta_{n-1} f\{r(a_1\cos\theta_1 + \ldots a_n\sin\theta_1 \ldots \sin\theta_{n-1})\}$$

$$= \int_A^B r^{n-1} dr \int \sin^{n-2}\theta'_1 f(ar\cos\theta'_1)\, d\theta'_1 \int \sin^{n-3}\theta'_2 d\theta'_2 \ldots \int \sin\theta'_{n-2} d\theta'_{n-2} \int d'\theta_{n-1}.$$

The limits for θ and θ' are the same.

That this equation may subsist for all values of A and B, it is necessary and sufficient that

$$\int \sin^{n-2}\theta_1\, d\theta_1 \ldots \int d\theta_{n-1} f\{r(a_1\cos\theta_1 + \ldots a_n\sin\theta_1 \ldots \sin\theta_{n-1})\}$$
$$= \int \sin^{n-2}\theta'_1 f(ar\cos\theta'_1)\, d\theta'_1 \int \sin^{n-3}\theta'_2\, d\theta'_2 \ldots \int d\theta'_{n-1} \ldots\ldots (2).$$

With respect to the limits of θ and θ', it is not difficult to perceive that if $\theta_1 \ldots \theta_{n-1}$ are taken between the limits 0 and π, $x_1 \ldots x_{n-1}$ will receive all the values of which they are capable, namely, all that included between $-r$ and $+r$; and that the same set of values cannot occur more than once. But in order that x_n may vary from 0 to $-r$, it is necessary to extend the superior limit of θ_{n-1} from π to 2π. Thus the limits of θ_{n-1} are 0 and 2π, while those of the other variables $\theta_1 \ldots \theta_{n-2}$ are 0 and π. And similarly for θ'.

On the second side of (2) we have the factor

$$\int_0^\pi \sin^{n-3}\theta'_2\, d\theta'_2 \int_0^\pi \sin^{n-4}\theta'_3 d\theta'_3 \ldots \int_0^{2\pi} d\theta'_{n-1} \ldots\ldots (A).$$

Let n be odd, then we shall have

$$\int_0^\pi \sin^{n-3}\theta'_2\,d\theta'_2 \int_0^\pi \sin^{n-4}\theta'_3\,d\theta'_3$$

$$= \frac{(n-4)\ldots 3.1}{(n-3)\ldots 4.2} \cdot \frac{(n-3)\ldots 4.2}{(n-4)\ldots 3.1}\, 2\pi = \frac{2\pi}{n-3}.$$

Similarly

$$\int_0^\pi \sin^{n-5}\theta'_4\,d\theta'_4 \int_0^\pi \sin^{n-6}\theta'_5\,d\theta'_5 = \frac{2\pi}{n-5}, \text{ \&c.} = \text{\&c.}$$

Lastly
$$\int_0^{2\pi} d\theta'_{n-1} = 2\pi.$$

Thus
$$(A) = 2\frac{\pi^{\frac{n-1}{2}}}{\left(\frac{n-1}{2}-1\right)\left(\frac{n-1}{2}-2\right)\ldots 2.1} = \frac{2\pi^{\frac{n-1}{2}}}{\Gamma\left(\frac{n-1}{2}\right)}.$$

Again, if n be even, we get

$$(A) = 2\frac{\pi^{\frac{n-2}{2}}}{\left(\frac{n-1}{2}-1\right)\left(\frac{n-1}{2}-2\right)\ldots \frac{3}{2}\frac{1}{2}}.$$

And this, multiplied by $\frac{\sqrt{(\pi)}}{\Gamma(\frac{1}{2})}$, or unity, gives, as before,

$$(A) = \frac{2\pi^{\frac{n-1}{2}}}{\left(\Gamma\frac{n-1}{2}\right)};$$

and thus (2) becomes, making $r = 1$,

$$\int_0^\pi \sin^{n-2}\theta_1\,d\theta_1 \int_0^\pi \sin^{n-3}\theta_2 d\theta_2 \ldots \int_0^{2\pi} d\theta_{n-1} f(a_1\cos\theta_1 + \ldots \text{\&c.})$$

$$= \frac{2\pi^{\frac{n-1}{2}}}{\Gamma\left(\frac{n-1}{2}\right)} \int_0^\pi \sin^{n-2}\theta'_1\,d\theta'_1 f(\alpha\cos\theta'_1) \quad\ldots\ldots (3).$$

This is the general result given at the end of Mr Boole's paper, and which includes the others.

NOTES ON MAGNETISM*.

No. I.

A GEOMETRICAL construction, by means of which the action of a small magnet on a distant particle of free magnetism may be readily determined, is mentioned, in a memoir by Weber, on the Bifilar Magnetometer (*Scientific Memoirs*, II. p. 270). It is due to Gauss, but I do not know where he has demonstrated it. A proof of it may be acceptable to some readers of the *Journal*.

I begin by enunciating the construction in question, which will be easily understood without a figure. Let AB be a small bar magnet, c its centre, P the particle of magnetism on which it acts. Join cP, draw PD perpendicular to it, meeting cP produced in D. Let $cQ = \frac{1}{3} cD$. Join PQ. Then PQ or QP (according to the sign of the magnetism of P and the direction of the poles of AB) is the direction of the action of AB on P; and $\frac{Mm}{cP^3} \frac{PQ}{cQ}$ is its magnitude, M being the measure of the magnetism of AB, m that of the magnetism of P.

The dimensions of the magnet being small, and its length in the direction of its axis being much greater than its breadth or thickness, we may proceed as follows.

Conceive the magnet to be composed of a series of intense particles ranged along its axis. Let s be the distance of any one of them from c, μds the measure of its magnetism. Also let $cP = r$, and call the angle cP makes with cD, θ.

* *Cambridge Mathematical Journal*, No. XX. Vol. IV. p. 90, February, 1844.

Then the distance of the particle in question from P is $(r^2+s^2-2rs\cos\theta)^{\frac{1}{2}}$, and consequently its action on P is

$$\frac{m\mu ds}{r^2+s^2-2rs\cos\theta},$$

magnetic attraction being supposed to follow the ordinary law. The component of this action along cD, is

$$\frac{m\mu ds}{(r^2+s^2-2rs\cos\theta)^{\frac{3}{2}}}(r\cos\theta-s).$$

This is approximately equal to

$$\frac{m\mu}{r^3}\left(1+3\frac{s}{r}\cos\theta\right)(r\cos\theta-s)\,ds,$$

or to
$$\frac{m\mu}{r^3}\{r\cos\theta+(3\cos^2\theta-1)\,s\}\,ds;$$

and therefore the total action of AB on P, parallel to cD, is

$$\frac{m\cos\theta}{r^2}\int\mu ds+m\frac{3\cos^2\theta-1}{r^3}\int\mu sds,$$

the integrals being taken along the whole length of the magnet. Consequently

$$\int\mu ds=0,$$

since the aggregate magnetism of a magnet, each element being taken with its proper sign, is zero. Again, $\int\mu sds$ being the *moment* of the magnetism of AB, is the measure of its magnetic power, or what we have called M.

Consequently the action parallel to cD is

$$\frac{Mm}{r^3}(3\cos^2\theta-1)\;\ldots\ldots\ldots\ldots\ldots\;(1).$$

The action of the element at s, perpendicular to cD, is

$$\frac{m\mu ds}{(r^2+s^2-2rs\cos\theta)^{\frac{3}{2}}}\,r\sin\theta,$$

or, approximately,

$$\frac{m\mu}{r^2}\left(1+3\frac{s}{r}\cos\theta\right)\sin\theta ds.$$

The total action perpendicular to cD is, therefore,

$$\frac{3Mm}{r^3}\sin\theta\cos\theta\;\ldots\ldots\ldots\ldots\ldots\;(2).$$

The equation to the resultant of these two forces is, since the line passes through P,

$$\frac{y - r\sin\theta}{3\sin\theta\cos\theta} = \frac{x - r\cos\theta}{3\cos^2\theta - 1}.$$

For $y = 0$, or at the point Q,

$$x_0 = r\cos\theta - \tfrac{1}{3}r\sec\theta\,(3\cos^2\theta - 1),$$

$$\text{or}\quad x_0 = \tfrac{1}{3}r\sec\theta.$$

Now $cD = r\sec\theta$, therefore

$$cQ = \tfrac{1}{3}cD,$$

which proves the first part of the construction.

Next, to find the magnitude of the whole action, square and add (1) and (2), then

$$\frac{Mm}{r^3}(1 - 6\cos^2\theta + 9\cos^4\theta + 9\sin^2\theta\cos^2\theta)^{\frac{1}{2}},$$

$$\text{or}\quad \frac{Mm}{r^3}(1 + 3\cos^2\theta)^{\frac{1}{2}}$$

is the magnitude sought. Now

$$\begin{aligned} PQ &= r\,\{\sin^2\theta + (\cos\theta - \tfrac{1}{3}\sec\theta)^2\}^{\frac{1}{2}} \\ &= r\,(1 - \tfrac{2}{3} + \tfrac{1}{9}\sec^2\theta)^{\frac{1}{2}} \\ &= \tfrac{1}{3}r\,(\sec^2\theta + 3)^{\frac{1}{2}}, \end{aligned}$$

and $\quad cQ = \tfrac{1}{3}r\sec\theta.$

Therefore $\quad \dfrac{PQ}{cQ} = (1 + 3\cos^2\theta)^{\frac{1}{2}}.$

Consequently, if R be the magnitude of the resultant sought,

$$R = \frac{Mm}{cP^3}\,\frac{PQ}{cP},$$

which was to be proved.

The preceding formulæ enable us to determine all the circumstances of the mutual action of two magnets, which are such as to fulfil the conditions of our hypothesis.

For instance, in the memoir *Intensitas vis Magneticæ Terrestris*, Gauss has shown that if a magnet and a needle be placed at right angles to one another, then the moment of rotation of the needle due to the action of the magnet is approximately twice as great when the line of the axis of the magnet passes through the centre of the needle, as it is when the line of the axis of the needle passes through the centre of the magnet.

In neither case has the transverse force (2) any tendency to produce rotation. Its effect is destroyed by the fixed centre of the needle.

Let y be the distance of any element of the needle from the centre, mdy its magnetism. Then, in the first case, we shall have $\theta = \frac{y}{R}$ approximately, (R being the distance between the centre of the magnet and that of the needle). Consequently the force (1) may be expressed by the formula

$$\frac{Mmdy}{(R^2+y^2)^{\frac{3}{2}}}\left(2-3\,\frac{y^2}{R^2}\right) \text{ since } \cos^2\theta = 1-\theta^2 \text{ nearly};$$

which, neglecting y^2, becomes

$$\frac{2Mmdy}{R^3};$$

the moment of this round the centre of the needle is $\frac{2Mmydy}{R^3}$, and the total moment is, therefore,

$$\frac{2MM'}{R^3} \text{ where } M' = \int mydy.$$

In the second case, $\theta = 0$ for every element of the needle, and the moment sought is, therefore,

$$\frac{MM'}{R^3}, \text{ since (1) then becomes } -\frac{Mm}{r^3},$$

and the sign is immaterial. The moment in this case is, therefore, one half of what it was before, which was to be proved.

This result is included in the following general investigation, in which we shall ascertain the moment of rotation due to the action of one magnet or needle upon another, whatever be their relative positions.

The data by which we shall suppose the position of the magnets to be determined, are the distance between their centres, the angles which the axis of each respectively makes with the line joining their centres, and the angle between the axes themselves. The last element may be readily replaced by the dihedral angle between two planes which intersect in the line joining the centres of the magnets, and one of which passes through the axis of one of the magnets while the other passes through that of the other.

Let the axis of the magnet, whose action on the other is to be calculated, be taken as axis of x, the centre of the magnet being the origin of co-ordinates. Let x, y, z be the co-ordinates of any point in the other magnet. Then, if $r^2 = x^2 + y^2 + z^2$,

$$3\cos^2\theta - 1 = \frac{3x^2 - r^2}{r^2}, \text{ and } \sin\theta\cos\theta = \frac{x\sqrt{(y^2+z^2)}}{r^2}.$$

Again, let a, b, c be the co-ordinates of the centre of the second magnet, α, β, γ the cos angles its axis makes with the co-ordinate axes, ρ the distance of any element $d\rho$ from the centre, X, Y, Z the forces on $d\rho$ parallel to the axes of co-ordinates, m the intensity of $d\rho$. Then

$$X = \frac{Mm}{r^5}(3x^2 - r^2)\,d\rho,$$

$$Y = \frac{3Mm}{r^5}xy\,d\rho,$$

$$Z = \frac{3Mm}{r^5}xz\,d\rho.$$

Let G, H, K be the moments of these forces about lines drawn through the centre of the second magnet parallel to the axes of co-ordinates,

$$G = \frac{3Mm}{r^5}x\{y(z-c) - z(y-b)\}\,d\rho,$$

$$H = \frac{Mm}{r^5}\{3xz(x-a) - (3x^2 - r^2)(z-c)\}\,d\rho,$$

$$K = \frac{Mm}{r^5}\{(3x^2 - r^2)(y-b) - 3xy(x-a)\}\,d\rho.$$

Now $\quad x = a + \alpha\rho, \quad y = b + \beta\rho, \quad z = c + \gamma\rho.$

Hence, neglecting the square, &c. of ρ,

$$G = \frac{3Mm}{r^5}a(b\gamma - c\beta)\,\rho d\rho,$$

$$H = \frac{Mm}{R^5}\{3ac\alpha - (3a^2 - R^2)\gamma\}\,\rho d\rho,$$

$$K = \frac{Mm}{R^5}\{(3a^2 - R^2)\beta - 3ab\alpha\}\,\rho d\rho,$$

where $R^2 = a^2 + b^2 + c^2$.

Integrating for ρ, we find, for the total moments,

$$(G) = \frac{3MM'}{R^5} a (b\gamma - c\beta),$$

$$(H) = \frac{3MM'}{R^5} a (c\alpha - a\gamma) + \frac{MM'}{R^3} \gamma,$$

$$(K) = \frac{3MM'}{R^5} a (a\beta - b\alpha) - \frac{MM'}{R^3} \beta,$$

M' being for the second magnet what M is for the first.

Let L be the resultant of these three moments, then

$$L^2 = (G)^2 + (H)^2 + (K)^2.$$

Consequently, since $\alpha^2 + \beta^2 + \gamma^2 = 1$, we shall have

$$L^2 = \left(\frac{3MM'}{R^5}\right)^2 a^2 \{R^2 - (a\alpha + b\beta + c\gamma)^2\}$$

$$+ 6 \left(\frac{MM'}{R^4}\right)^2 \{a\alpha (a\alpha + b\beta + c\gamma) - a^2\}$$

$$+ \left(\frac{MM'}{R^3}\right)^2 (\beta^2 + \gamma^2).$$

Now, let θ, θ' be the angles which the axes of the two magnets make respectively with the line joining their centres, and let ϕ be the angle between the axes themselves. Then

$$a = R \cos \theta \text{ and } a\alpha + b\beta + c\gamma = R \cos \theta', \text{ also } \alpha = \cos \phi.$$

Consequently

$$L^2 = \left(\frac{MM'}{R^3}\right)^2 \{9 \cos^2\theta \sin^2\theta' + 6 \cos \theta (\cos \theta' \cos\phi - \cos\theta) + \sin^2\phi\},$$

or $$L = \frac{MM'}{R^3} \{1 + 3 \cos^2 \theta - (3 \cos \theta \cos \theta' - \cos \phi)^2\}^{\frac{1}{2}}.$$

Let χ be the dihedral angle already mentioned, then

$$\cos \phi = \cos \theta \cos \theta' + \sin \theta \sin \theta' \cos \chi.$$

Consequently the last equation becomes

$$L = \frac{MM'}{R^3} \{1 + 3 \cos^2 \theta - (2 \cos \theta \cos \theta' - \sin \theta \sin \theta' \cos \chi)^2\}^{\frac{1}{2}},$$

the required expression.

ON A MULTIPLE DEFINITE INTEGRAL*.

IN the eighteenth number of the *Journal*, I pointed out the mode in which Fourier's theorem may be employed in the evaluation of certain definite multiple integrals. The theorem generally known as Liouville's, and another of the same degree of generality, were readily deduced from the considerations then suggested. I proceed to another application of the same method.

THEOR.
$$\int dx \int dy \dots \frac{f(mx+ny+\dots)}{(a^2+x^2)(b^2+y^2)\dots}$$
$$= \pi^{\nu-1}\frac{ma+nb+\dots}{ab\dots}\int_h^{h'}\frac{fu\,du}{u^2+(ma+nb+\dots)^2};$$

the integral being supposed to involve ν variables x, y, &c., and the limits being given by the inequalities

$$mx+ny+\dots \geqq h \text{ and } \leqq h'.$$

(Negative as well as positive values of the variables are admissible.)

DEM. Recurring to the general theorem stated at the commencement of the paper already mentioned, we see that the integral whose value is sought is equal to

$$\frac{1}{\pi}\int_h^{h'} fu\,du \int_0^{\infty} d\alpha \int_{-\infty}^{+\infty} dx \int_{-\infty}^{+\infty} dy \dots \frac{\cos\alpha\,(mx+ny+\dots-u)}{(a^2+x^2)(b^2+y^2)\dots}.$$

Develope the cosine in a series of products of sines and cosines of simple arcs. Every term involving a sine disappears on

* *Cambridge Mathematical Journal*, No. XXI. Vol. IV. p. 116, May, 1844.

integration, as the limits extend from $-\infty$ to $+\infty$. Consequently the last written expression becomes

$$\frac{1}{\pi}\int_h^{h'} fu\, du \int_0^\infty \cos \alpha u\, d\alpha \int_{-\infty}^{+\infty} \frac{\cos \alpha m x}{a^2+x^2}\, dx \int_{-\infty}^{+\infty} \frac{\cos \alpha n y}{b^2+y^2}\, dy \ldots$$

Now $$\int_{-\infty}^{+\infty} \frac{\cos \alpha m x}{a^2+x^2}\, du = \frac{\pi}{a} e^{-m a \alpha} \text{ \&c.} = \text{\&c.},$$

and thus the integral becomes

$$\frac{\pi^{\nu-1}}{ab\ldots}\int_h^{h'} fu\, du \int_0^\infty \cos \alpha u e^{-(ma+nb+\ldots)\alpha}\, d\alpha,$$

$$\text{or,}\quad \pi^{\nu-1}\frac{ma+nb+\ldots}{ab\ldots}\int_h^{h'} \frac{fu\, du}{u^2+(ma+nb+\ldots)^2};$$

which was to be proved.

It may be well to verify this result in a particular case. Let $\nu = 2$; then we have to prove that

$$\int dx \int dy\, \frac{f(mx+ny)}{(a^2+x^2)(b^2+y^2)} = \pi\, \frac{ma+nb}{ab} \int_h^{h'} \frac{fu\, du}{u^2+(ma+nb)^2};$$

for simplicity, we will suppose that

$$m^2+n^2 = 1.$$

Let $mx + ny = u$, and therefore $x = mu + nv$,

$nx - my = v$, $\qquad y = nu - mv$,

u and v being two new variables. As $u^2+v^2 = x^2+y^2$, $dx\, dy$ is to be replaced by $du\, dv$, and thus

$$\int dx \int dy\, \frac{f(mx+ny)}{(a^2+x^2)(b^2+y^2)} \text{ becomes}$$

$$\int fu\, du \int \frac{dv}{\{a^2+(mu+nv)^2\}\{b^2+(nu-mv)^2\}}.$$

The limits are easily seen to be $+\infty$ $-\infty$ for v; h' and h for u. For the integral expresses the volume of that portion of a solid bounded by the surface, whose equation is

$$z = \frac{f(mx+ny)}{(a^2+x^2)(b^2+y^2)};$$

which is included between the plane of (xy), the bounding surface, and two planes parallel to one another and perpendicular to (xy). The equations of these planes are respectively

$$mx + ny = h, \quad \text{and} \quad mx + ny = h',$$

and the introduction of u and v is equivalent to changing the axes of co-ordinates, so that one of the new axes, that of u, is perpendicular to these planes, while the other is parallel to them.

In order to find the value of

$$\int_{-\infty}^{+\infty} \frac{dv}{\{a^2 + (mu + nv)^2\}\{b^2 + (nu - mv)^2\}}, \text{ assume}$$

$$\frac{1}{\{a^2 + (mu + nv)^2\}\{b^2 + (nu - mv)^2\}}$$

$$= \frac{A(mu + nv) + Bna}{a^2 + (mu + nv)^2} + \frac{C(nu - mv) + Dmb}{b^2 + (nu - mv)^2}.$$

It is evident that the terms in A and C will disappear on integration between infinite limits: those in B and D become respectively πB and πD, and the integral in question is therefore

$$\pi(B + D).$$

Now it may be shown that

$$B = \frac{n}{a} \frac{u^2 - m^2a^2 + n^2b^2}{\{u^2 + (ma + nb)^2\}\{u^2 + (ma - nb)^2\}};$$

$$D = \frac{m}{b} \frac{u^2 + m^2a^2 - n^2b^2}{\{u^2 + (ma + nb)^2\}\{u^2 + (ma - nb)^2\}}.$$

Consequently $B + D = \dfrac{ma + nb}{ab} \dfrac{1}{u^2 + (ma + nb)^2}$;

and thus the integral sought is seen to be equal to

$$\pi \frac{ma + nb}{av} \int_h^{h'} \frac{fudu}{u^2 + (ma + nb)^2},$$

which was to be proved.

Similar considerations apply in the case of more variables, and doubtless by induction our general result might be established. But the method we have followed, besides being more analytical, is also very much simpler.

Another result of this same kind of analysis I shall indicate without a demonstration, which there will be no difficulty in supplying.

$$\int dx \int dy \dots e^{-a^2x^2-b^2y^2\dots} f(max + nby + \dots)$$

$$= \frac{\pi^{\frac{n-1}{2}}}{ab\dots} \frac{1}{\{m^2+n^2+\dots\}^{\frac{1}{2}}} \int_h^{h'} e^{-\frac{u^2}{m^2+n^2+\dots}} fu\, du,$$

the limits being given by

$$max + nby + \dots \overline{\overline{>}}\, h \text{ and } \overline{\overline{<}}\, h'.$$

In conclusion, it may be well to remark, that the analysis of which we have made use is not unfrequently applicable to questions which though not difficult in principle are nevertheless somewhat perplexing in practice.

ON A QUESTION IN THE THEORY OF PROBABILITIES*.

THE following question affords a good illustration of the methods employed in the more difficult parts of the theory of probabilities. In a paper presented to the Philosophical Society, I applied the kind of analysis we are about to make use of, to the celebrated *Rule of Least Squares.* There is, in fact, a close analogy between the two investigations. Laplace's solution of the present question is obtained by a process similar to that which he had employed when treating of the best method of combining discordant observations.

What is the probability that the sum of the times which each of n persons has respectively yet to live will amount to a given time T?

Let $\phi_p x_p \, dx_p$ be the probability that the pth person will live precisely a time x_p longer, ϕ_p denoting some function of x_p, which is necessarily such that

$$\int_0^\infty \phi_p x_p \, dx_p = 1,$$

as it is certain that he will die at some time or other.

Let $x_1, x_2, \ldots x_n$ be so related that

$$x_1 + x_2 + \ldots x_n = T.$$

The probability of this particular combination is

$$\phi_1 x_1 \, \phi_2 x_2 \ldots \phi_n x_n \, dx_1 \, dx_2 \ldots dx_n,$$

or $$\phi_1 x_1 \, \phi_2 x_2 \ldots \phi_n (T - x_1 - \ldots x_{n-1}) \, dx_1 \ldots dx_n,$$

and the aggregate probability sought is the integral of this expression obtained by giving all possible positive values to

* *Cambridge Mathematical Journal,* No. XXI. Vol. IV. p. 127. May, 1844.

$x_1 \ldots x_{n-1}$, which do not make $T-x_1-\ldots x_{n-1}$ negative. Thus we have

$$P=dx_n\int_0\ldots\int_0\phi_1x_1\,\phi_2x_2\ldots\phi_n(T-x_1-\ldots x_{n-1})\,dx_1\ldots dx_{n-1}.$$

Now, by Fourier's theorem,

$$\phi_n(T-x_1-\ldots x_{n-1})=\frac{1}{\pi}\int_0^\infty d\alpha\int_0^\infty \phi_nx_n\cos\alpha\,(T-x_1-\ldots x_{n-1}-x_n)\,dx_n,$$

$T-x_1\ldots x_{n-1}$ being supposed to lie between 0 and ∞; for all negative values of this quantity, the second member of the equation is equal to zero. Consequently, all the integrations may now be taken from zero to infinity; and thus, since as T and x_n vary together, $dT=dx_n$

$$P=\frac{dT}{\pi}\int_0^\infty d\alpha\int_0^\infty dx_1\ldots\int_0^\infty dx_n\phi_1x_1\ldots\phi_nx_n\cos\alpha\,(T-\Sigma x).$$

Now it may be shown that the greatest value of

$$\int_0^\infty dx_1\ldots\int_0^\infty dx_n\phi_1x_1\ldots\phi_nx_n\cos\alpha\,(T-\Sigma x)\ldots(\alpha)$$

corresponds to $\alpha=0$, and is, therefore, unity; and that when n is large (α) diminishes rapidly as α increases. Consequently the value of $\int_0^\infty(\alpha)\,d\alpha$ depends, when n is very large, on the elements for which α is very small. This consideration enables us to employ an approximate value of (α).

Let $T=t+m$, m being a disposable quantity; then

$$(\alpha)=\cos\alpha t\int_0^\infty dx_1\ldots\int_0^\infty dx_n\phi_1x_1\ldots\phi_nx_n\cos\alpha\,(m-\Sigma x)$$
$$+\sin\alpha t\int_0^\infty dx_1\ldots\int_0^\infty dx_n\,\phi_1x_1\ldots\phi_nx_n\sin\alpha\,(m-\Sigma x),$$

which may be thus written,

$$(\alpha)=\cos\alpha t\,G+\sin\alpha t\,H.$$

In order to obtain approximate values of G and H, expand $\cos\alpha\,(m-\Sigma x)$ and $\sin\alpha\,(m-\Sigma x)$; they become respectively, α being very small,

$$1-\tfrac{1}{2}\alpha^2(m-\Sigma x)^2 \text{ and } \alpha\,(m-\Sigma x).$$

Now let $\int_0^\infty \phi x \, x dx = K \int_0^\infty \phi x \, x^2 dx = k^2$;

then, since $\int_0^\infty \phi x dx = 1$, we shall have

$$G = 1 - \tfrac{1}{2}\alpha^2 (m^2 - 2m \, \Sigma K + 2\Sigma K_1 K_2 + \Sigma k^2),$$

$$H = \alpha \, (m - \Sigma K) \text{ approximately.}$$

Let $m = \Sigma K$, then

$$m^2 - 2m \, \Sigma K = - \{\Sigma K\}^2 = - \Sigma K^2 - 2\Sigma K_1 K_2,$$

and thus
$$G = 1 - \tfrac{1}{2}\alpha^2 \Sigma \, (k^2 - K^2),$$
$$H = 0;$$

and therefore, while α is very small,

$$(\alpha) = \cos \alpha t \, \{1 - \tfrac{1}{2}\alpha^2 \Sigma \, (k^2 - K^2)\}.$$

We have next to show that $\Sigma \, (k^2 - K^2)$ is a positive quantity.

Consider the definite integral

$$\int_0^\infty \int_0^\infty \phi x \, \phi z \, (z - x)^2 dx \, dz;$$

it is necessarily positive, since every element is so, $\phi x \, dx$ being the expression of a probability, and therefore essentially positive.

Expanding $(z - x)^2$, we find for the value of this integral

$$k^2 - 2K^2 + k^2 \text{ or } 2 \, (k^2 - K^2).$$

Hence $k^2 - K^2$, and therefore $\Sigma \, (k^2 - K^2)$, is positive.

(This demonstration is due to Poisson, *Con. des Tems*, 1827.)

Returning to the value we have found for (α), we see that we may in all cases represent (α) by $\cos \alpha t \, e^{-\frac{1}{2}\alpha^2 \Sigma (k^2 - K^2)}$, since when α is very small, the two expressions tend to coincide; and when α is not so, both are sensibly zero, $\Sigma \, (k^2 - K^2)$ being a large quantity of the order n. Consequently

$$\int_0^\infty (\alpha) \, d\alpha = \sqrt{\left(\frac{\pi}{2}\right)} \frac{1}{\{\Sigma \, (k^2 - K^2)\}^{\frac{1}{2}}} e^{-\frac{t^2}{2\Sigma (k^2 - K^2)}}$$

and $$\Gamma = \frac{dt}{\sqrt{(2\pi)}} \frac{1}{\{\Sigma \, (k^2 - K^2)\}^{\frac{1}{2}}} e^{-\frac{t^2}{2\Sigma (k^2 - K^2)}} \ldots (p),$$

P being the probability that the required sum T shall be precisely equal to $\Sigma K + t$. The greatest value of P corresponds to $t = 0$; consequently the most probable value of T is ΣK.

It is to be remarked that the approximate formula (p) is independent of the law of probability expressed by the function ϕx: it depends merely on the two definite integrals

$$\int_0^\infty x\phi\, x dx \text{ and } \int_0^\infty x^2 \phi\, x dx.$$

We have stopped the approximation to the value of G at the second power of α. Had we gone farther, and retained only the *principal* term in the coefficient of each power of α, a similar result, viz. one which may be assumed as coincident with the exponential function, would, there is little doubt, have been obtained; while the coefficient of each power of α in H would be negligible in comparison of the corresponding power in G. Some remarks on this, or at least on a cognate question, will be found in the paper already mentioned.

As a verification of the approximation we have employed, which is in effect the same as that of Laplace, let us suppose that the functions $\phi_1, \phi_2 \dots$ are all of the same form ϕ, and that $\phi x = e^{-x}$. Then, as we have seen, the required probability is obtained by integrating

$$e^{-x_1 - x^2 \dots -(T - x_1 - x_2 - \dots)} dx_1 \dots dx_{n-1},$$

for all positive values of $x_1, x_2 \dots x_{n-1}$ which do not transgress the limits

$$x_1 + x_2 \dots x_{n-1} = T.$$

Now $$e^{-x_1 - x_2 \dots -(T - x_1 - x_2 - \dots)} = e^{-T},$$

and thus we have $P = e^{\ T} dT \int_0 dx_1 \dots \int_0 dx_{n-1}$,

the limits being given by

$$x_1 + x_2 \dots x_{n-1} \leqq T.$$

Hence, it is easily seen that

$$P = \frac{e^{-T}}{\Gamma(n)} T^{n-1} dT.$$

In order to compare this with the approximate expression (p), I remark that $T = n - 1$ renders P a maximum; assume, therefore,

$$T = n - 1 + t,$$

$$\text{then } P = \frac{e^{-(n-1)}(n-1)^{n-1}}{\Gamma(n)} e^{-t}\left(1+\frac{t}{n-1}\right)^{n-1} at.$$

Now, by Stirling's theorem, or by that which Binet proposes to substitute for it (vide *Journal de l'Ecole Polytechnique*, XVI. p. 226), we have, when n is very large,

$$\Gamma(n) = \sqrt{(2\pi)}\, e^{-(n-1)} (n-1)^{n-\frac{1}{2}}.$$

Consequently

$$\frac{e^{-(n-1)}(n-1)^{n-1}}{\Gamma(n)} = \frac{1}{\sqrt{\{2\pi(n-1)\}}} = \frac{1}{\sqrt{(2\pi n)}} \ldots q.p.$$

$$\text{Again } \left\{1+\frac{t}{n-1}\right\}^{n-1} = 1+t+1.\left(1-\frac{1}{n-1}\right)\frac{t^2}{1.2}$$

$$+\left(1-\frac{1}{n-1}\right)\left(1-\frac{2}{n-1}\right)\frac{t^3}{1.2.3} + \&c.$$

$$= e^t + ft,$$

ft being a certain function of t and n. Therefore

$$e^{-t}\left(1+\frac{t}{n-1}\right)^{n-1} = 1 + e^{-t} ft.$$

Now the coefficient of t^2 in ft is

$$-\frac{1}{(n-1).1.2},$$

$$\text{that of } t^3 \text{ is } \left\{-\frac{3}{n-1}+\frac{2}{(n-1)^2}\right\}\frac{1}{1.2.3},$$

and that of t^4 is

$$\left\{-\frac{6}{n-1}+\frac{11}{(n-1)^2}-\frac{6}{(n-1)^3}\right\}\frac{1}{1.2.3.4}.$$

Hence the coefficient of t^2 in $e^{-t} ft$ is

$$-\frac{1}{2(n-1)},$$

that of t^3 is

$$\frac{1}{2(n-1)} - \frac{1}{2(n-1)} + \frac{1}{3(n-1)^2} \text{ or } \frac{1}{3(n-1)^2};$$

and, lastly, that of t^4 is

$$-\frac{1}{4(n-1)}+\frac{2}{4(n-1)}-\frac{2}{1.2.3(n-1)^2}-\frac{1}{4(n-1)}$$
$$+\frac{11}{1.2.3.4}\frac{1}{(n-1)^2}-\frac{1}{4(n-1)^3},$$

$$\text{or } \frac{1}{1.2}\left\{\frac{1}{2(n-1)}\right\}^2-\frac{1}{4(n-1)^3}.$$

Now if, in forming the approximate expression, we reject all terms of the form $\left\{\frac{t}{\sqrt{(n)}}\right\}^p \frac{1}{n^q}$, where q is different from zero, *i.e.* if we look on $\frac{t}{\sqrt{(n)}}$ as a quantity all whose powers are to be retained, except when divided by any power of n, the value of $e^{-t}\left(1+\frac{t}{n-1}\right)^{n-1}$ may be taken as equal to

$$1-\frac{t^2}{2(n-1)}+\frac{1}{1.2}\cdot\frac{t^4}{4(n-1)^2}-\&c.$$

which, as similar results would have been obtained had we pursued the investigation farther, might be shown to be equal to

$$e^{-\frac{t^2}{2(n-1)}},$$

and thus $$P=\frac{1}{\sqrt{(2\pi n)}}e^{-\frac{t^2}{2(n-1)}}dt,$$

P is the probability that T is equal to $n-1+t$: writing $t+1$ for t in $\frac{t^2}{n-1}$ and reducing, we find that, within the limits of the approximation, it may also be assumed as the probability that t is equal to $n+t$: also $\frac{t^2}{n-1}=\frac{t^2}{n}$ ... *q.p.*, and thus

$$P=\frac{1}{\sqrt{(2\pi n)}}e^{-\frac{t^2}{2n}}dt;$$

which, as in our case, $k^2=2$ and $K=1$ is precisely equivalent to the result deduced from the general formula (p).

The legitimacy of some parts of the preceding approximation may be questioned; as quantities which are neglected may, under certain conditions, be larger than those which are retained: and, as the result coincides with that of the general method, the doubt thus suggested appears to extend to the latter. The subject of approximation by means of definite integrals is certainly not free from obscurity.

The method of this paper extends *m.m.* to the case in which we seek to determine the degree of improbability that the average length of the reigns of a series of kings shall exceed by a given quantity the average deduced from authentic history. The application of considerations of this nature to historical criticism appears to have been first made in Sir Isaac Newton's Chronology. They are doubtless entitled to much attention; but any attempt to evaluate their legitimate influence, would, for more than one reason, be unsatisfactory.

ON THE BALANCE OF THE CHRONOMETER*.

It is well known that a common watch goes more slowly when its temperature is raised, and *versâ vice*. The reason of this is that the elasticity of the balance-spring decreases with the increment of temperature and increases with its decrement. Neglecting the mass of the spring and the connection of the balance with the other parts of the watch, we may take as the equation for determining the oscillations of the balance,

$$\frac{d^2\theta}{dt^2}+\frac{e\theta}{I}=0,$$

where e depends on the form and elasticity of the spring, and I is the moment of inertia of the balance. The time of oscillation depends, of course, on the ratio $\frac{e}{I}$, e being, as we have said, a function of t the temperature. In order, therefore, to the equable rate of the watch, it would be necessary that I should be such a function of t, that $\frac{e}{I}$ may be constant. In the balance of a common watch I is sensibly constant. Hence the inequality of which we have spoken.

In the chronometer the balance is so constructed that its figure alters when the temperature varies. The figure repre-

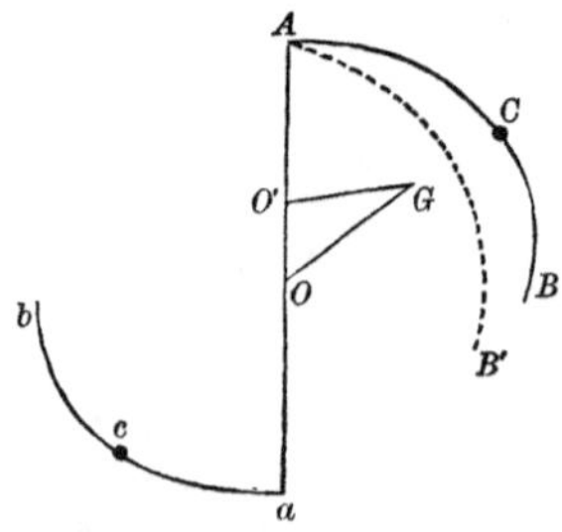

* *Cambridge Mathematical Journal*, No. XXI. Vol. IV. p. 133, May, 1844.

sents a common form of the chronometer balance. The arc AB, which carries a weight at C, is formed of two concentric laminæ of different kinds of metal, the outer lamina being the most expansible. These two laminæ are securely united in their whole length, so that an increase of temperature necessarily distorts the arc AB into some form like AB'. Similarly for ab. Contrary effects are produced by a decrease of temperature. Thus, the moment of inertia I decreases as t increases; as e also does. And thus we are enabled, by suitable adjustments, to make $\frac{e}{I}$, at least approximately, constant.

It would, I believe, be impossible, without some hypothesis, to determine the form which AB assumes under the influence of a change of temperature. The following suppositions are probably sufficiently near the truth to be applicable when the variations of t are not excessive.

Let us suppose the laminæ to be cylindrical and concentric, and bounded by four plane surfaces, two of which are perpendicular to the axis of the cylinder, while the other two, which form the boundaries at A and B, pass through the axis. These conditions being fulfilled, whatever the value of t may be, it is clear that the variation of form can depend on two elements only, namely, the radius of the cylinder, and the angle which AB subtends at its centre. To determine these, we assume that the middle filament of each lamina expands as it would do if free.

In the normal state, let 2ϵ, $2\epsilon'$ be the thicknesses of the outer and inner laminæ respectively, r the radius of the boundary of the two laminæ, μ, μ' the coefficients of expansibility of the outer and inner laminæ ($\mu > \mu'$), θ the angle subtended at the centre.

The radii of the middle filaments are, therefore, $r+\epsilon$, $r-\epsilon'$; let their lengths be l and l', then

$$l=(r+\epsilon)\,\theta, \quad l'=(r-\epsilon')\,\theta.$$

For an increase of temperature t, let r and θ become r_1 and θ_1: then we shall have

$$l\,(1+\mu t)=(r_1+\epsilon)\,\theta_1, \quad l'\,(1+\mu' t)=(r_1-\epsilon')\,\theta_1;$$

ϵ and ϵ' being so small that their variations may be neglected.

Hence $$\frac{r_1 + \epsilon}{r_1 - \epsilon'} = \frac{l}{l'} \frac{1 + \mu t}{1 + \mu' t},$$

and therefore

$$r_1 - \epsilon' = (\epsilon + \epsilon') \frac{l' (1 + \mu t)}{l (1 + \mu t) - l' (1 + \mu' t)},$$

which, as μ and μ' are very small, is approximately

$$r_1 - \epsilon' = (\epsilon + \epsilon') \frac{l'}{l - l'} \left\{1 - (\mu - \mu') \frac{l'}{l - l'} t\right\}.$$

When $t = 0$, $$r - \epsilon' = (\epsilon + \epsilon') \frac{l'}{l - l'}.$$

Consequently, for a first approximation,

$$\Delta r = - \Delta\mu \frac{r^2}{\tau} t \text{ .... (1)},$$

where $\Delta r = r_1 - r$, $\Delta\mu = \mu - \mu'$, and $\tau = \epsilon + \epsilon_1$.

Again $$\tau\theta_1 = l - l' + (\mu l - \mu' l') t.$$

When $t = 0$, $$\tau\theta = l - l';$$

and therefore $$\tau\Delta\theta = (\mu l - \mu' l') t.$$

But $$\mu l - \mu' l' = r\theta\Delta\mu + (\mu\epsilon - \mu'\epsilon') \theta,$$

and the last term is negligible. Therefore

$$\Delta\theta = \frac{r}{\tau} \theta\Delta\mu t \text{ .... (2)}.$$

We distinctly perceive from (1) and (2) why the effects of *distortion* are so considerable in comparison with those of simple expansion; it is because the expressions of Δr and $\Delta\theta$ have the small quantity τ in the denominator. *AB* becomes a larger arc of a smaller circle.

To apply these results: we suppose that when $t = 0$ the centre of *AB* coincides with the central point *O*; and *AB*, being securely fastened at *A*, continues perpendicular at that point to the line *OA*, consequently its centre remains in that line. Let O' be its new position, then $OO' = -\Delta r$. If m_1 be the mass of *AB*, its moment of inertia about *O* was $m_1 r^2$;

about O' it is $m_1 r^2 - 2m_1 \Delta\mu \frac{r^3}{\tau} t$ nearly. Let G be the centre of gravity of AB'; then, in the triangle $OO'G$, we have $OG^2 = OO'^2 + (O'G)^2 + 2OO'.O'G \cos \frac{1}{2}\theta_1$, since $\angle GO'A = \frac{1}{2}\theta$. OO'^2 or $(\Delta r)^2$ may be neglected, then, approximately,

$$(OG)^2 = (O'G)^2 - 2\Delta r . O'G \cos \tfrac{1}{2}\theta,$$

and, as $$O'G = r_1 \frac{\sin \frac{1}{2}\theta_1}{\frac{1}{2}\theta_1} = 2r \frac{\sin \frac{1}{2}\theta}{\theta} \text{ nearly},$$

we have $$(OG)^2 - (O'G)^2 = -2r \frac{\sin\theta}{\theta} \Delta r.$$

Now the moment of inertia round O is equal to that round O' increased by $m\{(OG)^2 - (O'G)^2\}$; hence, finally,

$$\Delta I_1 = -2m_1 \Delta\mu \frac{r^3}{\tau} t \left(1 - \frac{\sin\theta}{\theta}\right) \text{ .... (3)},$$

I_1 being the moment of inertia of the arc AB.

(In accordance with the rest of the approximation the expansion of AO is neglected.)

Again, we will suppose the weight at C to be a material particle, and that the angle AOC is equal to ϕ. Then, I_2 being the moment of inertia of this weight, whose mass we will denote by m_2, we shall have

$$\Delta I_2 = -2m_2 \Delta\mu \frac{r^3}{\tau} t (1 - \cos\phi) \text{ .... (4)}.$$

Consequently, as the inertia of the bar OA does not undergo any sensible alteration, and as every thing which has been proved of OAB is true of Oab, we have, finally,

$$\Delta I = -4\Delta\mu \frac{r^3}{\tau} t \left\{m_1 \left(1 - \frac{\sin\theta}{\theta}\right) + m_2 (1 - \cos\phi)\right\} \text{ ...... (5)}.$$

It appears that the variation of e is exactly proportional to t: so that e becomes $e(1 - \nu t)$, ν being some constant. Consequently we must have, in order that $\frac{e(1-\nu t)}{I + \Delta I} = \frac{e}{I}$,

$$\nu I = 4\Delta\mu \frac{r^3}{\tau} \left\{m_1 \left(1 - \frac{\sin\theta}{\theta}\right) + m_2 (1 - \cos\phi)\right\} \text{ ...... (6)}.$$

In calculating the value of I we may take into account the moment of inertia of aOA; moreover, instead of the approximate expression m_1r^2 for the moment of inertia of AB, we may employ a more accurate one involving the quantities ϵ and ϵ'; the approximate expression is sufficiently accurate for the determination of ΔI.

The adjustment for compensation is effected by shifting the weight m_2 along AB; that is, by altering the value of ϕ until (6) is fulfilled. On the hypothesis we have made, the value of I for $t=0$ is not affected by the change of ϕ.

In determining the approximate expressions (5) and (6), we have neglected all terms in which $\Delta\mu$ occurs not divided by τ; all terms involving $\Delta\mu$ multiplied by ϵ or ϵ'; all terms into which any power of μ or μ' enters. In consequence of the last restriction t can only rise to the first power in the result. If this were absolutely correct it would follow that, if the compensation were effected for a particular value of t, it would subsist accurately for all values of t. For instance, if we give t equal values, positive and negative, the decrease of I in the one case ought to be equal to its increase in the other. But when t is considerable, it is found that there is a sensible deviation from this result; and, assuming that the expression for Δe does not in any perceptible manner involve powers of t, it follows that that of ΔI must do so. Any term involving t^2 (and, *a fortiori*, any higher powers of that quantity), must be very small, since t always occurs multiplied by μ or μ'; but it may, nevertheless, sensibly affect the chronometer's daily rate. On the usual construction, the balance oscillates 216,000 times in twenty-four hours. Consequently a very slight change in the moment of inertia of the balance will become perceptible in that period.

In order to obviate the consequent error, it has been proposed by Mr Dent, a distinguished chronometer-maker of the present

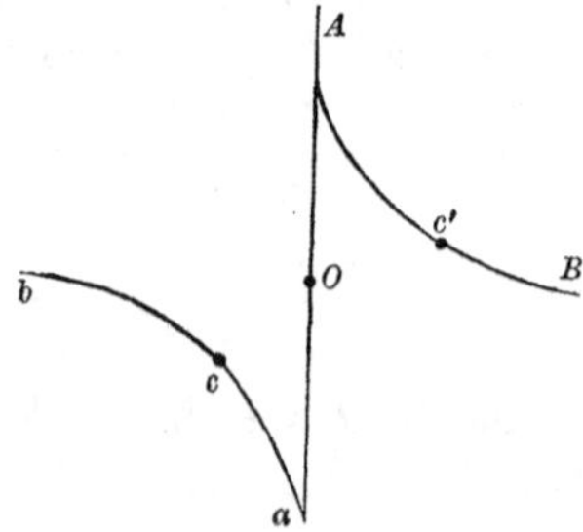

day, to alter the form of the balance. The figure represents one of those which he proposes to substitute for that in common use. It would be easy to determine the corresponding expression for ΔI, to the degree of approximation of our previous results. As, however, the comparison of the merits of the two forms must depend on the terms involving t^2, it may be well to reserve it for another opportunity. If there appears reason to believe that our hypotheses represent the facts with sufficient accuracy to encourage us to proceed farther, I hope to resume the subject in the next number of the *Journal.*

NOTES ON MAGNETISM *.

No. II.

In order to a distinct understanding of the results obtained in the last number of the *Journal*, it will be desirable to consider the established conventions with respect to the signs of the symbols which we had occasion to employ.

North magnetism is assumed to be positive; hence, of course, south magnetism must be considered as negative.

The measure M of the magnetic power of a bar magnet is, as we have seen, equal to $\int \mu s ds$, μ being the magnetism of the element ds, which is situated at a distance from the origin equal to s. The limits of the integral are such as to include the whole length of the magnet.

The position of the origin is arbitrary: we may conveniently place it at the centre of the magnet, but the value of $\int \mu s ds$ is the same whether this be done or any other point be taken. For let the origin be shifted through a distance a, so that $s = s' - a$, then

$$\int \mu s ds = \int \mu (s' - a)\, ds' = \int \mu s' ds' - a \int \mu ds' :$$

and as all the integrals extend throughout the length of the magnet $\int \mu ds' = 0$, and therefore

$$\int \mu s ds = \int \mu s' ds' \quad \text{or} \quad M = M',$$

which was to be proved.

But the value of $\int \mu s ds$ changes its sign if the direction in which s is measured changes. Let l be the length of the magnet; then, s being measured in one direction, say from left to right, we have

$$M = \int_0^l \mu s ds.$$

* *Cambridge Mathematical Journal*, No. XXI. Vol. IV. p. 139, May, 1844.

Now suppose that $s' = l - s$, then the limits are interchanged and $ds' = -ds$; consequently

$$M = \int_0^l \mu\,(l - s')\,ds' = -\int_0^l \mu s' ds' = -M',$$

s' being measured in the direction opposite to that of s, or from right to left.

The magnetism of a magnet may thus be always represented by a positive quantity.

Any two points in the axis of a magnet may be taken as its poles. But although the position of the poles is matter of convention, yet relatively to one another, one is the north and the other the south pole.

The physical character by which they are distinguished is this: if a particle of north magnetism be placed in the prolongation of the axis from south to north, it is repelled from the magnet. Contrariwise, if it be placed in the prolongation of the axis towards the south. Further, we must integrate $\mu s ds$ from south to north, *i. e.* s must be taken as positive when μds lies to the north of the origin, in order that M may be positive. This may be shown by supposing a particle of north magnetism m placed in the prolongation of the axis towards the north, and at a distance r from the centre of the magnet. If we assume that from south to north is positive, the action of the magnet on m is $\frac{2Mm}{r^3}$; and as this action is repulsive its expression will be positive, and therefore M is so. If we had assumed from north to south to be positive, the action of the magnet would have been represented by $-\frac{2Mm}{r^3}$, and as this is positive, M will necessarily be negative. So that, in order to make the measure of the magnet's power positive, we must take the direction $S \ldots N$ as positive.

Consequently the angle θ must be measured from it. We suppose it measured in the usual manner, viz. in the *unscrew* direction.

The general expression for the moment of rotation due to the action of one magnet on another is much simplified when the two magnets are supposed to lie in one plane.

The dihedral angle χ is then zero, and consequently the equation

$$L = \frac{MM'}{R^3}\{1 + 3\cos^2\theta - (2\cos\theta\cos\theta' - \sin\theta\sin\theta'\cos\chi)^2\}^{\frac{1}{2}}$$

becomes

$$L = \frac{MM'}{R^3}\{1 + 3\cos^2\theta - 4\cos^2\theta\,(1 - \sin^2\theta')$$
$$+ 4\sin\theta'\cos\theta\sin\theta\cos\theta' - \sin^2\theta\sin^2\theta'\}^{\frac{1}{2}}.$$

The quantity between the brackets is equal to

$$4\sin^2\theta'\cos^2\theta + 4\sin\theta'\cos\theta.\sin\theta\cos\theta' + \sin^2\theta\cos^2\theta'.$$

Consequently

$$L = \frac{MM'}{R^3}(\sin\theta\cos\theta' + 2\sin\theta'\cos\theta).$$

This result may be readily established by an independent process, which the reader will find no difficulty in supplying. The last result may be put in the following form:

$$L = \frac{MM'}{2R^3}\{3\sin(\theta + \theta') - \sin(\theta - \theta')\}.$$

Professor Lloyd, in the 19th volume of the *Memoirs of the Royal Irish Academy*, has investigated this case of the mutual action of two magnets. His result is (mutatis mutandis)

$$L = \frac{MM'}{2R^3}\{\sin(\theta + \theta') - 3\sin(\theta - \theta')\}.$$

This differs from the last written result, merely because, in the Professor's analysis, θ and θ' are measured in opposite directions. If we replace θ in Prof. Lloyd's result by $2\pi - \theta$, it becomes

$$L = \frac{MM'}{2R^3}\{3\sin(\theta + \theta') - \sin(\theta - \theta')\},$$

as before.

The general formula affords a simple solution of the following problem. The position of a magnet, and that of the centre of a needle being given, to place the needle in the position in which the moment of rotation due to the action of the magnet is a maximum.

By the formula established in the last number of the *Journal*, we have

$$L = \frac{MM'}{R^3}\{1 + 3\cos^2\theta - (3\cos\theta\cos\theta' - \cos\phi)^2\}^{\frac{1}{2}}.$$

We suppose the magnet and needle to be in the same plane. In the figure let O be the centre, SON the line of the axis of

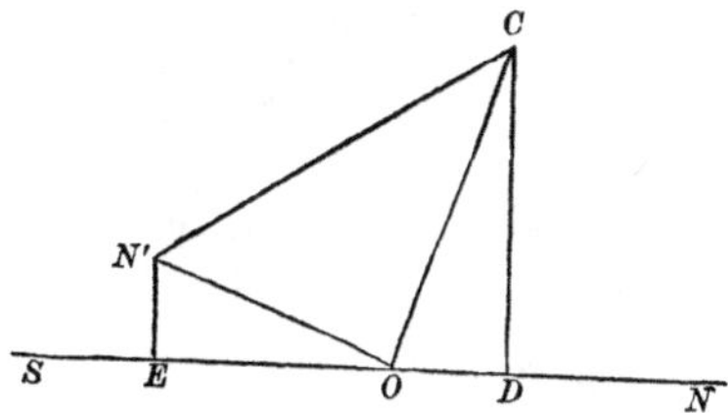

the magnet, C the centre of the needle. Project C on ON in D, take $OE = 2\,OD$, draw EN' perpendicular to SN meeting ON', which is at right angles to OC in N', CN' is the line in which the axis of the needle must be placed, its north pole being turned towards N'.

In order to prove this, we have only to remark that the angle θ or CON is constant, the position of C being given; consequently the condition to be fulfilled, in order that the moment of rotation L may be a maximum, is

$$3\cos\theta\cos\theta' - \cos\phi = 0.$$

Now as ϕ is the angle between $N'C$ and SN, we have

$$ED = CN'\cos\phi,$$

and therefore $$OD = \tfrac{1}{3}\,CN'\cos\phi.$$

Again, $OC = CN'\cos\theta'$ and $OD = OC\cos\theta$.

Hence $$OD = CN'\cos\theta\cos\theta'.$$

Consequently $$3\cos\theta\cos\theta' - \cos\phi = 0,$$

or the required condition is fulfilled.

There are three particular cases worth noticing:

(1) $\theta = 0$. In this case C lies in the axis ON, D coincides with it, and ON' is perpendicular to ON, and therefore parallel to EN'. Consequently the point N' is removed to an infinite distance, and CN' is therefore perpendicular to ON. The corresponding value of L is $\frac{2MM'}{R^3}$, and is the maximum maximorum.

(2) $\theta = \theta'$. In this case the magnet and needle are parallel

to one another. The quadrilateral $CDEN'$ is a parallelogram, EN' is equal to DC, and consequently

$$\tan COD : \tan N'OE :: EO : OD :: 2 : 1.$$

But $$COD = \theta \text{ and } N'OE = \frac{\pi}{2} - \theta,$$

since CON' is a right angle. Consequently

$$\tan\theta = 2\cot\theta,$$

$$\text{or } \tan\theta = \sqrt{2}.$$

The corresponding value of $3\cos^2\theta$ is therefore unity; and consequently we have, in this case,

$$L = \sqrt{2}\frac{MM'}{R^3} (\theta = 54^\circ\, 44' 8'').$$

(3) $\theta = \frac{\pi}{2}$. Here D (and therefore E) coincides with O, while ON' lies in the axis OS'. Consequently N' is at O, and the needle is therefore again perpendicular to the magnet. In this case

$$L = \frac{MM'}{R^3}.$$

The first and third cases were noticed in the last number of the *Journal.* In the second, the value of L is a mean proportional between what is in the other two cases, in the last of which it is a minimum maximorum.

If we were required, for a given position of C, to find the position in which the needle would be in equilibrium, or the moment L equal to zero, we might have recourse to Gauss's construction already mentioned; for if the needle be placed along the line in which the magnet tends to attract or repel C, as the dimensions of the needle are small, every element would approximately be attracted or repelled along this line, and therefore the total action would be destroyed by the resistance of C.

Thus there are always two directions for every position of C'; one of maximum moment and the other of equilibrium: these two directions are at right angles to one another.

MATHEMATICAL NOTE*.

$$\int_0^1 \frac{\log(1+x)}{1+x^2}\,dx.$$

THIS definite integral is evaluated in a curious manner by M. Bertrand in *Liouville's Journal*. The demonstration I am about to give of his result, is somewhat different in form from that which he made use of.

The method employed in an ingenious paper which appeared in the last volume of the *Journal* (III. p. 168), will apply to the integral we are about to consider.

Let $$fu = \int_0^1 \frac{\log(1+ux)}{1+x^2}\,dx.$$

Then $$\frac{d}{du} fu = \int_0^1 \frac{xdx}{(1+ux)(1+x^2)}.$$

Now $$\frac{x}{1+x^2} + \frac{u}{1+u^2} = (1+ux)\frac{x+u}{(1+x^2)(1+u^2)}.$$

Consequently

$$\frac{x}{(1+ux)(1+x^2)} = \frac{x+u}{(1+x^2)(1+u^2)} - \frac{u}{(1+ux)(1+u^2)},$$

and therefore

$$\frac{d}{du} fu = \frac{1}{1+u^2}\int_0^1 \frac{xdx}{1+x^2} + \frac{u}{1+u^2}\int_0^1 \frac{dx}{1+x^2} - \frac{u}{1+u^2}\int_0^1 \frac{dx}{1+ux}$$

$$= \tfrac{1}{2}\log 2\,\frac{1}{1+u^2} + \frac{\pi}{4}\,\frac{u}{1+u^2} - \frac{\log(1+u)}{1+u^2}.$$

Integrate for u, from 0 to 1,

$$f(1) - f(0) = \frac{\pi}{8}\log 2 + \frac{\pi}{8}\log 2 - f(1).$$

* *Cambridge Mathematical Journal*, No. XXI. Vol. IV. p. 143, May, 1844.

But $f(0)=0$, since $\log 1 = 0$. Therefore

$$f(1) = \frac{\pi}{8} \log 2.$$

But
$$f(1) = \int_0^1 \frac{\log(1+x)}{1+x^2}\, dx.$$

Therefore
$$\int_0^1 \frac{\log(1+x)}{1+x^2}\, dx = \frac{\pi}{8} \log 2.$$

The singularity of this method, and its applicability in other cases give it interest: but, as the writer of the paper already noticed pointed out to me, the integral may be got by assuming $x = \tan y$; it then becomes

$$\int_0^{\frac{\pi}{4}} \log(1+\tan y)\, dy,$$

and, by his fundamental equation,

$$\int_0^{\frac{\pi}{4}} \log(1+\tan y)\, dy = \int_0^{\frac{\pi}{4}} \log\left\{1 + \tan\left(\frac{\pi}{4} - y\right)\right\} dy:$$

$$1 + \tan\left(\frac{\pi}{4} - y\right) = 1 + \frac{1-\tan y}{1+\tan y} = \frac{2}{1+\tan y};$$

and therefore

$$2\int_0^{\frac{\pi}{4}} \log(1+\tan y)\, dy = \frac{\pi}{4} \log 2,$$

whence the truth of M. Bertrand's result is obvious.

MEMOIR OF THE LATE D. F. GREGORY, M.A. FELLOW OF TRINITY COLLEGE, CAMBRIDGE*.

THE subject of the following memoir died in his thirty-first year. He had, nevertheless, accomplished enough not only to justify high expectations of his future progress in the science to which he had principally devoted himself, but also to entitle his name to a place in some permanent record.

Duncan Farquharson Gregory was born at Edinburgh in April 1813. He was the youngest son of Dr James Gregory, the distinguished professor of Medicine, and was thus of the same family as the two celebrated mathematicians James and David Gregory. The former of these, his direct ancestor, is familiarly remembered as the inventor of the telescope which bears his name; he lived in an age of great mathematicians, and was not unworthy to be their contemporary.

Of the early years of Mr Gregory's life but little need be said. The peculiar bent of his mind towards mathematical speculations does not appear to have been perceived during his childhood; but, in the usual course of education, he shewed much facility in the acquisition of knowledge, a remarkably active and inquiring mind, and a very retentive memory. It may, perhaps, be mentioned here, that his father, whom he lost before he was seven years old, used to predict distinction for him; and was so struck with his accurate information and clear memory, that he had pleasure in conversing with him, as with an equal, on subjects of history and geography. In his case, as in many others, ingenuity in little mechanical contrivances seems to have preceded, and indicated the developement of a taste for abstract science.

* *Cambridge Mathematical Journal*, No. I. Vol. IV. p. 145, November, 1844.

Two years of his life were passed at the Edinburgh Academy; when he left it, being considered too young for the University, he went abroad and spent a winter at a private academy in Geneva. Here his talent for mathematics attracted attention; in geometry, as well as in classical learning, he had already made distinguished progress at Edinburgh.

The following winter he attended classes at the University of Edinburgh, and soon became a favourite pupil of Professor Wallace's, under whose tuition he made great advances in the higher parts of mathematics. The Professor formed the highest hopes of Mr Gregory's future eminence: those who long afterwards saw them together in Cambridge, speak with much interest of the delighted pride he shewed in his pupil's success and increasing reputation.

In 1833, Mr Gregory's name was entered at Trinity College in the University of Cambridge, and shortly afterwards he went to reside there. He brought with him a very unusual amount of knowledge on almost all scientific subjects: with Chemistry he was particularly well acquainted, so much so that he had been at Cambridge but a few months when it was proposed to him by one of the most distinguished men in the University to act as assistant to the professor of Chemistry; which for some time he did. Indeed, it is impossible to doubt that, had not other pursuits engaged his attention, he might have achieved a great reputation as a chemist. He was one of the founders of the Chemical Society in Cambridge, and occasionally gave lectures in their rooms.

He had also a very considerable knowledge of botany, and indeed of many subjects which he seemed never to have studied systematically: he possessed in a remarkable degree the power of giving a regular form, and, so to speak, a unity to knowledge acquired in fragments.

All these tastes and habits of thought Mr Gregory cultivated, to a certain extent, during the first years of his residence in Cambridge, of course in subordination to that which was the end principally in view in his becoming a member of the University, namely, the study of mathematics and natural philosophy.

He became a Bachelor of Arts in 1837, having taken high mathematical honours: more, however, might, we may believe, have been effected in this respect, had his activity of mind permitted him to devote himself more exclusively to the prescribed course of study.

From henceforth he felt himself more at liberty to follow original speculations, and, not many months after taking his degree, turned his attention to the general theory of the combination of symbols.

It may be well to say a few words of the history of this part of mathematics.

One of the first results of the differential notation of Leibnitz, was the recognition of the analogy of differentials and powers. For instance, it was readily perceived that

$$\frac{d^{m+n}}{dx^{m+n}} y = \frac{d^m}{dx^m} \frac{d^n}{dx^n} y,$$

or, supposing the y to be *understood*, that

$$\left(\frac{d}{dx}\right)^{m+n} = \left(\frac{d}{dx}\right)^m \left(\frac{d}{dx}\right)^n,$$

just as in ordinary algebra we have, a being any quantity,

$$a^{m+n} = a^m a^n.$$

This, and one or two other remarks of the same kind, were sufficient to establish an analogy between $\frac{d}{dx}$ the symbol of differentiation and the ordinary symbols of algebra. And it was not long afterwards remarked that a corresponding analogy existed between the latter class of symbols and that which is peculiar to the calculus of finite differences. It was inferred from hence that theorems proved to be true of combinations of ordinary symbols of quantity, might be applied by analogy to the differential calculus and to that of finite differences. The meaning and interpretation of such theorems would of course be wholly changed by this kind of transfer from one part of mathematics to another, but their form would remain unchanged. By these considerations many theorems were suggested, of which it was thought almost impossible to obtain direct demonstrations. In this point of view the subject was developed by Lagrange, who left undemonstrated the results to which he was led, intimating, however, that demonstrations were required. Gradually, however, mathematicians came to perceive that the analogy with which they were dealing, involved an essential identity; and thus results, with respect to which, if the expression may be used, it had only been felt that they must be true, were now actually seen to be so. For, if the algebraical theorems by which these results were suggested, were true, *because* the symbols they involve represented quantities, and such operations as may be performed on quantities, then indeed the analogy would be altogether precarious. But if, as is really the case, these theorems are true, in virtue of certain fundamental laws of combination, which hold both for algebraical symbols, and for those peculiar to the higher branches of mathematics, then each algebraical theorem and its analogue constitute, in fact, only one and the same theorem, except *quoad* their distinctive interpretations, and therefore a demonstration of either is in reality a demonstration of both*.

The abstract character of these considerations is doubtless the reason why so long a time elapsed before their truth was distinctly perceived. They would almost seem to require, in order that they may be readily apprehended, a peculiar faculty—a kind of mental *disinvoltura* which is by no means common.

Mr Gregory, however, possessed it in a very remarkable degree. He at once perceived the truth and the importance of the principles of which we have been speaking, and proceeded to apply them with singular facility and fearlessness.

It had occurred to two or three distinguished writers that the analogy, as it was called, of powers, differentials, &c., might be made available in the solution of differential equations, and of equations in finite differences.

This idea, however, probably from some degree of doubt as to the legitimacy of the methods which it suggested, had not been fully or clearly developed: it seems to have been chiefly employed as affording a convenient way of expressing solutions already obtained by more familiar considerations.

To this branch of the subject Mr Gregory directed his

* The values of certain definite integrals are to be looked upon as merely arithmetical results; in such cases we are not at liberty to replace the constants involved in the definite integrals by symbols of operation. In other cases we are at liberty to do so, and this remarkable application of the principles stated in the text, has already led Mr Boole of Lincoln, with whom it seems to have originated, to several curious conclusions.

attention, and from the general views of the laws of combination of symbols already noticed, deduced in a regular and systematic form, methods of solution of a large and important class of differential equations (linear equations with constant coefficients, whether ordinary or partial) of systems of such equations existing simultaneously, of the corresponding classes of equations in finite and mixed differences; and lastly, of many functional equations. The steady and unwavering apprehension of the fundamental principle which pervades all these applications of it, gives them a value quite independent of that which arises from the facility of the methods of solution which they suggest.

The investigations of which I have endeavoured to illustrate the character and tendency, appeared from time to time in the *Cambridge Mathematical Journal.*

In this periodical publication Mr Gregory took much interest. He had been active in establishing it, and continued to be its editor, except for a short interval, from the time of its first appearance in the autumn of 1837, until a few months before his death. For this occupation he was for many reasons well qualified; his acquaintance with mathematical literature was very extensive, while his interest in all subjects connected with it was not only very strong, but also singularly free from the least tinge of jealous or personal feeling. That which another had done or was about to do, seemed to give him as much pleasure as if he himself had been the author of it, and this even when it related to some subject which his own researches might seem to have appropriated.

This trait, as the recollections of those who knew him best will bear me witness, was intimately connected with his whole character, which was in truth an illustration of the remark of a French writer, that to be free from envy is the surest indication of a fine nature.

To the *Cambridge Mathematical Journal*, Mr Gregory contributed many papers beside those which relate to the researches already noticed. In some of these he developed certain particular applications of the principles he had laid down in an Essay on the Foundations of Algebra, presented to the Royal Society of Edinburgh in 1838, and printed in the fourteenth volume of their Transactions. I may particularly mention a paper on the curious question of the logarithms of negative quantities, a question which, it is well known, has often been discussed among mathematicians, and which even now does not appear to be entirely settled.

In 1840, Mr Gregory was elected Fellow of Trinity College; in the following year he became Master of Arts, and was appointed to the office of moderator, that is, of principal mathematical examiner. His discharge of the duties of this office (which is looked upon as one of the most honourable of those which are accessible to the younger members of the University) was distinguished by great good sense and discretion.

In the close of the year 1841, Mr Gregory produced his "Collection of Examples of the Processes of the Differential and Integral Calculus;" a work which required, and which manifests much research, and an extensive acquaintance with mathematical writings. He had at first only wished to superintend the publication of a second edition of the work with a similar title, which appeared more than twenty-five years since, and of which Messrs. Herschel, Peacock, and Babbage were the authors. Difficulties, however, arose, which prevented the fulfilment of this wish, and it is not perhaps to be regretted that Mr Gregory was thus led to undertake a more original design. It is well known that the earlier work exercised a great and beneficial influence on the studies of the University, nor was it in any way unworthy of the reputation of its authors. The original matter contributed by Sir John Herschel is especially valuable. Nevertheless, the progress which mathematical science has since made, rendered it desirable that another work of the same kind should be produced, in which the more recent improvements of the calculus might be embodied.

Since the beginning of the century, the general aspect of mathematics has greatly changed. A different class of problems from that which chiefly engaged the attention of the great writers of the last age has arisen, and the new requirements of natural philosophy have greatly influenced the progress of pure analysis. The mathematical theories of heat, light, electricity, and magnetism, may be fairly regarded as the achievement of the last fifty years. And in this class of researches an idea is prominent, which comparatively occurs but seldom in purely dynamical enquiries. This is the idea of discontinuity. Thus, for instance, in the theory of heat, the conditions relating to the surface of the body whose variations of temperature we are considering, form an essential and peculiar element of the problem; their peculiarity arises from the discontinuity of the transition from the temperature of the body to that of the space in which it is placed. Similarly, in the undulatory theory of light, there is much difficulty in determining the conditions which belong to the bounding surfaces of any portion of ether; and although this difficulty has, in the ordinary applications of the theory, been avoided by the introduction of proximate principles, it cannot be said to have been got rid of.

The power, therefore, of symbolizing discontinuity, if such an expression may be permitted, is essential to the progress of the more recent applications of mathematics to natural philosophy, and it is well known that this power is intimately connected with the theory of definite integrals. Hence the principal importance of this theory, which was altogether passed over in the earlier collection of examples.

Mr Gregory devoted to it a chapter of his work, and noticed particularly some of the more remarkable applications of definite integrals to the expression of the solutions of partial differential equations. It is not improbable that in another edition he would have developed this subject at somewhat greater length. He had long been an admirer of Fourier's great work on heat, to which this part of mathematics owes so much; and once, while turning over its pages, remarked to the writer,—"All these things seem to me to be a kind of mathematical paradise."

In 1841, the mathematical Professorship at Toronto was offered to Mr Gregory: this, however, circumstances induced him to decline. Some years previously he had been a candidate for the Mathematical Chair at Edinburgh.

His year of office as moderator ended in October 1842. In the University Examination for Mathematical Honours in the following January, he, however, in accordance with the usual routine, took a share, with the title of examiner,—a position little less important, and very nearly as laborious, as that of moderator. Besides these engagements in the University, he had been for two or three years actively employed in lecturing and examining in the College of which he was a Fellow. In the fulfilment of these duties, he shewed an earnest and constant desire for the improvement of his pupils, and his own love of science tended to diffuse a taste for it among the better order of students. He had for some time meditated a work on Finite Differences, and had commenced a treatise on Solid Geometry, which, unhappily, he did not live to complete. In the midst of these various occupations, he felt the earliest approaches of the malady which terminated his life.

The first attack of illness occurred towards the close of 1842. It was succeeded by others, and in the spring of 1843, he left Cambridge never to return again. He had just before taken part in a college examination, and notwithstanding severe suffering, had gone through the irksome labour of examining with patient energy and undiminished interest.

Many months followed of almost constant pain. Whenever an interval of tolerable ease occurred, he continued to interest himself in the pursuits to which he had been so long devoted; he went on with the work on Geometry, and, but a little while before his death, commenced a paper on the analogy of differential equations and those in finite differences. This analogy it is known that he had developed to a great length; unfortunately, only a portion of his views on the subject can now be ascertained.

At length, on the 23rd February 1844, after sufferings, on which, notwithstanding the admirable patience with which they were borne, it would be painful to dwell, his illness terminated in death. He had been for a short time aware that the end was at hand, and, with an unclouded mind, he prepared himself calmly and humbly for the great change; receiving and giving comfort and support from the thankful hope that the close of his suffering life here, was to be the beginning of an endless existence of rest and happiness in another world. He retained to the last, when he knew that his own connection with earthly things was soon to cease, the unselfish interest which he had ever felt in the pursuits and happiness of those he loved.

A few words may be allowed about a character where rare and sterling qualities were combined. His upright, sincere, and honourable nature secured to him general respect. By his intimate friends, he was admired for the extent and variety of his information, always communicated readily, but without a thought of display,—for his refinement and delicacy of taste and feeling,—for his conversational powers and playful wit; and he was beloved by them for his generous, amiable disposition, his active and disinterested kindness, and steady affection. And in this manner his high-toned character acquired a moral influence over his contemporaries and juniors, in a degree remarkable in one so early removed.

To this brief history, little more is to be added; for though it is impossible not to indulge in speculations as to all that Mr Gregory might have done in the cause of science and for his own reputation, had his life been prolonged, yet such speculations are necessarily too vague to find a place here; and even were it not so, it would perhaps be unwise to enter on a subject so full of sources of unavailing regret.

ON THE SOLUTION OF EQUATIONS IN FINITE DIFFERENCES*.

THE partial differential equations which occur in various branches of mathematical physics are, for the most part, of such forms that solutions of them may be obtained without much difficulty. As is well known, the great difficulty in almost all such cases consists in the necessity of determining which of all possible solutions satisfies the particular conditions of the problem on which we are engaged. It seems that before the time of Fourier's researches on heat, the course which mathematicians had uniformly followed was, first to obtain the general solution of the equation of the problem, and then to determine by particular considerations the arbitrary functions which it involved. This course undoubtedly would be the most direct and analytical, were there any general method for determining the form of the functions in question: as, however, there is none, the analytical generality of the first part of the process is in many cases sterile and useless.

Fourier's methods, which depend essentially on the linearity of the partial differential equations which occur in the theory of heat, consist in assuming some simple solution of the equation of the problem, in deducing from hence a more general solution of it, and in determining successively and by means of particular considerations the arbitrary quantities thus introduced in such a manner as to satisfy all the conditions of the question. The general solution with arbitrary functions does not make its appearance in his process; and the reason why it is so much more manageable than the other appears to be, that it is far easier to determine arbitrary constants in accordance with certain

* *Cambridge Mathematical Journal*, No. XXII. Vol. IV. p. 182, November, 1844.

conditions than arbitrary functions. There will, generally speaking, be an infinite number of arbitrary constants, and it is therefore necessary to treat them in *classes*. The ingenious synthesis by which this is effected by Fourier, in the different problems discussed by him in the *Théorie de la Chaleur*, forms one of the most interesting parts of that admirable work. The same kind of reasoning is made use of by Poisson, in his researches on similar subjects: and there can be little doubt that the methods of Fourier, developed and extended as they have been by subsequent writers, will long continue to be an essential element in the application of mathematics to physical researches. Similar methods may be made available in the solution of equations in partial finite differences. Such equations do not, it is true, present themselves very often, as the continuity of the causes to which natural phenomena are due, leads rather to differential equations than to those in finite differences. In fact, I am not aware of any subject, except the theory of probabilities, in which we meet with problems whose solution depends on that of an equation in partial finite differences.

In this theory, however, such problems are not uncommon. One of the most interesting of them, both in its own nature and historically, may serve as an illustration of the application of the methods of Fourier to finite differences. This problem, which has engaged the attention of several writers on the subject of probabilities, and of which a solution was among the earliest efforts of Ampère, is that of the *duration of play*. Professor De Morgan has spoken of this solution and of that of Laplace, as being of the highest order of difficulty: that which I am about to enter on has, I think, a decided advantage in this respect.

The problem itself may be thus stated:—Two persons, M and N, have between them a number a of counters: they play at a game at which M's chance is p, and N's q. The losing player gives one counter to the other, and they are to play on until one or other have lost all his counters. What is the probability that the party will terminate in M's favour after any assigned number of games, N being supposed to have originally x of the a counters?

Let y_{xz} be the probability that M will win the party at the $(z+1)^{\text{th}}$ game. If he win the next game (of which the probability is p), this becomes $y_{x-1.z-1}$; if he lose it (of which the probability is q), it becomes $y_{x+1.z-1}$, and therefore

$$y_{xz} = py_{x-1.z-1} + qy_{x+1.z-1} \quad\ldots\ldots\ldots\ldots\ldots\ldots (1).$$

This is the equation of the problem. It is clear that

$$y_{0z} = 0, \quad y_{az} = 0 \quad\ldots\ldots\ldots\ldots\ldots\ldots (2),$$

as the party ceases as soon as M or N has a counters. Again,

$$y_{x0} = 0 \text{ unless } x = 1, \text{ and } y_{1.0} = p \quad\ldots\ldots\ldots\ldots (3);$$

for if N have more than one counter he cannot lose them all at the next game; and if he have only one, his chance of his being left without any is p.

Let us assume $\qquad y_{xz} = \alpha^z v_x \quad\ldots\ldots\ldots\ldots\ldots\ldots (4),$

α being arbitrary. Then

$$\alpha v_x = pv_{x-1} + qv_{x+1} \quad\ldots\ldots\ldots\ldots\ldots\ldots (5).$$

Of this a solution is

$$v_x = C\left(\frac{p}{q}\right)^{\frac{x}{2}} \sin(\mu x + \omega) \quad\ldots\ldots\ldots\ldots\ldots (6),$$

where C and ω are arbitrary, and μ such that

$$2\sqrt{(pq)} \cos \mu = \alpha \quad\ldots\ldots\ldots\ldots\ldots\ldots (7):$$

this form of solution is therefore real if α^2 is less than $4pq$. In order that (4) may satisfy the conditions (2), we must have

$$v_0 = C \sin \omega = 0, \quad v_a = C\left(\frac{p}{q}\right)^{\frac{a}{2}} \sin(\mu a + \omega) = 0 \quad\ldots\ldots (8).$$

It is impossible to satisfy these two conditions without making $C = 0$, which would give a nugatory result, unless $\sin \mu a = 0$ or $\mu = \frac{r\pi}{a}$, r being an integer. Let us therefore assume this value for μ; and then, by (7),

$$\alpha = 2\sqrt{(pq)} \cos \frac{r\pi}{a} \quad\ldots\ldots\ldots\ldots\ldots\ldots (9).$$

In order to satisfy (8), we have now only to make $\omega = 0$, and then, substituting the values of v_x and α in (4), we get

$$y_{xz} = (4pq)^{\frac{z}{2}} \left(\frac{p}{q}\right)^{\frac{x}{2}} C \sin \frac{r\pi}{a} x \left(\cos \frac{r\pi}{a}\right)^z \quad\ldots\ldots\ldots (10).$$

This value, in which C and r are arbitrary, satisfies (1) and (2), and in consequence of the linearity of these equations they will be satisfied by a sum of similar values, and we shall thus have a more general solution, viz.

$$y_{xz} = (4pq)^{\frac{z}{2}} \left(\frac{p}{q}\right)^{\frac{x}{2}} \Sigma C \sin \frac{r\pi}{a} x \left(\cos \frac{r\pi}{a}\right)^z \text{......... (11).}$$

If in this we put $z = 0$, we have

$$y_{x0} = \left(\frac{p}{q}\right)^{\frac{x}{2}} \Sigma C \sin \frac{r\pi}{a} x \text{.................. (12).}$$

Now, by (3), this is to be equal to p for $x = 1$, and to 0 for the $a - 2$ values of x, $2 . 3 \ldots\ldots a - 1$. There are thus $a - 1$ conditions for (12) to fulfil, and therefore we have, extending the summation Σ from $r = 1$ to $r = a - 1$, the following system of equations:

$$\left.\begin{aligned} \sqrt{(pq)} &= C_1 \sin \frac{\pi}{a} + \ldots\ldots + C_{a-1} \sin \frac{a-1}{a} \pi \\ 0 &= C_1 \sin \frac{2\pi}{a} + \ldots\ldots + C_{a-1} \sin 2 \frac{a-1}{a} \pi \\ 0 &= \ldots\ldots\ldots\ldots\ldots\ldots \\ 0 &= C_1 \sin \frac{a-1}{a} \pi + \ldots + C_{a-1} \sin \frac{(a-1)^2}{a} \pi \end{aligned}\right\} \text{...... (13).}$$

From these $a - 1$ equations we have to determine the $a - 1$ quantities $C_1 \ldots C_{a-1}$. In order to do this, multiply the first equation by $\sin \frac{r}{a} \pi$, the second by $\sin 2 \frac{r}{a} \pi$, and so on, (r being an integer less than a), and add. Then, as may be easily shown, the coefficient of every one of the quantities C, except C_r, will in the resulting sum be equal to zero, while that of C_r will be $\frac{a}{2}$. Consequently (13) is equivalent to the system of equations included in the general formula

$$C_r = \frac{2}{a} \sqrt{(pq)} \sin \frac{r}{a} \pi \text{.................. (14),}$$

and consequently (11) becomes

$$y_{xz} = \frac{1}{a} (4pq)^{\frac{z+1}{2}} \left(\frac{p}{q}\right)^{\frac{x}{2}} \Sigma_1^{a-1} \sin \frac{r}{a} \pi \sin \frac{rx}{a} \pi \left(\cos \frac{r}{a} \pi\right)^z \text{...... (15),}$$

which is the required probability.

We may deduce from this formula, by indirect considerations, one or two analytical theorems. For it is obviously impossible that the party should terminate in M's favour in less than x games, as x is the number of counters he must win from N. Consequently

$$\Sigma_1^{a-1} \sin \frac{r}{a}\pi \sin \frac{rx}{a}\pi \left(\cos \frac{r}{a}\pi\right)^z = 0 \text{ .... (16)},$$

for all integer values of z less than $x-1$.

Again, M may win the party at the xth game, if he win x games in succession, the probability of which is p^x. Hence, putting $z = x-1$, we have

$$p^x = \frac{1}{a}(4pq)^{\frac{x}{2}} \left(\frac{p}{q}\right)^{\frac{x}{2}} \Sigma_1^{a-1} \sin \frac{r}{a}\pi \sin \frac{rx}{a}\pi \left(\cos \frac{r}{a}\pi\right)^{x-1},$$

$$\text{or } \Sigma_1^{a-1} \sin \frac{r}{a}\pi \sin \frac{rx}{a}\pi \left(\cos \frac{r}{a}\pi\right)^{x-1} = \frac{a}{2^x} \text{ ...... (17)}.$$

These formulæ may undoubtedly be established by other methods, but I have thought it worth while to point out this way of deducing them, from the analogy it bears to that in which many remarkable theorems are obtained by Poisson, in his *Théorie de la Chaleur*, namely by considering the nature of the quantities which his formulæ represent. This mode of establishing analytical theorems by considerations founded on the interpretation of our results, is one of the most curious features of the more recent methods of treating physical questions.

To (16) and (17) another theorem may be added, by the following consideration. M, if he win, must win the party either in x games or in $x+$ and even number of games. For if he lose k games he must win back k games and x more or there must have been $x+2k$ games in the party. Hence his chance is zero whenever $z+1 = x+2k+1$, and therefore

$$\Sigma_1^{a-1} \sin \frac{r}{a}\pi \sin \frac{rx}{a}\pi \left(\cos \frac{r}{a}\pi\right)^{x+2k} = 0 \text{ .... (18)},$$

k being any positive integer whatever.

When a is infinite, the sums contained in the last three equations become definite integrals. Let

$$\frac{r}{a}\pi = \phi, \quad \text{then } \frac{\pi}{a} = d\phi, \quad \text{and } \frac{a-1}{a}\pi = \pi.$$

Consequently (16), (17), (18), become respectively

$$\int_0^\pi \sin\phi \sin x\phi (\cos\phi)^z d\phi = 0 \text{ .... } (19),$$

(z being integral and less than $x-1$),

$$\int_0^\pi \sin\phi \sin x\phi (\cos\phi)^{x-1} d\phi = \frac{\pi}{2^x} \text{ .... } (20),$$

$$\int_0^\pi \sin\phi \sin x\phi (\cos\phi)^{x+2k} d\phi = 0 \text{ .... } (21).$$

If, instead of seeking the probability that M will win the party *at* the $(z+1)^{\text{th}}$ game, we wished to find that of his winning it *after* z or more games shall have been played, we should only have to sum (15) for z from z to infinity. Calling this new probability u_{xz}, we should thus get

$$u_{xz} = \frac{1}{a}(4pq)^{\frac{z+1}{2}} \left(\frac{p}{q}\right)^{\frac{x}{2}} \Sigma_1^{a-1} \frac{\sin\frac{r}{a}\pi \sin\frac{rx}{a}\pi}{1-2\sqrt{(pq)}\cos\frac{r}{a}\pi} \left(\cos\frac{r}{a}\pi\right)^z \text{ .... } (22).$$

If in (22) we put z equal to zero, we have then the probability of M's winning the party at the first, second, &c. games, *i. e.* of his winning it at all. Writing simply u_x for u_{x0}, we shall thus get

$$u_x = \frac{2}{a}\sqrt{(pq)} \left(\frac{p}{q}\right)^{\frac{x}{2}} \Sigma_1^{a-1} \frac{\sin\frac{r}{a}\pi \sin\frac{rx}{a}\pi}{1-2\sqrt{(pq)}\cos\frac{r}{a}\pi} \text{ .... } (23).$$

Now of this probability we can obtain, as is well known, a much simpler expression. For it is easily seen that we shall have

$$u_x = pu_{x-1} + qu_{x+1} \text{ .... } (24)$$

for every value of x, provided that, instead of considering u_x as the probability that M *will* win the party, we make it denote the probability that he either *has* won or *will* win it. As it is impossible that he can have already won it while x differs from zero, this alteration does not affect the value represented by u_x except for the case of $x=0$. In this case the value of u_x, as expressed by (23), will be zero, as the party is at an end, M having already won it. But according to the proposed modification, the new value of u_0 will be unity, and therefore we have for the initial and final values of u_x,

$$u_0 = 1, \quad u_a = 0 \text{ .... (25).}$$

The necessity of this modification arises from this, that otherwise the relation expressed by (24) would not be in all cases true. For when $x = 1$, we should have $u_1 = qu_2$, whereas the true value is of course

$$u_1 = p + qu_2.$$

From (24) we have (introducing the relation $p + q = 1$),

$$u_x = \alpha + \beta\left(\frac{p}{q}\right)^x \text{ .... (26),}$$

α and β being arbitrary constants: and thence, by (25), we get

$$1 = \alpha + \beta,$$

$$0 = \alpha + \beta\left(\frac{p}{q}\right)^a,$$

and consequently $$u_x = \frac{\left(\frac{p}{q}\right)^x - \left(\frac{p}{q}\right)^a}{1 - \left(\frac{p}{q}\right)^a} \text{ .... (27).}$$

This expression is therefore, except for $x = 0$, equivalent to (23), into which however the relation already mentioned, viz. that $p + q = 1$ has not as yet been introduced.

When p and q are equal, (27) becomes

$$u_x = \frac{a - x}{a} \text{ .... (28),}$$

while (23) similarly becomes

$$u_x = \frac{1}{a}\Sigma_1^{a-1}\frac{\sin\frac{r}{a}\pi\sin\frac{rx}{a}\pi}{1 - \cos\frac{r}{a}\pi},$$

$$\text{or } u_x = \frac{1}{a}\Sigma_1^{a-1}\cot\frac{r}{a}\frac{\pi}{2}\sin\frac{rx}{a}\pi \text{ .... (29).}$$

Comparing (28) and (29), we have the following theorem: writing x for $a - x$,

$$x = \Sigma_1^{a-1} \pm \cot\frac{r}{a}\frac{\pi}{2}\sin\frac{rx}{a}\pi \text{ .... (30),}$$

the upper sign to be taken when r is odd.

This theorem, like the preceding ones (16), (17), &c., requires x not to transgress the limits $x = 1$, $x = a - 1$. In the case supposed (viz. when p and q are each equal to $\frac{1}{2}$), (22) becomes

$$u_{xz} = \frac{1}{a} \Sigma_1^{a-1} \cot \frac{r}{a} \frac{\pi}{2} \sin \frac{rx}{a} \pi \left(\cos \frac{r}{a} \pi\right)^z \ldots\ldots\ldots (31).$$

But as the party cannot be won in less than x games, $u_{x0} = u_{xz}$ while z is less than x, and therefore

$$x = \Sigma_1^{a-1} \pm \cot \frac{r}{a} \frac{\pi}{2} \sin \frac{rx}{a} \pi \left(\cos \frac{r}{a} \pi\right)^z \ldots\ldots\ldots (32),$$

of which (30) is a particular case.

If, instead of seeking the probability that at the $(z+1)^{\text{th}}$ game N would lose the party, by losing the last of his x counters, we had sought that of his having at the termination of this game any assigned number of counters k, the following method might have been made use of.

Let y_{xz} be the probability in question. It is clear that it will satisfy, as before, equations (1) and (2). But instead of (3), we shall in this case have

$$y_{x0} = 0, \text{ unless } x = k \pm 1, \text{ and } y_{k\pm 1.0} = \tfrac{1}{2} \{p + q \pm (p - q)\} \ldots (3).$$

Equation (11) therefore, which depends merely on (1) and (2), will still obtain; but instead of the system of equations (13), we shall have the following:

$$\left.\begin{aligned} 0 &= C_1 \sin \frac{\pi}{a} + \ldots + C_{a-1} \sin \frac{a-1}{a} \pi \\ \&\text{c.} &= \&\text{c.} \\ \left(\frac{q^{k+1}}{p^{k-1}}\right)^{\frac{1}{2}} &= C_1 \sin \frac{k-1}{a} \pi + \ldots + C_{a-1} \sin \frac{(k-1)(a-1)}{a} \pi \\ 0 &= \&\text{c.} \\ \left(\frac{q^{k+1}}{p^{k-1}}\right)^{\frac{1}{2}} &= C_1 \sin \frac{k+1}{a} \pi + \ldots + C_{a-1} \sin \frac{(k+1)(a-1)}{a} \pi \\ \&\text{c.} &= \&\text{c.} \\ 0 &= C_1 \sin \frac{a-1}{a} \pi + \ldots + C_{a-1} \sin \frac{(a-1)^2}{a} \pi \end{aligned}\right\} \ldots (13').$$

From whence, by the same system of factors as. before, we deduce the general formula

$$C_r = \frac{4}{a}\left(\frac{q^{k+1}}{p^{k-1}}\right)^{\frac{1}{2}} \sin\frac{kr}{a}\pi \cos\frac{r}{a}\pi \ldots\ldots\ldots\ldots (14');$$

for the factors corresponding to the two equations whose first members are different from zero, are $\sin\frac{(k-1)r}{a}\pi$ and $\sin\frac{(k+1)r}{a}\pi$, and the sum of these is $2\sin\frac{kr}{a}\pi\cos\frac{r}{a}\pi$. Consequently the expression of the probability sought will be (accenting the y for distinctness),

$$y'_{xz} = \frac{2}{a}(4pq)^{\frac{z+1}{2}}\left(\frac{p}{q}\right)^{\frac{x-k}{2}} \Sigma_1^{a-1} \sin\frac{kr}{a}\pi \sin\frac{rx}{a}\pi\left(\cos\frac{r}{a}\pi\right)^{z+1} \ldots (15').$$

It is an obvious consequence of the discontinuity of the limiting conditions of the problem, that this expression does not reduce itself to (15) when k is taken equal to zero. For the same reason it is not applicable when k is equal to unity: and on the other hand, it is not to be greater than $a-2$.

It is unnecessary to trace the different corollaries deducible from the last written equation, as it has been introduced merely to illustrate the facility with which our method discusses any proposed modification of the question of the duration of play.

One point, which is perhaps worth notice, is the symmetrical manner in which x and p, k and q, enter into (15′): the result, however, which is the interpretation of this symmetry may probably be obtained by general considerations.

A more general question would arise from supposing it possible for M to win or lose at each game any number of counters not greater than a. The method we have been illustrating would apply to this question, but the solution of it involves that of an algebraical equation of a degree superior to the second.

Another part of the subject, namely, the numerical calculation of the expressions already obtained, would not be consistent with the design of this paper. When z is sufficiently large, all the summations with respect to r may be reduced to their first and last terms, unless a is extremely large, in which case other methods of approximating (those, namely, of Laplace), may be made use of.

Enough has probably been said to show the facility which the method I have proposed is capable of giving to questions of acknowledged difficulty. I am not aware that it has been before pointed out; but as I am not at present able to refer to any work on the subject, I cannot speak confidently on this point. [x, a are integral throughout.]

GENERAL THEOREMS ON MULTIPLE INTEGRALS*.

IN Liouville's theorem for the reduction of a certain class of definite multiple integrals, the integrations comprise all positive values of the variables which do not transgress a limiting inequality, which either is of, or may easily be reduced to, a linear form. Take for illustration the case of two variables, and let $mx + ny < h$ be the limiting inequality in question, m, n and h being positive. Then, geometrically, $mx + ny = h$ is the equation of a straight line which forms the base of a triangle of which the intercepts of the positive half axes of co-ordinates are the sides, and our integration extends over the whole surface of this triangle. A similar interpretation may of course be given in the case of three variables. But to return to that of two. Let $mx + ny = h$ cut the axis of x in the point M and that of y in the point N: conceive another straight line $m'x + n'y = h'$; m', n', h' being also all positive; and let it cut the axes in M', N' respectively. Let us suppose for distinctness that $\frac{m}{m'}$ is greater than $\frac{n}{n'}$. Then, if the value of $\frac{h}{h'}$ be intermediate between those of the two fractions $\frac{m}{m'}$ and $\frac{n}{n'}$, it will be easily seen that the two lines must intersect in some point A, lying in the positive quadrant of co-ordinates, and that we shall have a quadrilateral $OMAN'$, (O being the origin of co-ordinates,) formed by the axes and by the two bounding lines. If now we integrate any function of x and y for all positive values of the variables

* *Cambridge and Dublin Mathematical Journal*, Vol. I. p. 1, 1846.

not transgressing the *two* inequalities $mx+ny \leqq h$, $m'x+n'y \leqq h'$, we shall in effect integrate over the surface of the quadrilateral $OMAN'$. But if the two lines did not intersect within the positive quadrant, then one or other bounding inequality would be inoperative, and we should in effect integrate over the surface, not of a quadrilateral, but of a triangle, as in the case contemplated by Liouville's theorem. It is manifest that we may have, instead of two limiting inequalities, any larger number we please, and that our integrations may thus be made to extend over an irregular polygon of a greater or less number of sides. I do not believe that any writer on multiple integrals has considered the case in which the limits are given by more than one inequality, but the restriction to that of one is clearly unnecessary.

Let us suppose there are r variables $x, y, \dots z$, and that we have to evaluate the integral

$$\int_0 dx \dots \int_0 dz\, e^{-ax \dots -cz}\, \phi\,(mx+\dots pz)\, \phi_{,}\,(m_{,}x+\dots p_{,}z) \;\dots\dots(1),$$

subject to the two inequalities

$$mx+\dots pz \leqq h, \quad m_{,}x+\dots p_{,}z \leqq h_{,},$$

$m \dots p,\, h;\; m_{,} \dots p_{,},\, h_{,}$ being all positive; and ϕ and $\phi_{,}$ any functions whose values may be represented within the limits of integration by Fourier's theorem.

Let the value of the integral in question be I; then, by considerations analogous to those of which I made use in a paper which appeared at the commencement of the last volume of the *Journal**, we shall have

$$I = \frac{1}{\pi^2}\int^h \phi u\, du \int^{h_{,}} \phi_{,} u_{,}\, du_{,} \int_0^\infty d\alpha \int_0^\infty d\alpha_{,} \,.\, G,$$

where

$$G = \int_0^\infty dx \dots \int_0^\infty dz\, e^{-ax \dots -cz} \cos\alpha\,(mx+\dots pz-u) \cos\alpha_{,}(m_{,}x+\dots p_{,}z-u_{,}),$$

and the lower limits of integration with respect to u and u' may be any negative quantities.

I remark in the first place, that

$$\int_0^\infty d\alpha \int_0^\infty d\alpha_1\, G = \tfrac{1}{4}\int_{-\infty}^\infty d\alpha \int_{-\infty}^\infty d\alpha_{,} H, \quad \text{where}$$

* Page 150 of this volume.

$$H=\int_0^\infty dx\ldots\int_0^\infty dz\, e^{-ax-\ldots-cz}\cos\{(am+\alpha_{,}m_{,})\,x+\ldots$$
$$(ap+\alpha_{,}p_{,})\,z-\alpha u-\alpha_{,}u_{,}\},$$

and therefore

$$I=\frac{1}{4\pi^2}\int^h\phi u\,du\int^{h'}\phi_{,}u_{,}\,du_{,}\int_{-\infty}^{\infty}d\alpha\int_{-\infty}^{\infty}d\alpha_{,}H.$$

Let $\quad H=K\cos(\alpha u+\alpha_{,}u_{,})+L\sin(\alpha u+\alpha_{,}u_{,}).$

Then it will easily be seen that

$$K=\frac{N}{D},\quad L=\frac{N'}{D};$$

where, if we take the case of three variables,

$$N=abc\left(1-\frac{am+\alpha_{,}m_{,}}{a}\,\frac{an+\alpha_{,}n_{,}}{b}-\frac{am+\alpha_{,}m_{,}}{a}\,\frac{ap+\alpha_{,}p_{,}}{c}\right.$$
$$\left.-\frac{an+\alpha_{,}n_{,}}{b}\,\frac{ap+\alpha_{,}p_{,}}{c}\right),$$

$$N'=abc\left(\frac{am+\alpha_{,}m_{,}}{a}+\frac{an+\alpha_{,}n_{,}}{b}\right.$$
$$\left.+\frac{ap+\alpha_{,}p_{,}}{c}-\frac{am+\alpha_{,}m_{,}}{a}\,\frac{an+\alpha_{,}n_{,}}{b}\,\frac{ap+\alpha_{,}p_{,}}{c}\right),$$

$$D=\{a^2+(am+\alpha_{,}m_{,})^2\}\,\{b^2+(an+\alpha_{,}n_{,})^2\}\,\{c^2+(ap+\alpha_{,}p_{,})^2\}.$$

(Precisely the same law of formation of these quantities would obtain if we were to take any number of variables. I have taken the case of three merely for distinctness of representation.)

Putting for $\cos(\alpha u+\alpha_{,}u_{,})$ and $\sin(\alpha u+\alpha_{,}u_{,})$ their exponential values, we find that

$$HD=a\ldots c\left\{1-\sqrt{(-1)}\,\frac{am+\alpha_{,}m_{,}}{a}\right\}\ldots\left\{1-\sqrt{(-1)}\,\frac{ap+\alpha_{,}p_{,}}{c}\right\}e^{(\alpha u+\alpha_{,}u_{,})\sqrt{(-1)}}$$
$$+a\ldots c\left\{1+\sqrt{(-1)}\,\frac{am+\alpha_{,}m_{,}}{a}\right\}\ldots\left\{1+\sqrt{(-1)}\,\frac{ap+\alpha_{,}p_{,}}{c}\right\}e^{-(\alpha u+\alpha_{,}u_{,})\sqrt{(-1)}}:$$

and as

$$a^2+(am+\alpha_{,}m_{,})^2=a^2\left\{1-\sqrt{(-1)}\,\frac{am+\alpha_{,}m_{,}}{a}\right\}\left\{1+\sqrt{(-1)}\,\frac{am+\alpha m}{a}\right\}$$

$$H = \frac{e^{(au+a_,u_,)\sqrt{(-1)}}}{\{a+\sqrt{(-1)}\,(am+\alpha_,m_,)\}\,\ldots\,\{c+\sqrt{(-1)}\,(ap+\alpha_,p_,)\}}$$

$$+\frac{e^{-(au+a_,u_,)\sqrt{(-1)}}}{\{a-\sqrt{(-1)}\,(am+\alpha_,m_,)\}\,\ldots\,\{c-\sqrt{(-1)}\,(ap+\alpha_,p)\}}.$$

Now, assume that

$$\frac{1}{(a+am+\alpha_,m_,)\,\ldots\,(c+ap+\alpha_,p_,)}$$

$$=\frac{F_{ab}}{(a+am+\alpha_,m_,)(b+an+\alpha_,n_,)}+\frac{F_{ac}}{(a+am+\alpha_,m_,)(c+ap+\alpha_,p_,)}+\&c.$$

...... (2);

where F_{ab}, F_{ac}, &c. are independent of α and $\alpha_,$. This assumption is justifiable because it introduces $\frac{r\,.\,r-1}{2}$ disposable quantities F, viz. as many as there are combinations two and two of the r quantities $a, b \ldots c$, and it will be easily seen that there are the same number of conditions to be satisfied.

Consequently as

$$e^{(au+a'u')\sqrt{(-1)}} = \cos\,(au+\alpha'u') + \sqrt{(-1)}\,\sin\,(au+\alpha'u'),$$

we shall have

$$\left.\begin{array}{l} H = F_{ab}\,\{ab-(am+\alpha_,m_,)\,(an+\alpha_,n_,)\}\cos\,(au+\alpha_,u_,) \\ \qquad +\{a\,(an+\alpha_,n_,)+b\,(am+\alpha_,m_,)\}\sin\,(an+\alpha_,n_,) \\ \text{divided by} \\ \qquad \{a^2+(am+\alpha'm')^2\}\,\{b^2+(an+\alpha'n')^2\} \end{array}\right\}+\&c.$$

Let us next assume $u = mx + ny$, $u_, = m_,x + n_,y$, x and y being here two new variables; also $\alpha' = am + \alpha_,m_,$, and $\beta' = an + \alpha_,n_,$: then the coefficient of F_{ab} in the expression of H will become

$$\frac{(ab-\alpha'\beta')\cos\,(\alpha'x+\beta'y)+(a\beta'+b\alpha')\sin\,(\alpha'x+\beta'y)}{(a^2+\alpha'^2)\,(b^2+\beta'^2)}.$$

Moreover $dudu_1d\alpha d\alpha_1$ will be replaced by $dxdyd\alpha'd\beta'$; and therefore, as we have

$$I = \frac{1}{4\pi^2}\int^h \phi u du \int^h \phi_, u_, du_, \int_{-\infty}^{+\infty} d\alpha \int_{-\infty}^{+\infty} d\alpha_,\, H, \text{ we shall have}$$

$$I = \frac{1}{4\pi^2}\,\Sigma F_{ab}\iint \phi\,(mx+ny)\,\phi_,\,(m_,x+n_,y)\,dxdy\,M,$$

where the sign of summation extends to all the quantities F, and where

$$M=\int_{-\infty}^{+\infty}d\alpha'\int_{-\infty}^{+\infty}d\beta'\,\frac{(ab-\alpha'\beta')\cos(\alpha'x+\beta'y)+(a\beta'+b\alpha')\sin(\alpha'x+\beta'y)}{(a^2+\alpha_{,}^2)(b^2+\beta_{,}^2)}.$$

From the known integrals

$$\int_{-\infty}^{+\infty}\frac{\cos\alpha x\,.\,d\alpha}{a^2+\alpha^2}=\frac{\pi}{a}\,e^{\mp ax}\int_{-\infty}^{+\infty}\frac{\alpha\sin\alpha x\,.\,d\alpha}{a^2+\alpha^2}=\pm\,\pi e^{\mp ax},$$

the upper signs to be taken when x is positive, it follows that

$$M=\pi^2e^{\mp ax\mp by}\,(1\pm1\pm1\pm1).$$

If x and y are both positive, the bracket becomes $1+1+1+1$ or 4; if x only be negative, it becomes $1-1-1+1$ or 0; if y only be negative, it becomes $1-1+1-1$ or 0; and similarly if both x and y are negative. Thus generally

$$M=4\pi^2e^{-ax-by}\text{ or }M=0.$$

There are, indeed, exceptional cases; as if y be zero, x being positive, when $M=2\pi^2e^{-ax}$, and similarly if x be zero, y being positive; and again, if x and y are both zero, when $M=\pi^2$: but of these, as we are about to multiply M by the element $dx\,dy$, it is unnecessary to take account. Therefore, in integrating for x and y, we include only positive values of the variables; and as u and u' are not to be greater than h and $h_{,}$ respectively, x and y must be such as not to transgress the inequalities

$$mx+ny\leqq h,\quad m_{,}x+n_{,}y\leqq h_{,}.$$

Thus we find that

$$I=\Sigma F_{ab}\textstyle\int_0 dx\int_0 dy\,\phi\,(mx+ny)\,\phi_{,}(m_{,}x+n_{,}y)\,e^{-ax-by},$$

the limits being given by the two above-written inequalities. It appears, therefore, that the integral (1), when there are two limiting inequalities, is reducible to the sum of a series of double integrals.

This result is analogous to that which is obtained in the case of the function $\phi\,(mx+\dots pz)\,e^{-ax\dots-cz}$, in the paper already referred to.

It remains to determine the form of the quantity F_{ab}. This is done at once by multiplying equation (2) by

$$(a+\alpha m+\alpha_{,}m_{,})\,(b+\alpha n+\alpha_{,}n_{,}),$$

and replacing α, $\alpha_{,}$ by values which make both these factors vanish. It hence appears that

$$F_{ab} = \frac{(mn_{,} - m_{,}n)^{r-2}}{\{c\,(mn_{,} - m_{,}n) + a\,(np_{,} - n_{,}p) + b\,(pm_{,} - p_{,}m)\}\,\ldots},$$

the denominator being the continued product of $r-2$ factors, each of the same form as the one written down. Of course the other quantities F are obtained in the same manner.

Let us now take the more general case in which there are s limiting inequalities, s being less than r, and in which the function to be integrated is

$$\phi_1\,(m_1x + \ldots p_1z) \ldots \phi_s\,(m_sx + \ldots p_sz)\; e^{-ax\ldots-cz},$$

the inequalities in question being

$$m_1x + \ldots p_1z \leqq h_1, \ldots m_sx + \ldots p_sz \leqq h_s.$$

We shall arrive at a perfectly analogous result in this more general case. In the first place the integral sought may be thus written,

$$\frac{1}{\pi^s}\int^{h_1}\phi_1 u_1\,du_1 \ldots \int^{h_s}\phi_s u_s\,du_s \int_0^\infty d\alpha_1 \ldots \int_0^\infty d\alpha_s G, \text{ where}$$

$$G = \int_0^\infty dx \ldots \int_0^\infty dz\; e^{-ax-\ldots cz}\cos\alpha_1(m_1x + \ldots p_1z - u_1) \ldots \cos\alpha_s(m_sx + \ldots p_sz - u_s$$

Now a little consideration will convince us that

$$\int_0^\infty d\alpha_1 \ldots \int_0^\infty d\alpha_s G = \frac{1}{2^s}\int_{-\infty}^{+\infty} d\alpha_1 \ldots \int_{-\infty}^{+\infty} d\alpha_s H, \text{ where}$$

$$H = \int_0^\infty dx \ldots \int_0^\infty dz\; e^{-ax-\ldots cz}\cos\,[x\Sigma\alpha m + \ldots + z\Sigma\alpha p - \Sigma\alpha u]:$$

for if we take the expression

$$\cos\,[x\,\Sigma\alpha m + \ldots + z\,\Sigma\alpha p - \Sigma\alpha u],$$

make α_1 negative, add the resulting expression to the original one: then in the two terms thus got make α_2 negative, and as before add the results, we shall, continuing this process, get in all 2^s terms, which will be found to be equal to 2^s times the continued product of the cosines involved in G.

Effecting the integrations indicated in H, we see that

$$H = \frac{N \cos \Sigma \alpha u + N' \sin \Sigma \alpha u}{D},$$

where $D = \{a^2 + (\Sigma \alpha m)^2\} \dots \{c^2 + (\Sigma \alpha p)^2\}$,

and N and N' follow the same law of formation as in the particular case already considered, except that for $\frac{\alpha m + \alpha' m'}{a}$, &c. we substitute $\frac{\Sigma \alpha m}{a}$, &c. With this remark we perceive that

$$H = \frac{e^{\Sigma \alpha u \sqrt{(-1)}}}{\{a + \sqrt{(-1)}\, \Sigma \alpha m\} \dots \{c + \sqrt{(-1)}\, \Sigma \alpha p\}} + \frac{e^{-\Sigma \alpha u \sqrt{(-1)}}}{\{a - \sqrt{(-1)}\, \Sigma \alpha m\} \dots \{c - \sqrt{(-1)}\, \Sigma \alpha p\}} \dots\dots (3).$$

The assumption now to be made is that

$$\frac{1}{(a + \Sigma \alpha m) \dots (c + \Sigma \alpha p)} = \Sigma \frac{F}{\Delta} \dots\dots\dots\dots\dots (2'),$$

where Δ is the product of every set of s factors taken out of the whole number of r factors

$$a + \Sigma \alpha m, \ \dots \ c + \Sigma \alpha p,$$

and F is independent of $\alpha_1 \dots \alpha_s$.

There will thus be $\frac{r \,.\, r-1 \dots r-s+1}{1 \,.\, 2 \dots s}$ disposable quantities F, which will be found to be the number required to make (2′) identically true. Consequently we shall have

$$H = \Sigma F \frac{\nu \cos \Sigma \alpha u + \nu' \sin \Sigma \alpha u}{\delta},$$

where δ is the product of s factors of the form $a^2 + (\Sigma \alpha m)^2$; and ν and ν' are formed just as in the case of $s = 2$: that is to say, we shall have

$$\nu = a \dots c\,(1 - C_2 + C_4 - \&c.), \quad \nu' = a \dots c\,(C_1 - C_3 + \&c.),$$

where C_t is the sum of the products of every combination that can be made of the s quantities $\frac{\Sigma \alpha m}{a}$, &c., taken t and t together.

In order to simplify the expression

$$\frac{\nu \cos \Sigma \alpha u + \nu' \sin \Sigma \alpha u}{\delta},$$

let us denote the s quantities $\Sigma \alpha m$, &c., which are involved in it, by the single symbols $\beta_1 \ldots \beta_s$, and assume

$$u_1 = \Sigma m_1 x, \quad u_2 = \Sigma m_2 x, \ldots u_s = \Sigma m_s x,$$

the sign of summation Σ extending only to that set of s out of the r quantities $x \ldots z$, which corresponds to the factors involved in the denominator δ. Of course x, y, &c. are here, as before, new variables. (In the case of $s = 3$, for instance, these assumptions will be of the form

$$\beta_1 = \alpha_1 m_1 + \alpha_2 m_2 + \alpha_3 m_3, \quad u_1 = m_1 x + n_1 y + p_1 z,$$
$$\beta_2 = \alpha_1 n_1 + \alpha_2 n_2 + \alpha_3 n_3, \quad u_2 = m_2 x + n_2 y + p_2 z,$$
$$\beta_3 = \alpha_1 p_1 + \alpha_2 p_2 + \alpha_3 p_3, \quad u_3 = m_3 x + n_3 y + p_3 z.)$$

It follows from this that $du_1 \ldots du_s d\alpha_1 \ldots d\alpha_s$ will be replaced by $dx \ldots dy \,.\, d\beta_1 \ldots d\beta_s$, and that the factors in δ will take the simpler form $a^2 + \beta_1^2$, $b^2 + \beta_2^2$, &c.; while $\Sigma \alpha u$ will become $\beta_1 x + \beta_2 y + \ldots$

The integrations with respect to β extend, like those for α, from $-\infty$ to $+\infty$. Let

$$\int_{-\infty}^{+\infty} d\beta_1 \ldots \int_{-\infty}^{+\infty} d\beta_s \, \frac{\nu \cos \Sigma \beta x + \nu' \sin \Sigma \beta x}{\delta} = M'.$$

Then, from the obvious analogy between the forms ν and ν', and those of the developments of $\cos \Sigma \beta x$ and $\sin \Sigma \beta x$ respectively, it follows that if x, y, &c. are all positive,

$$M' = \pi^s e^{-ax - by \ldots} (1 + 1 + \ldots),$$

there being twice as many units within the brackets as there are terms in the development of $\sin (f_1 + \ldots f_s)$, or of $\cos (f_1 + \ldots f_s)$, that is to say, twice 2^{s-1} or 2^s.

Moreover, if any one, as x, of the quantities x, y, &c., is negative, $M' = 0$; and this, whether it alone is negative or any others, are so too. For if $x = -x'$, let its coefficient β_1 be assumed equal to $-\beta_1'$, when the expression of M' becomes of the same form as if x were positive, except that ν and ν' are changed by having $-\beta_1'$ wherever β_1 occurred previously. Now none of the quantities β can occur raised to any power, and therefore every term involving β_1 will change sign when β_1 is replaced by $-\beta_1'$. Hence we shall have

$$M_{,} = \pi^s e^{-ax'-by-\dots}(1 \pm 1 \dots),$$

there being as many negative units as positive within the brackets, since in the development of sin $(f_1 + \dots f_s)$ or cos $(f_1 + \dots f_s)$ there are 2^{s-2} terms independent of the sine of f_1 and 2^{s-2} terms which involve that quantity, and which therefore change sign when f_1 does so. Hence the quantity within the bracket, and consequently $M_{,}$, is equal to zero if x be negative; and so, of course, for the other variables $y \dots z$.

M will, in particular cases analogous to those already noticed, assume exceptional or limiting values, but of these we need not take account. And thus we arrive at the following remarkable theorem:

The definite integral of r *variables* x ... z

$$\int_0 dx \dots \int_0 dz\, \phi_1(m_1x + \dots p_1z) \dots \phi_s(m_sx + \dots p_sz)e^{-ax-\dots cz},$$

whose limits are given by s *inequalities*

$$m_1x + \dots p_1z \leqq h_1, \dots m_sx + \dots p_sz \leqq h_s,$$

can generally be expressed as a linear function of

$$\frac{r.(r-1)\dots(r-s+1)}{1.2\dots s}$$

integrals of s *variables each. The form of each of these integrals may be deduced from the original integral by omitting from it any set of* r − s *of the variables, and similarly the form of the limiting inequalities may be got by omitting the same set of variables from the original inequalities* (r > s).

In certain cases, however, when the constants a, m, &c. have particular values, the theorem fails because the assumption (2′) becomes illegitimate. This failure is indicated by certain of the quantities F becoming infinite. To determine the form of F, we have merely to multiply (2′) by Δ, and then to equate to zero all the s factors of which Δ is composed. All the quantities F, except the particular one under consideration, will then disappear, and we have s equations determining the s quantities α.

Hence it will appear that F is equal to a fraction whose numerator is unity, and denominator equal to the value assumed by the product of the remaining $r-s$ factors, when the values already assigned for the quantities α, &c. have been substituted for them; a result which it is obvious can be immediately expressed in the notation of determinants. F will therefore become infinite if our equating the s factors by which it is divided in (2′) to zero will make one or more of the remaining $r-s$ factors vanish. Let it make t of these factors vanish; then equating these t factors also to zero, we get in all $s+t$ equations, which are equivalent to s independent ones. Therefore any set of s out of these $s+t$ equations will satisfy the remaining t equations. Hence

$$\frac{(s+t)\,(s+t-1)\,\ldots\,(t+1)}{1\,.\,2\,\ldots\,t}$$

of the quantities F will become infinite, and therefore the second side of (2′) will consist of finite terms and of a finite quantity expressed in the form of the sum of that number of infinite terms. This indetermination of course indicates a change in the form of the function, the general character of which the reader will have little difficulty in perceiving. But the consideration of these particular cases, some of which are interesting, must be deferred to another occasion.

I am inclined to believe that the process developed in this paper will admit both of simplification and extension. For the exponential function we may substitute with certain modifications any function of $ax+\ldots cz$, in accordance with a result given by Mr Boole in his very interesting Memoir on a new Method in Analysis, which is published in the *Transactions of the Royal Society*. (This result would include the one which I obtained in the last volume of the *Journal*, from which however it might be deduced.)

Thus, if in the theorem established in this paper we replace $a, b, \ldots c$ by $ka, kb \ldots kc$, k being a wholly arbitrary quantity, we may, comparing the coefficients of its powers, deduce new theorems from the given one. Developing the first side of the equation, the coefficient of k^n will be

$$\frac{1}{1\,.\,2\,\ldots\,n}\int_0 dx\ldots\int_0 dz\,\phi_1(m_1x+\ldots p_1z)\ldots\phi_s(m_sx+\ldots p_sz)(ax+\ldots cz)^n,$$

and in the second it will be the sum of a series of terms of the form

$$\frac{F}{1 . 2 \ldots (n+r-s)} \int_0 dx \int_0 dy \ldots \phi \, (m_1 x + n_1 y + \ldots) \\ \ldots \phi_s \, (m_s x + n_s y + \ldots) \, (ax + by + \ldots)^{n+r-s},$$

as it is manifest that F will become $\frac{F}{k^{r-s}}$. Hence, if

$$\psi \, (ax + \ldots cz)$$

be such a function that its development may be substituted for it in the integrations, we shall have

$$\int_0 dx \ldots \int_0 dz \, \phi_1 \, (m_1 x + \ldots p_1 z) \ldots \phi_s \, (m_s x + \ldots p_s z) \, \psi \, (ax + \ldots cz) \\ = \Sigma F \int_0 dx \int_0 dy \ldots \phi_1 \, (m_1 x + n_1 y + \ldots) \ldots \phi_s \, (m_s x + n_s y + \ldots) \\ \psi_1 \, (ax + by + \ldots),$$

where $\frac{d^{r-s}}{dt^{r-s}} \psi_1 t = \psi t$ and all the differential coefficients of $\psi_1 t$ of an order lower than the $(r-s)^{\text{th}}$ vanish for $t = 0$. This is, I believe, in the case of s equal to unity, precisely equivalent to one of Mr Boole's results. It might also, I imagine, be obtained without having recourse to developments.

MATHEMATICAL NOTE*.

Solution of a Functional Equation.

To solve the functional equation

$$\int_{-\infty}^{+\infty} \phi_m x \phi_n (\alpha - x)\, dx = \phi_{m+n}\alpha,$$

provided $$\phi x = \phi(-x).$$

Let $\int_{-\infty}^{+\infty} \phi\alpha \cos \alpha z \, d\alpha = \psi z$. Then

$$\psi_{m+n} z = \int_{-\infty}^{+\infty} d\alpha \int_{-\infty}^{+\infty} dx \phi_m x \phi_n (\alpha - n) \cos \alpha z.$$

Write for α, $\alpha + x$, then $\cos \alpha z$ becomes

$$\cos \alpha z \cos xz - \sin \alpha z \sin xz,$$

and as the sines change sign with their arcs, we get

$$\psi_{m+n} z = \int_{-\infty}^{+\infty} d\alpha \int_{-\infty}^{+\infty} dx \phi_m x \phi_n \alpha \cos \alpha z \cos xz.$$

Or $$\psi_{m+n} z = \psi_m z \psi_n z,$$

whence $$\psi_m z = \chi z^m,$$

χ being independent of m, and therefore, by Fourier's theorem,

$$\phi_m x = \frac{1}{\pi} \int_0^{\infty} \chi z^m \cos xz \, dz.$$

The required solution.

* *Cambridge and Dublin Mathematical Journal*, Vol. VII. p. 103, 1852.

ON THE AREA OF THE CYCLOID*.

To the Editor of the Cambridge and Dublin Mathematical Journal.

SIR,—The determination of the area of the Cycloid, so easily effected by modern analysis, was regarded by the geometricians of the seventeenth century as a problem of no small difficulty. Mersenne was the first who attempted a solution: he was however unsuccessful. It was proposed by him in despair to Roberval in 1628, who also failed in his attempt at that time. About seven years afterwards, however, Roberval overcame the difficulty, and communicated his good fortune to Mersenne.

In a letter to Descartes, Mersenne made mention of Roberval's discovery of the area of the Cycloid as a great feat in geometry, simply stating the result obtained by Roberval, without giving any clue to the method. Descartes, solving the problem himself with little difficulty, communicated his method in reply to Mersenne, with some supercilious remarks about the supposed difficulty of the problem. Fermat and other mathematicians of that day exercised their ingenuity in the same question. A solution of the problem by pure geometry, which was some time ago communicated to me by Mr R. L. Ellis of Trinity College, possesses so great a superiority over any of the geometrical methods of these early mathematicians which I have seen, that I think it may be acceptable to those readers of your *Journal* who take an interest in the history of mathematics.

"The motion of the generating circle may be resolved into two uniform motions, a motion of translation parallel to the

* *Cambridge and Dublin Mathematical Journal*, Vol. IX. p. 263, 1854.

directrix and of rotation round its own centre. The area generated by the describing point may be considered as generated by these two motions: that of translation nowise affects the motion of rotation, and the area due to the latter is the same as if the former did not exist, that is, it is equal to the area of the generating circle. Contrariwise the motion of rotation does affect the area due to that of translation, inasmuch as in virtue of it the distance of the describing point from the directrix is varied: the mean distance, viewed as depending on the motion of rotation, is equal to the radius of the generating circle, and the corresponding area is therefore a rectangle, the base of which is the space slided over and altitude that radius; and, as this space is the circumference of the generating circle, the area in question is equal to twice the area of that circle: on the whole, therefore, the area of the cycloid is equal to three times that of the generating circle.

"The reason is just the same as that by which what are called Guldinus's properties are established. We here resolve the motion of a describing point into motions parallel and perpendicular to the abscissa; the latter generates no area, the former generates a rectangular area having for its base the abscissa and for its altitude the mean value of the ordinates; that is, the ordinate of the centre of gravity of the arc, which is a known result. The only difference to be attended to in the two cases relates to the mode in which the average is to be taken."

Mr Ellis has remarked, that the same method may be extended to the determination of the areas of the hypocycloid and epicycloid.

I am, Sir,

Your obedient Servant,

WILLIAM WALTON.

Cambridge, July 31, 1854.

SUR LES INTÉGRALES AUX DIFFÉRENCES FINIES *.

On peut évaluer l'intégrale

$$(1) \qquad \int dx \int dy \ldots \int dz \phi(x, y, \ldots, z),$$

dans laquelle les variables $x, y, \ldots, z$ doivent prendre toutes les valeurs positives qui satisfont à l'inégalité

$$(2) \qquad \psi(x, y, \ldots, z) < h,$$

en remplaçant dans la formule (1) la fonction ϕ par une fonction discontinue, qui devient égale à zéro pour toutes les valeurs des variables non comprises dans la formule (2). On peut alors étendre les intégrations depuis zéro jusqu'à l'infini, ce qui simplifie beaucoup les calculs.

Je crois que c'est à M. Lejeune-Dirichlet qu'est due l'idée de cette manière d'évaluer les intégrales multiples ; c'est ainsi qu'il a obtenu, il y a quelques années, une généralisation très-remarquable d'un théorème dû à Euler.

La théorie des intégrales définies nous fournit plusieurs moyens d'exprimer les fonctions discontinues ; je me suis servi, pour cet objet, du théorème de Fourier. Au moyen de ce théorème, j'ai déterminé, dans un petit Mémoire inséré dans le *Journal de Mathématiques* de Cambridge †, les valeurs de deux intégrales multiples. La première de ces intégrales revient à la généralisation qu'a donnée M. Liouville du résultat de M. Dirichlet ; mais je crois que la seconde est nouvelle.

* Extrait du *Journal de Mathématiques pures et appliquées*, Tome IX. 1844.

† Page 150 of this volume.

La facilité avec laquelle j'avais obtenu ces résultats me fit penser qu'on pourrait peut-être appliquer une méthode semblable aux différences finies ; les résultats auxquels je suis parvenu par cette considération font le sujet de ce qui va suivre.

En suivant l'analogie qui existe entre les différences finies et les différences infiniment petites, on voit qu'à l'intégration multiple, il faut substituer des sommations par rapport à toutes les variables qui entrent dans la fonction donnée.

Soit $\phi(x, y)$ cette fonction. Je désigne par $\Sigma^b \Sigma_c^d \phi(x, y)$ la quantité suivante ($b-a$ et $d-c$ étant des nombres entiers et positifs),

$$\begin{aligned}&\phi(a, c)+\phi(a+1, c)+\ldots \phi(b, c)\\ &+\phi(a, c+1)+\ldots\ldots\ldots\ldots \phi(b, c+1)\\ &+\ \ldots\ldots\ldots\ldots\ldots\ldots\ldots\ldots\ldots\ldots\\ &+\phi(a, d)+\ldots\ldots\ldots\ldots\ldots\ldots \phi(b, d).\end{aligned}$$

Il est visible que cette notation pourrait s'étendre à un nombre quelconque de variables.

Le théorème de Fourier se remplacera par la formule suivante, dans laquelle $b-x$ et $x-a$ sont des nombres entiers et positifs,

$$fx=\frac{1}{\pi}\int_0^\pi d\alpha\, \Sigma_a^b fu \cos\alpha(x-u). \tag{3}$$

On peut donc poser

$$x=a, \ =a+1, \ldots =b;$$

mais si l'on donne à x (qui doit toujours être un nombre entier) une valeur quelconque non comprise dans ces limites, on aura

$$\int_0^\pi d\alpha\, \Sigma_a^b fu \cos\alpha(x-u)=0.$$

La démonstration de ce théorème est si facile, qu'il n'est pas nécessaire de s'y arrêter ; je ferai seulement observer en passant qu'elle suppose que la fonction fu ne devienne infinie pour aucune des valeurs

$$u=a, \quad a+1, \ldots, \ b.$$

Désignons par $\{x\}^p$ la fonction $\dfrac{\Gamma(x+p)}{\Gamma(x)}$: toutes les fois que p est un nombre entier et positif, nous aurons

$$\{x\}^p=x.x+1\ldots x+p-1.$$

Cela posé, entrons en matière.

Je vais chercher la valeur de la somme multiple

$$(4)\quad \Sigma_0 \Sigma_0 \dots \Sigma_0 \{x\}^{p-1} \{y\}^{q-1} \dots \{z\}^{r-1} f(x+y+\dots+z),$$

dans laquelle l'étendue de la sommation est donnée par l'inégalité

$$(5)\qquad x+y+\dots+z \overline{\overline{<}} h.$$

D'après l'idée fondamentale de notre analyse, je remplace dans la formule (4) la fonction $f(x+y+\dots+z)$ par

$$\frac{1}{\pi}\int_0^\pi d\alpha\, \Sigma_k^h fu \cos\alpha\,(x+y+\dots+z-u).$$

Donc nous aurons, en changeant l'ordre des sommations,

$$(6)\quad \frac{1}{\pi}\Sigma_k^h fu \int_0^\pi d\alpha \Sigma_0 \Sigma_0 \dots \Sigma_0 \{x\}^{p-1}\{y\}^{q-1}\dots\{z\}^{r-1}\cos\alpha\,(x+y+\dots+z-u)$$

(k est un nombre négatif quelconque).

Les sommations par rapport à x, y, etc., peuvent à présent s'étendre jusqu'à l'infini.

Nous allons déterminer les valeurs de

$$\Sigma_0^\infty \{x\}^{p-1}\cos\alpha x, \quad \text{et de } \Sigma_0^\infty \{x\}^{p-1}\sin\alpha x.$$

Soit $z = e^{\alpha\sqrt{-1}}$; nous aurons, par un théorème connu,

$$2\,\Sigma_0^\infty a^{-x}\cos\alpha x = \frac{1}{1-\frac{z}{a}} + \frac{1}{1-\frac{1}{az}},$$

puisqu'on a

$$1+\frac{z}{a}+\text{etc.} = \frac{1}{1-\frac{z}{a}}, \quad \text{et } 1+\frac{1}{za}+\text{etc.} = \frac{1}{1-\frac{1}{za}}.$$

Pareillement on a

$$1+\frac{p}{1}\frac{z}{a}+\frac{p\,.\,p+1}{1\,.\,2}\frac{z^2}{a^2}+\text{etc.} = \frac{1}{\left(1-\frac{z}{a}\right)^p},$$

et de là

$$(7)\quad \frac{\Gamma(p)}{\Gamma(1)}a^{-p}z + \frac{\Gamma(p+1)}{\Gamma(2)}a^{-(p+1)}z^2 + \text{etc.} = \Gamma(p)\frac{z}{(a-z)^p}.$$

En écrivant dans cette équation z^{-1} au lieu de z, elle deviendra

$$\frac{\Gamma(p)}{\Gamma(1)} a^{-p} z^{-1} + \text{etc.} = \Gamma(p) \frac{z^{-1}}{(a - z^{-1})^p}.$$

Ajoutant cette équation à la dernière, nous aurons, à cause de $\{x\}^p = \dfrac{\Gamma(x+p)}{\Gamma(x)}$,

$$(8) \quad 2\,\Sigma_0^\infty \{x\}^{p-1} a^{-(x+p-1)} \cos \alpha x = \Gamma(p) \left\{ \frac{z}{(a-z)^p} + \frac{z^{-1}}{(a-z^{-1})^p} \right\}.$$

(On doit remarquer que $\{0\}^p = 0$, puisque $\Gamma(0)$ a une valeur infinie.)

Ensuite, à cause de

$$(a - z)(a - z^{-1}) = 1 - 2a \cos \alpha + a^2,$$

nous aurons

$$\frac{z}{(a-z)^p} + \frac{z^{-1}}{(a-z^{-1})^p} = \frac{1}{(1-2a \cos \alpha + a^2)^p} \{z (a-z^{-1})^p + z^{-1} (a-z)^p\}.$$

Mais, puisque $z = \cos \alpha + \sqrt{-1} \sin \alpha$,

$$a - z = \sin \alpha \, (a \operatorname{cosec} \alpha - \operatorname{cotang} \alpha - \sqrt{-1}).$$

Posons donc

$$\operatorname{cotang} \phi = a \operatorname{cosec} \alpha - \operatorname{cotang} \alpha,$$

la valeur de $(a - z)^p$ deviendra

$$\left(\frac{\sin \alpha}{\sin \phi}\right)^p (\cos p\phi - \sqrt{-1} \sin p\phi),$$

tandis que celle de $(a - z^{-1})^p$ sera

$$\left(\frac{\sin \alpha}{\sin \phi}\right)^p (\cos p\phi + \sqrt{-1} \sin p\phi);$$

par conséquent

$$z (a - z^{-1})^p + z^{-1} (a - z)^p = 2 \left(\frac{\sin \alpha}{\sin \phi}\right)^p \cos (p\phi + \alpha).$$

Mais la valeur de $\sin \phi$ est égale à $\dfrac{\sin \alpha}{(1 - 2a \cos \alpha + a^2)^{\frac{1}{2}}}$. Donc, nous aurons finalement

$$(9) \quad \Sigma_0^\infty \{x\}^{p-1} a^{-(x+p-1)} \cos \alpha x = \Gamma(p) \frac{\cos (p\phi + \alpha)}{(1 - 2a \cos \alpha + a^2)^{\frac{p}{2}}}.$$

On trouvera de la même manière que

$$(10)\quad \begin{cases} \Sigma_0^\infty \{x\}^{p-1} a^{-(x+p-1)} \sin \alpha x = \Gamma(p) \dfrac{\sin(p\phi+\alpha)}{(1-2a\cos\alpha+a^2)^{\frac{p}{2}}}. \\ (a>1). \end{cases}$$

A présent faisons $a=1$. Alors nous aurons

$$\operatorname{cotang}\phi = \frac{1-\cos\alpha}{\sin\alpha} = \operatorname{tang}\frac{\alpha}{2},$$

ç'est-à-dire

$$\phi = \frac{\pi}{2} - \frac{\alpha}{2},$$

et les équations (9) et (10) deviendront

$$(11)\quad \Sigma_0^\infty \{x\}^{p-1} \cos\alpha x = \Gamma(p) \frac{\cos\left\{\dfrac{p}{2}(\pi-\alpha)+\alpha\right\}}{\left(2\sin\dfrac{\alpha}{2}\right)^p},$$

$$(12)\quad \Sigma_0^\infty \{x\}^{p-1} \sin\alpha x = \Gamma(p) \frac{\sin\left\{\dfrac{p}{2}(\pi-\alpha)+\alpha\right\}}{\left(2\sin\dfrac{\alpha}{2}\right)^p}.$$

Au moyen de ces équations, on prouvera facilement que

$$(13)\quad \begin{cases} \Sigma_0^\infty \Sigma_0^\infty \ldots \Sigma_0^\infty \{x\}^{p-1}\{y\}^{q-1}\ldots\{z\}^{r-1}\cos\alpha(x+y+\ldots+z-u) \\ =\Gamma(p)\Gamma(q)\ldots\Gamma(r) \dfrac{\cos\left\{\dfrac{p+q+\ldots r}{2}(\pi-\alpha)+n\alpha-\alpha u\right\}}{\left(2\sin\dfrac{\alpha}{2}\right)^{p+q+\ldots r}}, \end{cases}$$

n étant le nombre des variables $x, y, \ldots, z$.

Mais en écrivant, dans l'équation (11), $p+q+\ldots r$ au lieu de p, on aura

$$\Sigma_0^\infty \{x\}^{p+q+\ldots r-1} \cos\alpha x = \Gamma(p+q+\ldots r) \frac{\cos\left\{\dfrac{p+q+\ldots r}{2}(\pi-\alpha)+\alpha\right\}}{\left(2\sin\dfrac{\alpha}{2}\right)^{p+q+\ldots r}},$$

et pareillement

$$\Sigma_0^\infty \{x\}^{p+q+\ldots r-1} \sin\alpha x = \Gamma(p+q+\ldots r) \frac{\sin\left\{\dfrac{p+q+\ldots r}{2}(\pi-\alpha)+\alpha\right\}}{\left(2\sin\dfrac{\alpha}{2}\right)^{p+q+\ldots r}}.$$

En combinant ces deux équations, on trouvera que

$$(14)\quad \begin{cases} \Sigma_0^\infty \{x\}^{p+q+\ldots r-1} \cos\alpha\,(x-\nu) \\ = \Gamma(p+q+\ldots r)\dfrac{\cos\left\{\dfrac{p+q+\ldots r}{2}(\pi-\alpha)+\alpha-\alpha\nu\right\}}{\left(2\sin\dfrac{\alpha}{2}\right)^{p+q+\ldots r}}. \end{cases}$$

Mettons donc

$$\nu = u - n + 1,$$

et comparons les équations (13) et (14), nous aurons

$$(15)\quad \begin{cases} \Sigma_0\, \Sigma_0^\infty \ldots \Sigma_0^\infty \{x\}^{p-1}\{y\}^{q-1}\ldots\{z\}^{r-1}\cos\alpha\,(x+y+\ldots z-u) \\ = \dfrac{\Gamma(p)\,\Gamma(q)\ldots\Gamma(r)}{\Gamma(p+q+\ldots+r)}\Sigma_0^\infty \{x\}^{p+q+\ldots r-1}\cos\alpha\,\{x-(u-n+1)\}. \end{cases}$$

En multipliant les deux membres de cette équation par $d\alpha$ et en intégrant depuis zéro jusqu'à π, nous aurons, par la formule (3),

$$(16)\quad \begin{cases} \displaystyle\int_0^\pi d\alpha\, \Sigma_0^\infty\, \Sigma_0^\infty \ldots \Sigma_0^\infty \{x\}^{p-1}\{y\}^{q-1}\ldots\{z\}^{r-1}\cos\alpha\,(x+y+\ldots z-u) \\ = \pi\dfrac{\Gamma(p)\,\Gamma(q)\ldots\Gamma(r)}{\Gamma(p+q+\ldots+r)}\{u-n+1\}^{p+q+\ldots r-1}, \end{cases}$$

pour toutes les valeurs positives de $u-n+1$.

Pour toutes les valeurs négatives de cette quantité, le second membre de l'équation (16) est égal à zéro.

Donc, en effectuant la sommation par rapport à u, il est inutile de donner à u des valeurs moindres que $n-1$. Cela posé, nous aurons finalement, en considérant la formule (4),

$$(17)\quad \begin{cases} \Sigma_0\Sigma_0\ldots\Sigma_0\{x\}^{p-1}\{y\}^{q-1}\ldots\{z\}^{r-1}f(x+y+\ldots+z) \\ = \dfrac{\Gamma(p)\,\Gamma(q)\ldots\Gamma(r)}{\Gamma(p+q+\ldots r)}\Sigma_{n-1}^{h} fu\,\{u-n+1\}^{p+q+\ldots r-1}, \end{cases}$$

l'étendue des sommations étant déterminée par l'inégalité

$$x+y+\ldots+z \leqq h.$$

Ce théorème est l'analogue pour les différences finies du théorème de M. Liouville, dont j'ai déjà parlé.

En effet, en supposant que l'inégalité qui détermine les limites des variables soit, comme ci-dessus,

$$x+y+\ldots+z \leqq h,$$

voici le théorème de M. Liouville:

$$\int_0 dx \int_0 dy \ldots \int_0 dz\, x^{p-1} y^{q-1} \ldots z^{r-1} f(x+y+\ldots+z)$$

$$= \frac{\Gamma(p)\Gamma(q)\ldots\Gamma(r)}{\Gamma(p+q+\ldots r)} \int_0^h fu\, u^{p+q+\ldots r-1} du.$$

Il est vrai que cette équation n'est qu'un cas particulier du résultat qu'a donné M. Liouville, mais malheureusement nous ne pouvons pas généraliser la formule (17) en supposant que l'étendue des sommations soit donnée par l'inégalité

$$ax + by + \ldots + cz \overset{=}{<} h,$$

sans au moins lui donner une forme beaucoup plus compliquée.

A présent, désignons, suivant la notation usitée, par $[x]^p$ la fonction $\frac{\Gamma(x+1)}{\Gamma(x-p+1)}$ (nous aurons, quand p sera un nombre entier,

$$[x]^p = x \,.\, x-1 \ldots x-p+1),$$

et tâchons d'évaluer la somme suivante,

$$\Sigma_{p-1} \Sigma_{q-1} \ldots \Sigma_{r-1} [x]^{p-1} [y]^{q-1} \ldots [z]^{r-1} f(x+y+\ldots z),$$

dans laquelle x peut prendre toutes les valeurs $p-1$, p, $p+1$, etc., tandis que y peut prendre toutes les valeurs $q-1$, q, $q+1$, etc., et ainsi de suite pour les autres variables. L'étendue des sommations est déterminée par l'inégalité

$$x + y + \ldots + z \overset{=}{<} h,$$

dans laquelle h est égale à $p+q+\ldots r+$un nombre entier.

Nous allons premièrement trouver les valeurs de

$$\Sigma_{p-1}^{\infty} [x]^{p-1} \cos ax, \quad \text{et de } \Sigma_{p-1}^{\infty} [x]^{p-1} \sin ax.$$

Puisque nous avons

$$1 + \frac{p}{1} az + \frac{p \,.\, p+1}{1 \,.\, 2} a^2 z^2 + \text{etc.} = \frac{1}{(1-az)^p},$$

il s'ensuit que

$$\frac{\Gamma(p-1+1)}{\Gamma(1)} z^{p-1} + \frac{\Gamma(p+1)}{\Gamma(2)} az^p + \text{etc.} = \Gamma(p) \frac{z^{p-1}}{(1-az)^p}.$$

Remplaçons z par z^{-1}, nous aurons, en ajoutant les deux résultats et en posant $a=1$,

$$[p-1]^{p-1}\cos\alpha(p-1)+[p]^{p-1}\cos\alpha p+\text{etc.}$$

$$=\frac{1}{2}\Gamma(p)\left\{\frac{z^{p-1}}{(1-z)^p}+\frac{z^{-p+1}}{(1-z^{-1})^p}\right\},$$

c'est-à-dire nous aurons

$$\Sigma_{p-1}^{\infty}[x]^{p-1}\cos\alpha x=\frac{1}{2}\Gamma(p)\left\{\frac{z^{p-1}}{(1-z)^p}+\frac{z^{-p+1}}{(1-z^{-1})^p}\right\},$$

et pareillement

$$\Sigma_{p-1}^{\infty}[x]^{p-1}\sin\alpha x=\frac{1}{2}\Gamma(p)\left\{\frac{z^{p-1}}{(1-z)^p}-\frac{z^{-p+1}}{(1-z^{-1})^p}\right\}.$$

Il est facile de voir, en suivant à peu près la même route qu'auparavant, que ces deux équations reviennent à celles-ci :

$$(18)\qquad \Sigma_{p-1}^{\infty}[x]^{p-1}\cos\alpha x=\Gamma(p)\frac{\cos\left\{\frac{p}{2}(\pi-\alpha)-\alpha\right\}}{\left(2\sin\frac{\alpha}{2}\right)^p},$$

$$(19)\qquad \Sigma_{p-1}^{\infty}[x]^{p-1}\sin\alpha x=\Gamma(p)\frac{\sin\left\{\frac{p}{2}(\pi-\alpha)-\alpha\right\}}{\left(2\sin\frac{\alpha}{2}\right)^p}.$$

Cela posé, on peut facilement s'assurer que la somme dont nous cherchons la valeur est égale à

$$\frac{1}{\pi}\Sigma^h fu\int_0^\pi d\alpha\,\Sigma_{p-1}^{\infty}\Sigma_{p-1}^{\infty}\ldots\Sigma_{p-1}^{\infty}[x]^{p-1}[y]^{q-1}\ldots[z]^{r-1}\cos\alpha(x+y\ldots z-u).$$

Par conséquent elle est égale, en vertu des formules (18) et (19), à

$$\frac{\Gamma(p)\Gamma(q)\ldots\Gamma(r)}{\pi}\Sigma^h fu\int_0^\pi d\alpha\frac{\cos\left\{\frac{p+q+\ldots r}{2}(\pi-\alpha)-n\alpha-\alpha u\right\}}{\left(2\sin\frac{\alpha}{2}\right)^{p+q+\ldots r}},$$

et de là nous aurons finalement

$$(20)\qquad \begin{cases}\Sigma_{p-1}\Sigma_{q-1}\ldots\Sigma_{r-1}[x]^{p-1}[y]^{q-1}\ldots[z]^{r-1}f(x+y+\ldots z)\\ =\dfrac{\Gamma(p)\Gamma(q)\ldots\Gamma(r)}{\Gamma(p+q+\ldots r)}\Sigma_{p+q+\ldots r-1}^{h}fu\,[u+n-1]^{p+q+\ldots r-1}.\end{cases}$$

Les deux résultats (17) et (20) suffisent pour montrer l'esprit de notre analyse, mais je vais encore l'appliquer à un autre exemple.

Évaluons l'expression

$$(21) \qquad \Sigma_0 \Sigma_0 \ldots \Sigma_0 a^x b^y \ldots c^z f(mx + ny + \ldots pz),$$

$mx + ny + \ldots pz \gtreqless h$ étant l'inégalité qui détermine l'étendue des sommations.

Je suppose que $m, n, \ldots, p$ soient des nombres entiers. En faisant

$$mx = x', \quad ny = y', \text{ etc.}; \quad a^{\frac{1}{2n}} = a', \quad b^{\frac{1}{n}} = b', \text{ etc.},$$

la formule (21) deviendra

$$\Sigma_0 \Sigma_0 \ldots \Sigma_0 a'^{x'} b'^{y'} \ldots c'^{z'} f(x' + y' + \ldots z').$$

Nous pouvons donc admettre que $m, n, \ldots, p$ soient égales à l'unité; le résultat général se déduira facilement de ce cas particulier.

Nous aurons d'abord

$$(22) \quad \begin{cases} \Sigma_0 \Sigma_0 \ldots \Sigma_0 a^x b^y \ldots c^z f(x + y + \ldots z) \\ = \dfrac{1}{\pi} \Sigma^h fu \displaystyle\int_0^\pi d\alpha\, \Sigma_0^\infty \Sigma_0^\infty \ldots \Sigma_0^\infty a^x b^y \ldots c^z \cos\alpha(x+y+\ldots z-u). \end{cases}$$

A présent, puisque

$$\Sigma_0^\infty a^x \cos\alpha x = \frac{1 - a\cos\alpha}{1 - 2a\cos\alpha + a^2},$$

$$\Sigma_0^\infty a^x \sin\alpha x = \frac{a\sin\alpha}{1 - 2a\cos\alpha + a^2},$$

nous pouvons effectuer les sommations indiquées dans le second membre de l'équation (22).

En effet, on verra, avec un peu d'attention, que nous aurons

$$(23) \quad \Sigma_0^\infty \Sigma_0^\infty \ldots \Sigma_0^\infty a^x b^y \ldots c^z \cos\alpha\,(x + y + \ldots z - u) = \frac{\mathrm{N}}{\mathrm{D}},$$

D étant égal à

$$(1 - 2a\cos\alpha + a^2) \ldots (1 - 2c\cos\alpha + c^2),$$

tandis que N est égal à

$$\cos\alpha u - \mathrm{S}a \,.\, \cos\alpha\,(u + 1) + \mathrm{S}\,ab \,.\, \cos\alpha\,(u + 2) - \ldots$$
$$\pm\, ab \ldots c \cos\alpha\,(u + \nu),$$

le signe de sommation S ayant rapport aux ν quantités $a, b, \ldots, c$.

Afin de donner à $\frac{1}{\mathrm{D}}$ une forme plus commode, posons l'équation

$$\frac{1}{\mathrm{D}} = \frac{\mathrm{A}}{1 - 2a\cos\alpha + a^2} + \ldots + \frac{\mathrm{C}}{1 - 2c\cos\alpha + c^2};$$

donc nous aurons

$$\mathrm{A} = \frac{a^{\nu-1}}{(a-b)\ldots(a-c)} \cdot \frac{1}{(1-ab)\ldots(1-ac)},$$

et ainsi de suite pour B, ..., C.

Or, nous savons que

$$\frac{1}{1 - 2a\cos\alpha + a^2} = \frac{1}{1-a^2}(1 + 2a\cos\alpha + \text{etc.}),$$

et de là il est visible que le terme de $\frac{\mathrm{NA}}{1 - 2a\cos\alpha + a^2}$, qui ne renferme pas α, est égal à

$$\frac{a^{\nu-1}}{(a-b)\ldots(a-c)}\,\frac{1}{(1-a^2)(1-ab)\ldots(1-ac)}(a^u - a^{u+1}\mathrm{S}a + a^{u+2}\mathrm{S}ab - \text{etc.}),$$

ou à

$$\frac{a^{u+\nu-1}}{(a-b)\ldots(a-c)},$$

puisque

$$\frac{1 - a\,\mathrm{S}a + a^2\,\mathrm{S}ab - \text{etc.}}{(1-a^2)(1-ab)\ldots(1-ac)} = 1.$$

Nous voyons donc que, pour toutes les valeurs positives de u et pour $u = 0$,

$$(24)\quad \left\{\begin{aligned} &\frac{1}{\pi}\int_0^\pi d\alpha\,\Sigma_0^\infty\,\Sigma_0^\infty\ldots\Sigma_0^\infty\,a^x b^y\ldots c^z\cos\alpha\,(x+y+\ldots z-u) \\ &= \frac{a^{u+\nu-1}}{(a-b)\ldots(a-c)} + \frac{b^{u+\nu-1}}{(b-a)\ldots(b-c)} + \text{etc.} \end{aligned}\right.$$

Si u est négatif, faisons $u = -u'$, nous aurons

$$\mathrm{N} = \cos\alpha u' - \mathrm{S}a \,.\, \cos\alpha\,(u'-1) + \text{etc.} \pm ab\ldots c\cos\alpha\,(u'-\nu),$$

et le terme de $\frac{\mathrm{NA}}{1-2a\cos\alpha+a^2}$, qui ne renfermera pas α, sera égal à

$$\frac{a^{u'+\nu-1}}{(a-b)\ldots(a-c)}\,\frac{1-a^{-1}\,\mathrm{S}a+a^{-2}\,\mathrm{S}ab-\text{etc.}}{(1-a^2)(1-ab)\ldots(1-ac)}.$$

En supposant que u' ne soit pas moindre que ν, il est visible que

$$1-a^{-1}\mathrm{S}a+a^{-2}\,\mathrm{S}ab-\text{etc.}=0.$$

Si u' est moindre que ν, alors la formule précédente n'est pas égale à zéro, mais l'ensemble des termes semblables tels que

$$\frac{b^{u+\nu-1}}{(b-a)\ldots(b-c)}\,\frac{1-b^{-1}\,\mathrm{S}a+\text{etc.}}{(1-b^2)(1-ba)\ldots(1-bc)}$$

disparaîtra ; c'est ce que nous allons démontrer.

Désignons par $\mathrm{S}_{u'}$ l'ensemble de toutes les combinaisons qu'on peut faire avec u' des ν quantités, $a, b, \ldots, c$. Si u' est moindre que ν, le terme de $\frac{\mathrm{NA}}{1-2a\cos\alpha+a^2}$, qui ne renfermera pas α, sera égal à

$$(25)\qquad \frac{\mathrm{A}}{1-a^2}\left(a^{u'}-a^{u'-1}\,\mathrm{S}_1+\ldots\pm\mathrm{S}_{u'}\mp a\mathrm{S}_{u'+1}\pm\ldots\pm a^{\nu-u'}\mathrm{S}_\nu\right);$$

or

$$a^{u'}-a^{u'-1}\,\mathrm{S}_1+\ldots\pm\mathrm{S}_{u'}\mp a^{-1}\,\mathrm{S}_{u'+1}\pm\ldots\pm a^{-\nu+u'}\,\mathrm{S}_\nu=0.$$

Il suit de là que la formule (25) est égale à

$$\mp\frac{\mathrm{A}}{1-a^2}\{\mathrm{S}_{u'+1}(a-a^{-1})-\mathrm{S}_{u'+2}(a^2-a^{-2})+\ldots\pm\mathrm{S}_\nu(a^{\nu-u'}-a^{-\nu+u'})\},$$

et, en ajoutant toutes les formules semblables, nous aurons

$$\mp\,\mathrm{S}_{u'+1}\left\{\frac{\mathrm{A}}{1-a^2}(a-a^{-1})+\frac{\mathrm{B}}{1-b^2}(b-b^{-1})+\ldots\right\},$$

$$\pm\,\mathrm{S}_{u'+2}\left\{\frac{\mathrm{A}}{1-a^2}(a^2-a^{-2})+\frac{\mathrm{B}}{1-b^2}(b^2-b^{-2})+\ldots\right\},$$

$\mp$ etc. ....

$$\pm\,\mathrm{S}_\nu\left\{\frac{\mathrm{A}}{1-a^2}(a^{\nu-u'}-a^{-\nu+u'})+\frac{\mathrm{B}}{1-b^2}(b^{\nu-u'}-b^{-\nu+u'})+\ldots\right\}.$$

Or, je dis que chacune de ces quantités est séparément égale à zéro. En remplaçant $2 \cos \alpha$ par $z + z^{-1}$, nous aurons

$$\frac{z^{\nu}}{(1-az)\ldots(1-cz)(z-a)\ldots(z-c)}$$
$$= \frac{Az}{1-a^2}\left(\frac{a}{1-az} - \frac{a^{-1}}{1-a^{-1}z}\right) + \text{etc.}$$

Si nous développons les deux membres de cette équation selon les puissances positives de z, la puissance la moins élevée qui se trouvera dans le premier membre sera z^{ν}. Par conséquent, nous voyons que

$$\frac{A}{1-a^2}(a^p - a^{-p}) + \frac{B}{1-b^2}(b^p - b^{-p}) + \text{etc.} = 0,$$

pour toutes les valeurs entières et positives de p qui sont moindres que ν, ce qu'il fallait démontrer.

Donc, finalement, pour toutes les valeurs négatives de u,

$$(26) \qquad \int_0^{\pi} \Sigma_0^{\infty} \Sigma_0^{\infty} \ldots \Sigma_0^{\infty} a^x b^y \ldots c^z \cos \alpha (x + y + \ldots z - u) = 0.$$

A présent, il est facile de voir que

$$(27) \qquad \begin{cases} \Sigma_0 \Sigma_0 \ldots \Sigma_0 a^x b^y \ldots c^z f(x + y + \ldots z) \\ = \Sigma_0^h fu \left\{\frac{a^{u+\nu-1}}{(a-b)\ldots(a-c)} + \ldots + \frac{c^{u+\nu-1}}{(c-a)\ldots(c-b)}\right\}, \end{cases}$$

ce qui est le résultat que nous cherchions.

J'ai démontré, dans le *Journal de Mathématiques* de Cambridge, une équation que je vais reproduire ici, afin qu'on puisse la comparer avec l'équation dernière. Les limites étant déterminées par l'inégalité

$$x + y + \ldots z \leqq h,$$

nous aurons

$$\int_0 dx \int_0 dy \ldots \int_0 dz \, e^{-ax-by-\ldots cz} f(x + y + \ldots z)$$
$$= \int_0^h fu \, du \left\{\frac{e^{-au}}{(b-a)\ldots(c-a)} + \ldots + \frac{e^{-cu}}{(a-c)(b-c)\ldots}\right\}.$$

Les résultats que nous avons obtenus sont, ce me semble, d'un genre nouveau : c'est pourquoi je pense qu'ils pourront peut-être intéresser les géomètres.

REPORT ON THE RECENT PROGRESS OF ANALYSIS (THEORY OF THE COMPARISON OF TRANSCENDENTALS)*.

1. THE province of analysis, to which the theory of elliptic functions belongs, has within the last twenty years assumed a new aspect. A great deal has doubtless been effected in other subjects, but in no other I think has our knowledge advanced so far beyond the limits to which it was not long since confined.

This circumstance would give a particular interest to a history of the recent progress of the subject, even did it now appear to have reached its full development. But on the contrary, there is now more hope of further progress than at the commencement of the period of which I have been speaking. When, in 1827, Legendre produced the first two volumes of his 'Théorie des Fonctions Elliptiques,' he had been engaged on the subject for about forty years; he had reduced it to a systematic form; and had with great labour constructed tables to facilitate numerical applications of his results. But little more, as it seemed, was yet to be done; nor does the remark of Bacon, that knowledge, after it has been systematized, is less likely to increase than before, seem less applicable to mathematical than to natural science. Nevertheless, almost immediately after the publication of Legendre's work, the earlier researches of Abel and Jacobi became known, and it was at once seen that what had been already accomplished formed but a part, and not a large one, of the whole subject.

To say this is not to derogate from the merit of Legendre. He created the theory of elliptic functions; and it is impossible

* *Report of the Sixteenth Meeting of the British Association;* held at Southampton, in September, 1846.

not to admire the perseverance with which he devoted himself to it. The attention of mathematicians was given to other things, and though the practical importance of his labours was probably acknowledged, yet scarcely any one seems to have entered on similar researches.* This kind of indifference was doubtless discouraging, but not long before his death he had the satisfaction of knowing that there were some by whom that which he had done would not willingly be let die.

The considerations here suggested have led me to select the theory of the integrals of algebraical functions as the subject of the report which I have the honour to lay before the Association.

2. The theory of the comparison of transcendental functions appears to have originated with Fagnani. In 1714, he proposed, in the 'Giornale de Litterati d'Italia,' the following problem: To assign an arc of the parabola whose equation is

$$y = x^4,$$

such that its difference from a given arc shall be rectifiable.

Of this problem he gave a solution in the twentieth volume of the same journal.

The principle of the solution consists in the transformation of a certain differential expression by means of an algebraical and rational assumption which introduces a new variable. The transformed expression is of the same form as the original one, but is affected with a negative sign. By integrating both we are enabled to compare two integrals, neither of which can be assigned in a finite form. It is difficult, however, to perceive how Fagnani was led to make the assumption in question: a remark which applies more or less to his subsequent researches on similar subjects.

The theorem which has made his name familiar to all mathematicians, appeared in the twenty-sixth volume of the 'Giornale.' In its application to the comparison of hyperbolic arcs we find some indications of a more general method. We have here a symmetrical relation between two variables, x and

* Those of M. Gauss, which would doubtless have been exceedingly valuable, have not, I believe, been published. They are mentioned in a letter from M. Crelle to Abel. Vide the introduction to the collected works of the latter, p. vii.

z, such that the differential expression $f(x)\,dx$ may be written in the form $z\,dx$. It follows at once that $f(z)\,dz = x\,dz$, and consequently that

$$\int f(x)\,dx + \int f(z)\,dz = \int \{x\,dz + z\,dx\} = xz + C.$$

The remarkable manner in which the idea of symmetry here presents itself, suggested to Mr Fox Talbot his 'Researches in the Integral Calculus.'

In applying his methods to the division of the arc of the lemniscate, Fagnani obtained some very curious results, and has accordingly taken for the vignette of his collected works a figure of this curve with the singular motto, 'Deo veritatis gloria.'

3. In MacLaurin's Fluxions, and in the writings of D'Alembert, instances are to be found where the solution of a problem is made to depend on the rectification of elliptic arcs, or, as we should now express it, is reduced to elliptic integrals. But of these instances Legendre has remarked that they are isolated results, and form no connected theory. MacLaurin is charged, in a letter appended to the works of Fagnani, with taking from the latter without acknowledgement, a portion of his discoveries with respect to the lemniscate and the elastic curve.

4. In 1761, Euler, in the 'Novi Commentarii Petropolitani' for 1758 and 1759, published his memorable discovery of the algebraical integral of the equation

$$\frac{m\,dx}{(A + Bx + Cx^2 + Dx^3 + Ex^4)^{\frac{1}{2}}} = \frac{n\,dy}{(A + By + Cy^2 + Dy^3 + Ey^4)^{\frac{1}{2}}},$$

m and n being any rational numbers.

He says he had been led to this result by no regular method, 'sed id potius tentando, vel divinando elicui,' and recommends the discovery of a direct method to the attention of analysts. In effect his investigations resemble those of Fagnani: he begins by assuming a symmetrical algebraical relation between the variables, and hence finds a differential equation which it satisfies. In this differential equation the variables are separated, so that each term may be considered as the differential of some function. With one form of assumed relation we are led to the differentials of circular, and with another to those of elliptic integrals, and so on. It is in this manner that Dr Gudermann, in the elaborate researches which he has published in Crelle's Journal, has commenced the discussion of the theory of elliptic functions.

5. In the fourth volume of the Turin Memoirs, Lagrange accomplished the solution of the problem suggested by Euler. He integrated the general differential equation already mentioned by a most ingenious method, which, with certain modifications, has remained ever since an essential element of the theory of elliptic functions. He proceeded to consider the more general equation

$$\frac{dx}{\sqrt{X}}=\frac{dy}{\sqrt{Y}},$$

where X and Y are any similar functions of x and y respectively, and came to the conclusion, that if they are rational and integral functions, the equation cannot, except in particular cases, be integrated, if they contain higher powers than the fourth. He also integrated this equation in a case in which X and Y involve circular functions of the variables. It had been already pointed out in the summary of Euler's researches, given in the *Nov. Com. Pet.* Tom. VI., that if X and Y are polynomials of the sixth degree, the last-written equation does not in general admit of an algebraical integral, since, if so, it would follow that the solution of the equation $\frac{dx}{1+x^3}=\frac{dy}{1+y^3}$, which (as the square of $1+x^3$ is a polynomial of the sixth degree) is a particular case of that which we are considering, could be reduced to an algebraical form. Now this solution involves both circular functions and logarithms, and therefore the required reduction is impossible. This acute remark* showed that Euler's result did not admit of generalisation in the manner in which it was natural to attempt to generalise it. It was reserved for Abel to discover the direction in which generalisation is possible.

6. The discovery of Euler, of which we have been speaking, is in effect the foundation of the theory of elliptic func-

* M. Richelot, in one of his memoirs on Abelian or hyper-elliptic integrals, quotes it, in a slightly modified form, from Euler's *Opuscula*.

tions, as the generalisation of it by Abel, or more properly speaking, the theory of which Euler's result is an isolated fragment, is the foundation of our knowledge of the higher transcendents. We may therefore conveniently divide the subject of this report into two portions, viz. the general theory of the comparison of algebraical integrals, and the investigations which are founded on it. Mathematicians have been led, by comparing different transcendents, to introduce new functions into analysis, and the theory of these functions has become an important subject of research.

The second portion may again be divided into two, viz. the theory of elliptic functions, and that of the higher transcendents.

This classification, though not perhaps unexceptional, will, I think, be found convenient.

7. About sixteen years after the publication of Lagrange's earlier researches on the comparison of algebraical integrals, he gave, in the New Turin Memoirs for 1784 and 1785, a method of approximating to the value of any integral of the form $\int \frac{Pdx}{R}$, where P is a rational function of x and R the square root of a polynomial of the fourth degree. I shall consider this important contribution to the theory of elliptic functions in connexion with the writings of Legendre. At present, in order to give a connected view of the first division of my subject, it will be necessary to go on at once to the works of Abel, and to those of subsequent writers. In the history of any branch of science the chronological order must be subordinate to that which is founded on the natural connexion of different parts of the subject.

I shall merely mention in passing, that in 1775, Landen published in the Philosophical Transactions a very remarkable theorem with respect to the arcs of a hyperbola. He showed that any arc of a hyperbola is equal to the difference of two elliptic arcs together with an algebraical quantity. In 1780 he published his researches on this subject in the first volume of his *Mathematical Memoirs*, p. 23. This theorem, as Legendre has remarked, might have led him to more important results. It contains the germ of the general theory of transformation, the eccentricities of the two ellipses being connected by the modular equation of transformations of the second order*. It is on this account that in a report on M. Jacobi's *Fundamenta Nova*, contained in the tenth volume of the Memoirs of the Institute, Poisson speaks of Landen's theorem as the first step made in the comparison of dissimilar elliptic integrals. Several writers have accordingly given Landen's name to the transformation commonly known as Lagrange's.

8. We have seen that even Lagrange failed in obtaining a result more general than that which had been made known by Euler, and yet, as we now know, Euler's theorem is but a particular case of a far more general proposition. But in order to further progress, it was necessary to introduce a wholly new idea. The resources of the integral calculus were apparently exhausted; Abel, however, was enabled to pass on into new fields of research, by bringing it into intimate connexion with another branch of analysis, namely, the theory of equations. The manner in which this was done shews that he was not unworthy to follow in the path of Euler and of Lagrange.

I shall attempt to state in a few words the fundamental idea of Abel's method.

Let us suppose that the variable x is a root of the algebraical equation $fx = 0$, and that the coefficients of this equation are rational functions of certain quantities $a, b, \dots c$, which we shall henceforth consider independent variables. Let us suppose also that in virtue of this equation we can express certain irrational functions† of x as rational functions of $x, a, b, \dots c$. For instance, if the equation were $x^2 + ax + \frac{1}{2}(a^2 - 1) = 0$, it follows that $\sqrt{1 - x^2} = a + x$. So that any irrational function of the form $F(x\sqrt{1 - x^2})$ can be expressed rationally (F being rational) in x and a.

* Vide *infra*, pp. 263 and 289.

† It must be remembered that an algebraical function is either explicit or implicit: explicit, when it can be expressed by a combination of ordinary algebraical symbols; implicit, when we can only define it by saying that it is a root of an algebraical equation whose co-efficients are integral functions of x. Thus y is an implicit function of x if $y^5 + xy + 1 = 0$. The remarks in the text apply to all algebraical functions, explicit or implicit.

From the given equation we deduce by differentiation the following,

$$dx = \alpha da + \beta db + \dots + \gamma dc,$$

where $\alpha, \beta, \dots \gamma$ are rational in $x, a, b, \dots, c$.

Let y be one of the functions which can be expressed rationally in x, &c., it follows that

$$ydx = Ada + Bdb + \dots + Cdc,$$

where $A, B, \dots C$ are also rational in x, &c.

The equation $fx = 0$ will have a number of roots, which we shall call $x_1, x_2, \dots x_\mu$. It follows that

$$y_1 dx_1 + \dots + y_\mu dx_\mu = (A_1 + \dots + A_\mu)\, da$$
$$+ (B_1 + \dots + B_\mu)\, db + \dots + (C_1 + \dots + C_\mu)\, dc,$$

where the indices affixed to y, A, &c. correspond to those affixed to x, so that y_1, for instance, is the same function of x_1 that y_2 is of x_2.

Now $A_1 + \dots + A_\mu$ is rational and symmetrical with respect to $x_1 \dots x_\mu$, therefore it can be expressed rationally in the coefficients of $f(x) = 0$, and therefore in $a, b \dots c$. We will call this sum R_a, and thus with a similar notation for b, &c. we get

$$y_1 dx_1 + \dots + y_\mu dx_\mu = R_a da + R_b db + \dots + R_c dc.$$

The second side of this equation is from the nature of the case a complete differential, and it is rational in a, b, c, &c.; it can therefore be integrated by known methods; and if we denote $\int^{x_1} ydx$ by $\psi(x_1)$, we get

$$\psi(x_1) + \dots + \psi(x_\mu) = M,$$

M being a logarithmic and algebraic function of a, b, &c., which we may suppose to include the constant of integration.

$\psi(x)$ is in general a transcendental function, while a, b, &c. are necessarily algebraical functions of $x_1, \dots, x_\mu$, and the result at which we have arrived is therefore an exceedingly general formula for the comparison of transcendental functions.

The simplicity and generality of these considerations entitle them to especial attention: it cannot be doubted that the application thus made of the properties of algebraical equations to the comparison of transcendents will always be a remarkable point in the history of pure analysis.

A very simple example may perhaps illustrate what has been said. Let us recur to the equation

$$x^2 + ax + \frac{1}{2}(a^2 - 1) = 0, \ldots\ldots\ldots\ldots\ldots(1),$$

and suppose that

$$y = \frac{1}{\sqrt{1 - x^2}}.$$

Differentiating the first of these equations, we find that

$$(2x + a)\,dx + (x + a)\,da = 0.$$

Comparing this with the general expression of dx, we perceive that

$$\alpha = -\frac{x + a}{2x + a}, \qquad \beta = \&c. = 0;$$

and as

$$y = \frac{1}{\sqrt{1 - x^2}} = \frac{1}{x + a} \text{ (vide } ante\text{, p. 243),*}$$

$$y\,dx = -\frac{da}{2x + a},$$

so that

$$A = -\frac{1}{2x + a}.$$

Let x_1 and x_2 be the two roots of our equation, we have thus to find the value of

$$A_1 + A_2 = -\left(\frac{1}{2x_1 + a} + \frac{1}{2x_2 + a}\right) = -\frac{2(x_1 + x_2 + a)}{(2x_1 + a)(2x_2 + a)} = 0,$$

since $$x_1 + x_2 = -a.$$

Hence $$y_1 dx_1 + y_2 dx_2 = 0,$$

and $$\psi x_1 + \psi x_2 = c.$$

Since $$x_1 + x_2 = -a,$$

and $$x_1 x_2 = \frac{1}{2}(a^2 - 1),$$

we see that $x_1^2 + x_2^2 = 1$, or $x_2 = \sqrt{1 - x_1^2}$.

* The ambiguous sign of the radical is to our purpose immaterial.

Hence, as $\psi x = \sin^{-1}x$, our result is merely this, that the sum of two arcs is constant if the sine of one is equal to the cosine of the other.

An infinity of analogous results may be obtained either by varying the form of y (*e.g.* by making $y = \sqrt{1-x^2}$), or by changing the equation (1). A formula applicable to all forms of y, and which, for each, includes all the results which can be established with respect to it, is, it will readily be acknowledged, one of the most general in the whole range of analysis. Abel's principal result is a formula of this nature; he developed at considerable length the various consequences which may be deduced from it.

Generally speaking, the number of independent variables $a, b, \dots c$ will be less than that of the different roots, $x_1 \dots x_\mu$; hence a certain number, say m, of the roots may be looked on as independent (viz. as many as there are quantities $a, b, \dots c$), and the rest will be functions of these. It may be shown that it will always be possible to make the *difference* $\mu - m$ constant, so that the sum of any number of the transcendents ψ is expressible by a fixed number of them, together with an algebraical and logarithmic function of the arguments, *i.e.* of $x_1, \dots x_m$. In the case of elliptic integrals, it had long been known that the sum of two may be thus expressed by a third; and Legendre pointed out that the sum of any number may similarly be expressed by means of one. Accordingly it appears from the general theory, that in this case $\mu - m$ may be made equal to unity.

9. The history of this important theory is curious. It was developed by Abel in an essay which he presented to the Institute in the autumn of 1826, when he had scarcely completed his twenty-fourth year.

In a letter to M. Holmboe, appended to the edition of his collected works, Abel writes, 'Je viens de finir un grand traité sur une certaine classe de fonctions transcendantes pour le présenter à l'Institut, ce qui aura lieu lundi prochain. J'ose dire sans ostentation que c'est un traité dont on sera satisfait. Je suis curieux d'entendre l'opinion de l'Institut la dessus. Je ne manquerai pas de t'en faire part.' Long before this memoir was published, Abel had become 'chill to praise or blame.' He died at Christiania in the spring of 1829.

M. Jacobi mentions in a note in Crelle's Journal, that while at Paris he represented, and as he believed not ineffectually, to Fourier, who was then one of the secretaries of the Institute, that the publication of this memoir would be very acceptable to mathematicians. A long period however was still to elapse before the publication took place. It was possibly retarded by the death of Fourier. In 1841 the memoir appeared in the seventh volume of the *Mémoires des Savans Etrangers.* It was prepared for publication by M. Libri.

Thus for about fifteen years Abel's general theory remained unpublished; but in the meanwhile Crelle's Journal was established, and to the third volume of this he contributed a paper which contains a theorem much less general than the researches he had communicated to the Institute, but far more so than anything previously effected in the theory of the comparison of transcendents. This is commonly known as Abel's Theorem. Legendre, in a letter to Abel, speaks thus of the memoir in which it appeared:—"Mais le mémoire...ayant pour titre *Remarques sur quelques propriétés générales*, &c., me parait surpasser tout ce que vous avez publié jusqu'à présent par la profondeur de l'analyse qui y règne ainsi que par la beauté et la généralité des résultats." In a previous letter, with reference I believe to the same subject, he had remarked, 'Quelle tête que celle d'un jeune Norvégien!'

Abel's theorem gives a formula for the comparison of all transcendental functions whatever whose differentials are irrational from involving the square root of a rational function of x.

In a very short paper in the fourth volume of Crelle's Journal, which must have been the last written of Abel's productions, the chief idea of his general theory is stated; and in the second volume of his collected works we find a somewhat fuller development of it, in a paper written before his visit to Paris, but not published during his life-time.

While Abel's great memoir remained unpublished at Paris, several mathematicians, developing the ideas which he had made known in his contributions to Crelle's Journal, succeeded in establishing results of a greater or less degree of generality. Researches of this kind may be presented in a variety of forms, because the algebraical function to be integrated, which we have called y, may be defined or expressed in different ways. For instance, if M and N are general symbols denoting any integral functions of x, the two suppositions $y = \frac{\sqrt{M}}{N}$ and $y = \frac{M}{\sqrt{N}}$ are precisely equivalent, since by an obvious reduction, and by changing the signification of M and N, the one may be transformed into the other; and so in more general cases. Thus the same function may assume a variety of aspects, and there will be a corresponding variety in the form of our final results.

In Crelle's Journal we find a good many essays on this part of the subject: of these I shall now mention several.

M. Broch is the author of a paper in the twentieth volume of Crelle's Journal, p. 178. It relates to the integration of certain functions irrational in consequence of involving a polynomial of any degree raised to a fractional power. For these functions he establishes formulæ of summation, which of course include Abel's theorem, since the latter relates to cases in which the fractional power in question is the $(\frac{1}{2})$th. Subsequently to the publication of this paper he presented to the Institute a memoir on the same subject, but gave to the functions to be integrated a different but not essentially more general form. This memoir, which was ordered to be printed among the *Savans Etrangers*, but which will be found in Crelle's Journal (XXIII. 145), may be divided into two portions: the first contains results analogous to Abel's theorem: the second relates to the discussion and reduction of the transcendents which they involve. In this part of his researches M. Broch has followed the method, and occasionally almost adopted the phraseology of a memoir of Abel, on the reduction and classification of Elliptic Integrals (Abel's Works, II. p. 93). MM. Liouville and Cauchy, in reporting on the memoir, conclude by remarking that the author 'n'a pas trop présumé de ses forces en se proposant de marcher sur les traces d'Abel.'

M. Jürgenson has contributed two papers to Crelle's Journal on the subject of which we are speaking. The first, which is very short, contains a general theorem for the summation of algebraical integrals* when the function to be integrated is expressed in a particular form. This paper appears in the

* I have used the expression "algebraical integrals," though perhaps not correctly, to denote the integrals of algebraical functions.

nineteenth volume, p. 113. In the second (Vol. XXIII. p. 126) the author reproduces the results he had already obtained, pointing out the equivalence of one of them to the theorem established in M. Broch's first essay. Besides this, he discusses a question connected with the reduction of algebraical integrals.

M. Ramus, in the twenty-fourth volume of Crelle's Journal, p. 69, has established two general formulæ of summation; from the second he deduces with great facility Abel's theorem, and also another result, which Abel mentions in a letter to Legendre, published in the sixth volume of Crelle's Journal, but which he left undemonstrated.

M. Rosenhain's researches (Crelle's Journal, XXVIII. p. 249, and XXIX. p. 1) embrace both the summation and reduction of algebraical integrals. His analysis depends on giving the function to be integrated a peculiar form, which he conceives leads to a simpler mode of investigation than any other.

A paper by Poisson will be found in the twelfth volume of Crelle's Journal, p. 89. It relates to the comparison of algebraical integrals, but is not I think so valuable as that great mathematician's writings generally are.

Beside the memoirs thus briefly noticed, I may mention two or three by M. Minding: that which appears in the twenty-third volume of Crelle's Journal, p. 255, is the one which is most completely developed.

There is also a very brief note by M. Jacobi in the eighth volume of Crelle's Journal.

10. To the *Philosophical Transactions* for 1836 and 1837 Mr Fox Talbot contributed two essays, entitled *Researches in the Integral Calculus*. These researches may be said to contain a development and generalisation of the methods of Fagnani. They are however far more systematic than the writings of the Italian mathematician, and if they had appeared in the last century would have placed Mr Talbot among those by whom the boundaries of mathematical science have been enlarged. But it cannot be denied that they fall far short of what had been effected at the time they were published, nor does it appear that they contain anything of importance not known before. I have assuredly no wish to speak disparagingly of

Mr Talbot; his mathematical writings bear manifest traces of the ability he has shown in so many branches of science*. But as in this country they seem to have been thought, and by men not apparently unqualified to judge, to contain great additions to our knowledge, I cannot avoid inquiring whether this be true.

Mr Talbot points out in the early part of his first paper, that if there are $n-1$ symmetrical relations among the n variables $x, y \ldots z$, then the identical equation

$$\{y \ldots z\}\, dx + (x \ldots z)\, dy + \ldots + \{xy \ldots\}\, dz = d\,\{xy \ldots z\}$$

will assume the form

$$\phi\,(x)\, dx + \phi\,(y)\, dy + \ldots + \phi\,(z)\, dz = d\,\{xy \ldots z\},$$

and thus give us

$$\int \phi\,(x)\, dx + \int \phi\,(y)\, dy + \ldots + \int \phi\,(z)\, dz = xy \ldots z + C.$$

Precisely the same remark, though expressed in a different notation, is the foundation of M. Hill's memoir, published in 1834, on what he calls 'functiones iteratæ.' It will be found in Crelle's Journal, XI. p. 193. A much more general theorem might be established by similar considerations: they are of course applicable whether the function ϕ be algebraical or transcendent.

In the course of his researches, Mr Talbot recognised the important principle, that the existence of $n-1$ symmetrical algebraical relations among n variables may be expressed by treating them as the roots of an equation, one of whose coefficients at least is variable, the others being either constant or functions of the variable one. Unfortunately he did not pass from hence to the more general view, that the existence of $n-p$ symmetrical relations may be expressed in a similar manner if we consider p of the coefficients of the equation as arbitrary quantities. Had he done so, it is possible, though not likely, that he would have rediscovered Abel's theorem; but as it is, he has never introduced, except once, and then as it were by accident, more than one arbitrary quantity. Thus only one of

* It must be remembered also that Mr Talbot admits himself to have been anticipated to a considerable extent by the publication of Abel's theorem.

his variables is independent, and consequently, in more than one instance, his results are unnecessarily restricted cases of more general theorems.

The character of his analysis will be perceived from what has been said. If $\int X dx$ be the transcendent to be considered, X being an algebraical function of x, he makes the following assumption—

$$X = f(xv),$$

v being a new variable, and f a rational function. From this assumption he deduces an algebraical equation in x, the coefficients of which are rational functions of v. This equation then is one of those of which we have spoken, by means of which the function to be integrated can be expressed in a rational form. Taking the sum with respect to the roots of this equation, we get

$$\Sigma (X dx) = \Sigma \{f(xv)\, dx\}.$$

It must be remarked that many forms might be assigned to the function f, which would give rise to a difficulty, of the means of surmounting which Mr Talbot has given no idea. If x and v are mixed up in $f(xv)$, it is manifest that we cannot integrate $f(xv)\, dx$, since v is a function of x, which if we eliminate we merely return to our function X. We must therefore express $\Sigma f(xv)\, dx$ in the form $V dv$, V being a function and, as Abel has shown, an integrable function of v. Abel has given formulæ by means of which this reduction may be effected in all possible cases. But there is nothing analogous to this in the writings of Mr Talbot, and consequently he could not, setting aside the defect already noticed, obtain results as general as many previously known. In Mr Talbot's investigations, $f(xv)\, dx$ is such that $\Sigma f(xv)\, dx$ may be put in the form—

$$V_1 \Sigma \{\phi'_1 x dx\} + V_2 \Sigma \{\phi'_2 x dx\} + \&c.,$$

$\phi_1 x$, $\phi_2 x$, &c. (of which $\phi'_1 x$, $\phi'_2 x$, &c. are the derived functions) being rational functions of x. Then $\Sigma \phi x =$ a rational function of v by a well-known theorem. Let the form of this function be ascertained, and let us denote it by χv. Then differentiating,

$$\Sigma \phi' x dx = \chi' v dv,$$

and hence

$$\Sigma X dx = \Sigma f(xv)\, dx = [V_1 \chi'_1 v + V_2 \chi'_2 v + \ldots]\, dv,$$

and the second side of this equation is of course rational and integrable. But the form of the function $f(xv)$ is unnecessarily restricted in order that this kind of reduction may be possible. Nevertheless, Mr Talbot's papers, from their fulness of illustration and the clear manner in which particular cases of the general theory are worked out by independent methods, will be found very useful in facilitating our conceptions of the branch of analysis which forms as it were the link between the theory of equations and the integral calculus.

In Mr Talbot's second memoir (*Phil. Trans.* 1837, part 2. p. 1) he has applied his method to certain geometrical theorems. Three of them relate to the ellipse, and are proved by the three following assumptions:

$$\left\{\frac{1-e^2x^2}{1-x^2}\right\}^{\frac{1}{2}} = 1 + vx, \text{ or } = \frac{1-ex}{\sqrt{v}}, \text{ or } = \frac{\sqrt{v}}{1-x}.$$

These assumptions are all cases of the following:

$$\left\{\frac{1-e^2x^2}{1-x^2}\right\}^{\frac{1}{2}} = \frac{c+c^1x}{a+a^1x},$$

where a, a^1, c, c^1 are arbitrary quantities. The results of this assumption are completely worked out by Legendre (*Théorie des Fonctions Elliptiques*, III. p. 192) in showing how the known formulæ of elliptic functions may be derived from Abel's theorem. Mr Talbot's first theorem is a case of the fundamental formula for the comparison of elliptic arcs. This remark has reference to an inquiry which Mr Talbot suggests as to the relation in which his theorems stand to the results obtained by Legendre and others.

In conclusion, it may be well to observe that Mr Talbot has remarked that, apparently, a solution discovered by Fagnani of a certain differential equation cannot be deduced from Abel's theorem; but as this solution may be easily derived from the ordinary formula for the addition of elliptic integrals of the first kind it is manifestly included in the theorem in question.

II.

11. I now come to the history of researches into the properties of particular classes of algebraical transcendents. The earliest, and still perhaps the most important class of these

researches relates to the transcendents which are commonly called elliptic functions or elliptic integrals. For a reason which will be mentioned hereafter the latter name seems preferable, and it is sanctioned by the authority of M. Jacobi, though the former was used by Legendre. Elliptic integrals then may be defined as those whose differentials are irrational in consequence of involving a radical of the form

$$\sqrt{\{\alpha + \beta x + \gamma x^2 + \delta x^3 + \epsilon x^4\}}.$$

But it may perhaps be more correct to say that all such integrals may be reduced to three standard integrals, to which the name of elliptic integrals has been given.

In the Turin Memoirs for 1784 and 1785, p. 218, Lagrange considered, as has been already mentioned, the theory of these transcendents. He showed that the integration of every function irrational in consequence of containing a square root may be made to depend on that of a function of the form $\frac{P}{R}$, P being rational, and R the radical in question; and that if under the sign of the square root x does not rise above the fourth degree, it may ultimately be made to depend on that of

$$\frac{N dx}{\sqrt{(1 \pm p^2x^2)\,(1 \pm q^2x^2)}},$$

where N is rational in x^2. He thus laid the foundation of that part of the theory of elliptic transcendents in which a proposed integral is reduced to certain canonical or standard forms, or to the simplest combination of such forms of which the case admits. In Legendre's earliest writings on elliptic functions there is nothing relating to this part of the subject. Having thus, in the simple manner which distinguishes his analysis, reduced the general case to that which admits of the application of his method, Lagrange proceeded to prove that if we introduce a new variable whose ratio to x is the subduplicate of the ratio of $1 \pm p^2x^2$ to $1 \pm q^2x^2$, the last written integral is made to depend on another of similar form, but in which p and q are replaced by new quantities p^1 and q^1. If p is greater than q, p^1 will be greater than p, and q^1 less than q, and thus by successive similar transformations we ultimately come to an integral in which q is so small that the factor $1 \pm q^1x^2$ may be replaced by unity, and the elliptic integral is therefore reduced to a circular or logarithmic form. Or by successive transformations in the opposite direction we come to an integral in which p^1 and q^1 are sensibly equal, in which case also the elliptic integral is reduced to a lower transcendent. This most ingenious method is the foundation of all that has since been effected in the transformation of elliptic integrals, or at least whatever has been done has been suggested by it. Thus it is to Lagrange that we owe the origin of two great divisions of the theory of these functions.

In the Memoirs of the French Academy for 1786, p. 616, we find Legendre's first essay on the subject to which he afterwards gave so much attention. We recognise in it what may I think be considered the principal aim of his researches in elliptic functions, namely to facilitate, by the tabulation of these functions, the numerical solution of mathematical and physical problems.

He begins, not with a general form as Lagrange had done, but with the integral $\int\sqrt{1-c^2\sin^2\phi}\,d\phi$, which as we know represents an elliptic arc, and shows how other functions, for instance the value of the hyperbolic arc, may be expressed by means of it, and of its differential coefficient with respect to the eccentricity c. The memoir does not contain much that is now of interest. After writing it he became aware of the existence of Landen's researches; and in a second memoir appended to the first gave a demonstration of Landen's principal theorem. This demonstration is founded on Legendre's own methods, and he deduces from it the remarkable conclusion, that if of a series of ellipses, whose eccentricities are connected by a certain law, we could rectify any two, we could deduce from hence the rectification of all the rest. The law connecting the eccentricities of the ellipses is that which would be obtained by making use of Lagrange's method of transformation, with which accordingly this result is closely allied.

Legendre's next work was an essay on transcendents*, presented to the Academy in 1792 and published separately the year after. It contains the same general view as that which is

* A translation of it appeared in Leybourne's *Mathematical Repository*, Vols. II. and III. The original I have not seen—it has long been scarce.

developed in the first volume of the *Exercises de Calcul Intégral*, which appeared in 1811.

12. The theory of elliptic functions, as it is presented to us by Legendre, may conveniently be considered under the following heads:

α. The reduction of the general integral,

$$\int \frac{P dx}{\sqrt{\alpha + \beta x + \gamma x^2 + \delta x^3 + \epsilon x^4}},$$

in which P is rational to three standard forms, since known as elliptic integrals of the first, second and third kinds *.

This classification, though the reduction of the general integral had, as we have seen, been already considered by Lagrange, is I believe entirely due to Legendre. If we consider how much it has facilitated all subsequent researches, we can hardly over-rate the importance of the step thus made. It may almost be said that Legendre, in thus showing us the primary forms with which the theory of elliptic integrals is conversant, created a new province of analysis: he certainly gave unity and a definite form to the whole subject.

For the three species of functions thus recognised Legendre suggested the names of *nome*, *epinome* and *paranome*, the name of the first being derived from the idea that it involves, so to speak, the law on which the comparison of elliptic integrals depends. But these names do not seem felicitous, nor have they, I believe, been adopted. To this part of the subject an important theorem relating to the reduction of elliptic integrals of the third kind, whose parameters are imaginary, seems naturally to belong.

* These three forms are

$$\int_0^x \frac{dx}{\sqrt{(1-x^2)(1-c^2x^2)}}; \quad \int_0^x \sqrt{\frac{1-c^2x^2}{1-x^2}}\, dx; \quad \int_0^x \frac{dx}{(1+nx^2)\sqrt{(1-x^2)(1-c^2x^2)}}.$$

Legendre always replaces x by $\sin \phi$, so that the integrals become

$$\int_0^\phi \frac{d\phi}{\sqrt{1-c^2\sin^2\phi}}; \quad \int_0^\phi \sqrt{1-c^2\sin^2\phi}\, d\phi; \quad \int_0^\phi \frac{d\phi}{(1+n\sin^2\phi)\sqrt{1-c^2\sin^2\phi}}.$$

The radical $\sqrt{1-c^2\sin^2\phi}$ is often denoted by Δ.

The constant c is called the *modulus;* the second constant n (in the third kind) is called the *parameter*. The modulus may always be supposed less than unity, and if $c = \sin \epsilon$, then ϵ is the *angle of the modulus*.

β. The comparison of elliptic integrals of the same form differing only in the value of the variable, or as it is often called, the amplitude of each. This part of the subject divides itself into three heads, corresponding to the three classes of integrals. The fundamental results are to be found in the memoirs of Euler, of which we have already spoken. By Legendre however they were more fully developed.

It is interesting to observe that Legendre suggested that the discovery of Euler (namely that the differential equation

$$\frac{dx}{\sqrt{f(x)}} + \frac{dy}{\sqrt{f(y)}} = 0$$

admits an algebraical integral, $f(x)$ being the polynomial

$$\alpha + \beta x + \gamma x^2 + \delta x^3 + \epsilon x^4)$$

might be generalised, if we consider the differential equation

$$\frac{dx}{\sqrt{f(x)}} + \frac{dy}{\sqrt{f(y)}} + \dots + \frac{dz}{\sqrt{f(z)}} = 0.$$

He remarks that this is perhaps the only way in which it can be generalised.

γ. Theorems relating to the comparison of different kinds of elliptic functions. One of the most remarkable of these is the relation between the complete integrals (those, namely, in which the variable x is unity) of the first and second kind, the moduli of which are complementary; that is, the sum of the squares of whose moduli is equal to unity. Legendre's demonstration of it is rather indirect, but many others have been since given. Another theorem may be mentioned,—that the complete integral of the third kind can always be expressed by means of the complete integrals of the first and second. A third and most important result shows that in elliptic integrals of the third kind we may distinguish two separate species, and that to one or other of these any such integral may be reduced. A memorable discovery of M. Jacobi has greatly increased the importance of this subdivision, of which we shall hereafter speak more fully. This part of the subject is, I imagine, entirely due to Legendre.

δ. The evaluation of elliptic integrals by means of expansions.

ϵ. The method of successive transformations. The idea of this method originated, as we have seen, with Lagrange. It is developed at great length by Legendre, with a special reference to the modifications required in applying it to the different species of integrals. As Lagrange had shown, the series of transformed integrals extending indefinitely both ways conducts us, in whichever direction we follow it, towards a transcendent of a lower kind than an elliptic integral, or in other words, towards a logarithmic or circular integral. There are thus two modes of approximation, one of which depends on a series of integrals with increasing moduli, and the other on a series whose moduli decrease. Thus for the three species of integrals there will be in all six approximative processes to be considered. In the case of the elliptic integral of the third kind, we have to determine the law of formation of the successive parameters n, n^1, &c.

ζ. Reductions of transcendents not contained in the general formula $\left(e.g. \int \frac{dx}{\sqrt{1-x^6}}\right)$ to elliptic integrals.

η. Lastly, applications to various mechanical and geometrical problems.

This analysis, however slight, will give an idea of the contents of that part of the *Exercices de Calcul Intégral* which relates to elliptic functions. In the third volume there are tables for facilitating the calculation of integrals of the first and second kind: they are accompanied with an explanation of the manner in which they were constructed. The ninth table is one with double entry, the two arguments being the angle of the modulus and the amplitude.

13. In 1825 Legendre presented to the Académie des Sciences the first volume of his *Traité des Fonctions Elliptiques.* A great part of this work is precisely the same as the *Exercices de Calcul Intégral.* By far the most important addition to the theory of elliptic functions consists in the discovery of a new system of successive transformations quite distinct from that of Lagrange.

In the earlier work Legendre had shown that a certain transcendent might be expressed in two ways by means of elliptic integrals of the first kind. Comparing the two results, he obtained a very simple relation between the two elliptic integrals. Their moduli are complementary; while the ratio of the Δ's in the two integrals can be expressed rationally in terms of the sine of the amplitude of one. This circumstance seems to have suggested to Legendre the possibility of generalising the result. He accordingly assumed a relation between the amplitudes of two integrals, of which the equation subsisting in the theorem of which we have been speaking is a particular case; and showed from hence that a simple relation perfectly similar to that which he had obtained in the particular instance existed between the two integrals, viz. that they bore to each other a ratio independent of their amplitudes. Their moduli are connected by an algebraical equation, but are not complementary. This circumstance therefore now appeared to be unessential, though in the *Exercices* the investigation is introduced for the sake of exhibiting a case in which an integral may be transformed into another with a complementary modulus.

Legendre thus obtained a new kind of transformation, which might be repeated any number of times or combined in an infinite variety of ways with that of Lagrange. To illustrate this he constructed a kind of table—a 'damier analytique.' In the central cell is placed the original modulus c. All the moduli contained in the same horizontal row are derivable from one another by Lagrange's scale of moduli; those in each vertical row by the newly discovered scale. He seems to have been very much struck by the infinite variety of transformations of which elliptic integrals admit. The integral of the first kind is especially remarkable, because of the simplicity of the relation which connects it with any of its transformations, viz. that their ratio is independent of the amplitudes.

Legendre's second work was, as we have remarked, presented to the Academy in 1825, but it was not published till 1827. In the summer of 1827 M. Jacobi announced in Schumacher's *Astronomischen Nachrichten*, No. 123, that he was in possession of a general method of transformation for elliptic integrals of the first kind. He was not acquainted with Legendre's discovery of a new scale, and as an illustration of the general theorem gave two cases of it, the first being equivalent to Legendre's method of transformation. Thus much was announced in a letter to M. Schumacher, dated June 13th;

but in one of a later date (August 2nd) he gave a formal enunciation of his theorem, but without demonstration. The two communications appear consecutively (*Ast. Nach.* VI. p. 33).

In No. 127 of the *Nachrichten*, VI. p. 133, M. Jacobi gave a demonstration of his theorem.

If we can so determine y in the terms of x as to satisfy the differential equation

$$\frac{dy}{\sqrt{(1-y^2)(1-\lambda^2 y^2)}} = \frac{1}{M}\frac{dx}{\sqrt{(1-x^2)(1-k^2x^2)}}$$

(M being constant), it is manifest that we shall have (F denoting the elliptic integral of the first kind) $F(kx) = MF(\lambda y)$, provided that y and x vanish together. The question therefore is, how may the differential equation be satisfied, for it is clear that by means of a solution of it we *transform* the elliptic integral $F(kx)$ into another, viz. into $F(\lambda y)$.

M. Jacobi shows that if y be equal to $\frac{U}{V}$, U and V being integral functions of x, the differential equation will be satisfied, provided U and V fulfil two general conditions, the second of which is found to be deducible from the first. He then makes an assumption which is equivalent to assigning particular forms to U and V, and thence shows, by a most ingenious method, that these forms of U and V are such as to fulfil the first of the required conditions, which, as has been said, implies the other. He thus verifies, *à posteriori*, the assumed value of the function y.

In proving that the forms assigned for U and V have the required property, it is necessary to pass from an expression of the value of $1-y$ in terms of x to one of $1-\lambda y$ in terms of the same quantity. This is done by means of a remarkable property of the functions U and V, namely, that if in both x be replaced by $\frac{1}{kx}$, $\frac{U}{V}$ or y will (the constants being properly adjusted) become $\frac{V}{\lambda U}$ or $\frac{1}{\lambda y}$. Therefore, in any form in which the relation connecting y and x can be put, we may replace x by $\frac{1}{kx}$, provided we at the same time replace y by $\frac{1}{\lambda y}$. This has been called the principle of double substitution, and by means of it we pass from the expression of $1-y$ to that of $1-\frac{1}{\lambda y}$, and thence obtain that of $1-\lambda y$. It is to be observed that this principle is used merely to prove a certain property of the functions U and V. Of course, as the change of x into $\frac{1}{kx}$ implies that of y into $\frac{1}{\lambda y}$ in the finite relation between these quantities, the same thing will be true in the differential equation by which they are connected, a remark which may very easily be verified. But, on the other hand, it by no means follows that because it is true in the differential equation therefore any assumed finite relation between y and x having this property is the integral required. The property in question therefore does not enable us to *verify* any assumed value of y.

This remark has reference to a communication from Legendre which appears in No. 130 of Schumacher's *Nachrichten*, VI. p. 201. In it he gives an account of M. Jacobi's researches, and an outline of the demonstration of which we have been speaking. I find it impossible to avoid the conclusion that this great mathematician mistook the character of the demonstration in question, and that to him it appeared to be in effect a mere verification of the assumed value of y by means of the principle of double substitution. He remarks that the direct substitution of the value of y in the differential equation is impracticable, but that M. Jacobi had avoided this substitution by means of 'une propriété particulière de cette équation qui doit être commune aux intégrales qui la représentent.' This property is the principle of double substitution; and after showing that it is true of the differential equation, the writer proceeds thus: 'Ce principe une fois posé, rien n'est plus facile que de vérifier l'équation trouvée $y=\frac{U}{V}$, car par la double substitution on obtient la même valeur de y à un coefficient près qui doit être égal à l'unité;' and, after a remark to our present purpose immaterial, concludes, 'Ainsi se trouve démontrée généralement l'équation $y=\frac{U}{V}$ ainsi que, etc.'

As we have seen, such a verification would be wholly inconclusive, nor is the essential point of M. Jacobi's reasoning, namely, that the assumed forms of U and V satisfy the general condition, laid down at the outset of his demonstration, here adverted to.

In 1828, Legendre published the first supplement to the *Traité des Fonctions Elliptiques*, &c. It contains an account of the researches of M. Jacobi, and of a memoir by Abel inserted in the third volume of Crelle's Journal. The account here given of M. Jacobi's demonstration is fuller and more explicit than that already noticed. It leaves, I think, no doubt of the error into which Legendre had fallen. No notice whatever is taken of the first part of M. Jacobi's reasoning: and after remarking that the differential equation is satisfied when the double substitution is made, he goes on, 'Tout se reduit donc à faire cette double substitution dans l'intégrale $y = \frac{x}{\mu}\frac{U}{V}$ et à examiner si elle est satisfaite.' After showing that it is so, he adds, 'Par ce procédé très simple il est constaté que l'équation $y = \frac{x}{\mu}\frac{U}{V}$ satisfait ... à l'équation différentielle dont l'intégrale est $F(k\phi) = \mu F(h\psi)$, etc.' (*Trait. des Fonct. Ell.* III. p. 10.)

Legendre remarks, that although M. Jacobi's demonstration rests on 'un principe incontestable et très ingénieux,' it is still desirable to have another verification of so important a theorem. He accordingly gives an original demonstration of it, which is however more nearly allied to M. Jacobi's than to him it seemed to be. This demonstration had already been hinted at in his communication to the *Nachrichten*. The principal difference is, that while M. Jacobi proved generally that if the first of the two required conditions were satisfied, the second would also be so, and then showed that the forms assigned to U and V satisfied the first condition; Legendre shows the assigned forms are such as to satisfy both conditions, on the connection between which it is therefore unnecessary for him to dwell. In the third supplement to the *Traité des Fonctions Elliptiques*, Legendre has given another demonstration of M. Jacobi's theorem, remarking that it is both more rigorous and more like M. Jacobi's than that which he had first given. I have thought it necessary to make these remarks, because it has been said that it was in the supplements to Legendre's work that the demonstration of this theorem received 'le dernier degré de rigueur*.'

14. In 1829 M. Jacobi's great work on elliptic functions, the *Fundamenta Nova Theoriæ Functionum Ellipticarum*, was published at Kœnigsberg. It contains his researches not merely on the theory of transformation, but also with respect to other parts of the subject. But the great problem of transformation is the fundamental idea of the whole work; the other parts are subordinate to it, or at least derived from it. The subject is treated with great fulness of illustration and in a manner not unlike that of Euler.

M. Jacobi begins by considering the possibility of transforming the general transcendent whose differential coefficient is unity divided by the square root of a polynomial of the fourth degree. Subsequently, having shown that this transcendent may be transformed by introducing a new variable y equal to the quotient of two integral functions of x, and also that the general transcendent may be reduced to one of the form

$$\int \frac{dy}{\sqrt{(1-y^2)(1-\lambda^2 y^2)}},$$

he proceeds to consider the latter in detail.

The first step of this reasoning, viz. the possibility of the transformation, depends on a comparison of the number of the disposable quantities in the assumed value of y with that of the conditions required, in order that the quantity under the radical in the transformed expression may be equal to the square of an integral function of x multiplied by four unequal linear factors. It is shown that the number of disposable quantities exceeds by three that of the required conditions. But, as Poisson has remarked in the report already mentioned (*Mem. de l'Institut.* x. p. 87), and as M. Jacobi himself intimates, this does not amount to an absolute *à priori* proof of the possibility of the transformation; *non constat* but that some of these conditions may be incompatible.

Granting however the possibility of putting the quantity under the radical in the required form, it is shown, as in Schumacher's Journal, that this condition is not only necessary

* Verhulst, *Traité Elémentaire des Fonctions Elliptiques.*

but also sufficient, or, in other words, that it involves the second condition already mentioned.

The transcendent $\int \frac{dy}{\sqrt{(1-y^2)(1-\lambda^2 y^2)}}$ may be transformed by assuming $y = \frac{U}{V}$, U being composed wholly of odd powers of x, and V of even powers of it. If the degree of U be greater than that of V, the transformation is said to be of an odd order, and of an even order in the contrary case.

This being premised, M. Jacobi discusses the particular cases of the transformations of the third and of the fifth order. The first is the same as that of Legendre. It is shown that if we put

$$y = \frac{(v + 2u^3)\,vx + u^6 x^3}{v^2 + v^3 u^2 (v + 2u^3)\,x^2},$$

where u and v are constants connected by the following equation—

$$u^4 - v^4 + 2uv\,\{1 - u^2 v^2\} = 0,$$

we shall get

$$\frac{dy}{\sqrt{(1-y^2)(1-\lambda^2 y^2)}} = \frac{v + 2u^3}{v} \cdot \frac{dx}{\sqrt{(1-x^2)(1-k^2 x^2)}},$$

in which $k = u^4$ and $\lambda = v^4$. The equation connecting u and v is called the modular equation.

The 'principle of double substitution' may be illustrated by writing $\frac{1}{u^4 x}$ for x in the expression for y, which then becomes, according to the principle in question, $\frac{1}{v^4 y}$.

If we seek to show that the assigned value of y actually satisfies the differential equation just stated, we begin by finding the value of $1 - y$. Reducing this value by means of the equation between u and v, we can put it in the form $(1 - x)\,\frac{R^2}{V}$, R being an integral function of x and V, as heretofore the denominator of the expression for y. The value of $1 + y$ is hence got by changing the sign of x and y, while that of $1 - v^4 y$ is obtained by simultaneously replacing x and y respectively by $\frac{1}{u^4 x}$ and $\frac{1}{v^4 y}$ and reducing. Similarly for $1 + v^4 y$. Hence it will appear that

$$(1-y^2)(1-v^8y^2) = (1-x^2)(1-u^8x^2)\frac{S^2}{V^4} \text{ .... } (\alpha),$$

where S, like R, is integral. By differentiating and reducing, we then show that

$$dy = \frac{v+2u^3}{v}\frac{S}{V^2}dx,$$

and combining these two results obtain the required verification.

The essence of M. Jacobi's demonstration consists in showing that if the value of y in terms of x is such that an equation of the form (α) subsists, then necessarily

$$\frac{dy}{dx} = \mu\frac{S}{V^2} \text{ .... } (\beta),$$

where μ is a constant; the existence of the two equations (α) and (β) being equivalent to the two conditions of which we have already spoken (p. 259). In the particular case we are now considering,

$$\mu = \frac{v+2u^3}{v}.$$

15. After considering the transformation of the fifth order (in which the modular equation is

$$u^6 - v^6 + 5u^2v^2\{u^2-v^2\} + 4uv\{1-u^2v^2\} = 0),$$

M. Jacobi prepares the way for a more general investigation by introducing a new notation. This step is one of the highest importance. We have been in the habit of calling ϕ the *amplitude* of the integral

$$\int_0^\phi \frac{d\phi}{\sqrt{1-k^2\sin^2\phi}}:$$

let this integral be called u. The new notation is contained in the equation $\phi = \text{am}\,u$; or if we call $\sin\phi$, x, so that

$$u = \int_0^x \frac{dx}{\sqrt{(1-x^2)(1-k^2x^2)}},$$

then $x = \sin\text{am}\,u$.

A new notation is in itself merely a matter of convenience: what gives it importance is its symbolizing a new mode of considering any subject. We had hitherto been accustomed to look on the value of the elliptic integral as a function of its amplitude, to make the amplitude (if the expression may so be used) the independent variable. But in reality a contrary course is on many accounts to be preferred. We have in the more advanced part of the theory more frequently occasion to consider the value of the amplitude as determined by the corresponding value of the integral than *vice versâ*; and it therefore becomes expedient to frame a notation by which the amplitude may be expressed as a function of the integral. In a paper in the ninth volume of Crelle's Journal by M. Jacobi, which, like many of his writings, contains in a short compass a philosophical view of a wide subject, he has made use of the analogy between circular and elliptic functions to illustrate the importance of the new notation for the latter. When the modulus of an elliptic integral of the first kind is equal to zero, the integral becomes

$$\int_0^x \frac{dx}{\sqrt{1-x^2}},$$

which, as we know, is equal to the arc whose sine is x, or to $\sin^{-1}x$. Now this is a function which we have much less often occasion to express than its inverse $\sin x$, and we accordingly always look on the latter as a *direct*, and on the former as an *inverse* function. Yet in the case of elliptic functions, the functional dependence for which we had an explicit and recognised notation, viz. that of the integral on the amplitude, corresponds to that which in circular functions has always and almost necessarily been treated merely as an inverse function. The origin of this discrepancy is obvious; our knowledge of the nature of circular functions is not derived from the algebraical integrals connected with them, and therefore these integrals are not brought so much into view as in the theory of elliptic functions the corresponding integrals necessarily are; but it is certain that while the discrepancy continued to exist the subject could never be fully or satisfactorily developed. The maxim "verba vestigia mentis" is as true of mathematical symbols as of the elements of ordinary language.

We shall see hereafter that Abel took the same step in his first essay on elliptic functions. At present I shall only remark, that one of the earliest consequences of the new notation was the recognition of a most important principle, viz. that the 'inverse function' $\sin \operatorname{am} u$, that is, the function corresponding to $\sin u$ in circular functions, is doubly periodic, or that it retains the same value when u increases by any multiple either of a certain real or of a certain imaginary quantity. Now

M. Jacobi has shown that no function* can be triply periodic, and therefore these inverse functions possess the most general kind possible of periodicity, a property which gives them great analytical importance.

Following M. Jacobi, we shall henceforth give the name of elliptic functions to those which are analogous to circular functions. It is on this account better to call Legendre's functions elliptic integrals than, as he has done, elliptic functions (vide *ante*, p. 253).

By the new notation we are led to consider a great variety of formulæ analogous to those of ordinary trigonometry. The sine or cosine of the amplitude of the sum of two quantities may be expressed in terms of the sines and cosines of the amplitudes of each, &c.†; and we have only to make the modulus equal to zero to pass from what has sometimes, though not with much propriety, been called elliptic trigonometry to the common properties of circular functions.

M. Jacobi gives a table of formulæ relating to the new elliptic functions, and proceeds to apply their properties to the problem of transformation. It was in this manner that he had treated the problem in the *Nachrichten*. As in his earlier essay, he assumes y equal to a rational function of x, whose coefficients

* *i. e.* no function of one variable.

† The fundamental formulæ are

$$\sin \text{am}\,(u+v)=\frac{\sin \text{am}\,u \cos \text{am}\,v \Delta \text{am}\,v+\sin \text{am}\,v \cos \text{am}\,u \Delta \text{am}\,u}{1-k^2 \sin^2 \text{am}\,u \sin^2 \text{am}\,v};$$

$$\cos \text{am}\,(u+v)=\frac{\cos \text{am}\,u \cos \text{am}\,v-\sin \text{am}\,u \sin \text{am}\,v \Delta \text{am}\,u \Delta \text{am}\,v}{1-k^2 \sin^2 \text{am}\,u \sin^2 \text{am}\,v};$$

$$\Delta \text{am}\,(u+v)=\frac{\Delta \text{am}\,u \Delta \text{am}\,v-k^2 \sin \text{am}\,u \sin \text{am}\,v \cos \text{am}\,u \cos \text{am}\,v}{1-k^2 \sin^2 \text{am}\,u \sin^2 \text{am}\,v};$$

being the modulus, and $\Delta \text{am}\,u=\sqrt{1-k^2 \sin^2 \text{am}\,u}$. If

$$K=\int_0^{\frac{\pi}{2}} \frac{d\phi}{\sqrt{1-k^2 \sin^2 \phi}} \text{ and } K'=\int_0^{\frac{\pi}{2}} \frac{d\phi}{\sqrt{1-k'^2 \sin^2 \phi}},$$

where $k^2+k'^2=1$, then it may be shown that

$$\sin \text{am}\,(u+4K)=\sin \text{am}\,u,$$

and

$$\sin \text{am}\,(u+2K'\sqrt{-1})=\sin \text{am}\,u,$$

so that $4K$ is the *real* and $2K'\sqrt{-1}$ the *imaginary* period of $\sin \text{am}\,u$. Hence it is obvious that we shall have generally

$$\sin \text{am}\,(u+4mK+2nK'\sqrt{-1})=\sin \text{am}\,u,$$

m and n being any integers.

are elliptic functions, and shows that this assumption satisfies the differential equation already mentioned. It may be asked what is gained by the introduction of elliptic functions into a problem of which, as we have seen, particular cases (*e.g.* the transformations of the third and fifth order) can be solved by algebraical considerations. The answer is, that the properties of these functions enable us to transform the assumed relation between y and x in a manner which would otherwise be impracticable. It is conceivable that any particular case might be solved by mere algebra, but it does not seem possible to discover in this way a general theorem for transformations of all orders, and practically the labour of obtaining the formulæ for the transformation of any high order would be intolerable.

Having proved the theorem for transformation in nearly the same manner as he had already done, M. Jacobi developes the demonstration which, as we have said, Legendre hinted at in No. 130 of Schumacher's Journal.

He then proceeds to consider the various transformations of any given order. We have seen that the modular equation for those of the third order rises to the fourth degree, that is to say, for a given value of the modulus of the original integral four new moduli exist, corresponding to four new integrals, into which the given one may be transformed. These four transformations are all included in the general formula for the third order; but it is to be remarked that in general only two of the roots of the modular equation are real. Thus there are two real transformations and no more. The same thing is true, *mutatis mutandis*, of the transformations of any prime order (to which M. Jacobi's attention is chiefly directed), that is to say, there will be $n+1$ transformations of the nth order, $n-1$ of which are imaginary. The two real transformations are called the first and the second; the second is sometimes called the impossible transformation, because it presents itself in an imaginary form*. Of the formulæ connected with these two transformations M. Jacobi gives copious tables.

* Mr Bronwin, in the *Cambridge Mathematical Journal* and in the *Phil. Mag.*, has made some objections to this transformation; but from a correspondence which I have recently had with him, I believe I am justified in stating that he does not object either to M. Jacobi's result or to the logical correctness of his reasoning, but only to the form in which the result is exhibited.

He next shows, in a very remarkable manner, that, corresponding to a transformation in which we pass from a modulus k to a modulus λ, there exists another, whose formulæ are derivable from those of the former, in which we pass from a modulus $\sqrt{1-k^2}$ to a modulus $\sqrt{1-\lambda^2}$, or which connects moduli *complementary* to λ and k. The latter is accordingly called, with reference to the former, the *complementary* transformation. The first real transformation of k corresponds to the second real transformation $\sqrt{1-k^2}$, and *vice versâ.*

The next theorem which M. Jacobi demonstrates is not less remarkable. It is that the combination of the first and second real transformations gives a formula for the multiplication of the original integral, or, in other words, that the modulus of the integral which results from this double transformation is the same as that of the original integral, so that the two integrals differ only in their amplitudes. Of this theorem he had in the earlier part of the work proved some particular cases*.

After fully developing this part of the subject, he next treats of the nature of the modular equation, and shows that it possesses several remarkable properties. One is, that all modular equations, of whatever order, are particular integrals of a differential equation of the third order, of which the general integral can be expressed by means of elliptic transcendents.

16. We now enter on the second great division of M. Jacobi's researches, the evolution of elliptic functions.

* It may be shown that if we pass from k to λ by the first transformation, we can pass from $\sqrt{1-\lambda^2}$ to $\sqrt{1-k^2}$ also by the first transformation. Also, as has been said, we derive from the transformation $\{k \text{ to } \lambda\}$ a transformation $\{\sqrt{1-k^2}$ to $\sqrt{1-\lambda^2}\}$, and similarly from $\{\sqrt{1-\lambda^2}$ to $\sqrt{1-k^2}\}$ a transformation $\{\lambda \text{ to } k\}$. The first and last of these transformations correspond respectively to the differential equations

$$\frac{dy}{\sqrt{(1-y^2)(1-\lambda^2y^2)}} = \frac{1}{M}\frac{dx}{\sqrt{(1-x^2)(1-k^2x^2)}},$$

$$\frac{dx'}{\sqrt{(1-x'^2)(1-k^2x'^2)}} = \frac{1}{M'}\frac{dy}{\sqrt{(1-y^2)(1-\lambda^2y^2)}}.$$

Hence, combining these equations and integrating,

$$F(kx') = \frac{1}{MM'}F(kx);$$

and it may also be shown that $\frac{1}{MM'}$ is an integer.

The evolution of elliptic functions into continued products with an infinite number of factors presents itself as the limit towards which M. Jacobi's theorem for the transformation of the nth order tends as n increases *sine limite.* It is for this reason that we may look on the problem of transformation as the leading idea in M. Jacobi's researches.

We may in some degree illustrate these evolutions by a reference to circular functions. A sine is, as we know, an elliptic function whose modulus is zero. Now if k is zero, λ is also zero. Thus if we apply a formula of transformation to a sine, we shall be led to another sine either of the same or of a multiple arc. Accordingly the first real transformation degenerates in the case in question into the known formula for the sine of a multiple arc; while the second, leading us merely to the sine of the same arc, becomes illusory. Thus in the case of a sine, transformation is merely multiplication; but from the formula for multiplication, viz.

$$\sin(2m+1)\theta=(2m+1)\sin\theta\left(1-\frac{\sin^2\theta}{\sin^2\dfrac{\pi}{2m+1}}\right)\ldots\left(1-\frac{\sin^2\theta}{\sin^2\dfrac{2m\pi}{2m+1}}\right),$$

we at once deduce, by making $(2m+1)\,\theta=\phi$ and $2m+1$ infinite, the common formula

$$\sin\phi=\phi\left(1-\frac{\phi^2}{\pi^2}\right)\left(1-\frac{\phi^2}{4\pi^2}\right)\ldots\ldots$$

This then is a formula of evolution deduced from the first real transformation. It is however only when k is zero that the first transformation will give such a formula. In all other cases it is, for a reason which we cannot here enter on, impossible to derive from it a formula of this kind. M. Jacobi's formulæ are accordingly derived from the second real transformation, and therefore are illusory when k is zero, or for the case of the sine. There is nothing therefore strictly analogous to them in the theory of angular sections. By means of them we express the function sin am x in terms of sin mx, m being a certain constant.

From the fundamental expressions in continued products, of which there are three, many important theorems may be derived. This part of the subject seems to admit of almost infinite increase, and it is difficult to give any general view of it.

I may, however, mention a remarkable transcendental function of the modulus k which is usually denoted by q, and which occurs perpetually in this part of the theory of elliptic functions. If for the moment we denote this function by Fk, so that $q = Fk$, then if for k we write k_n, which we suppose to represent the modulus of the first real transformation of the nth order, we find that $q^n = Fk_n$, so that if q_n is the same function of k_n that q is of k,

$$q_n = q^n.$$

This singular property, and others of an analogous character, are of great use in establishing various formulæ*.

Before discussing the evolution of integrals of the third kind, M. Jacobi has premised some important theorems. He proves that the elliptic integral of the third kind, though it involves three elements, viz. the amplitude, the modulus and the parameter, can yet be expressed in terms of other quantities severally involving but two. In order to this we introduce either a new transcendent† or a definite elliptic integral of the third kind, whose amplitude is a certain function of its modulus and parameter. It is almost impossible to tabulate the values of a function of three elements, on account of the enormous bulk of a table with *triple entry;* we therefore see the importance of the step thus made. M. Jacobi announced this discovery as generally true of elliptic integrals of the third kind, but his demonstration applies to that subdivision already mentioned, which was designated by Legendre 'Fonctions du troisième ordre à paramètre logarithmique,' and not to functions 'à paramètre circulaire‡.' It is probable that this limitation was in

* A method of calculating elliptic integrals by means of q was suggested by Legendre. Vide *Verhulst*, p. 252, and M. Jacobi in Crelle.

† This transcendent is denoted by Υ, and is defined by the equation

$$\Upsilon = \int E\,(c\phi)\,\frac{d\phi}{\Delta\,(c\phi)},$$

where $E\,(c\phi)$ is the elliptic integral of the second kind. If we introduce the inverse notation, and make $\phi = \text{am}\,u$, we can readily establish the following result,

$$\Upsilon = \frac{1}{2}\,u^2 - c^2 \iint \sin^2 \text{am}\,u du^2.$$

The function Υ, which is the logarithm of Ω (vide *infra*, p. 288), has many remarkable properties.

‡ In the former species $(1+n)\left(1+\frac{c^2}{n}\right)$ is negative, and in the latter positive

M. Jacobi's mind, but he does not seem to have expressed it. Further on, in the *Fundamenta Nova*, we find another mode of expressing integrals of the third kind in terms of functions of two elements, but this method also applies only to 'fonctions du troisième ordre à paramètre logarithmique,' the two methods being in fact closely allied.

Legendre appreciated the importance of this discovery of M. Jacobi. He speaks of it in a letter to Abel, as a 'découverte majeure,' but adds that his attempts to extend M. Jacobi's demonstration to the other class of integrals of the third kind had been unsuccessful. The same remarks occur in his second supplement (*Traité des Fonct. Ell.* III. p. 141). The distinction thus made between the two classes of integrals of the third kind appeared to Legendre sufficient to make it desirable to recognise in all four classes of elliptic integrals, so as to make the division between the two species of the third class coordinate with that between either and the first or second. Legendre says explicitly that M. Jacobi had announced, in making known his discovery, that it applied to functions 'à paramètre circulaire.' This however possibly arose from some misconception of M. Jacobi's meaning. Dr Gudermann, in the fourteenth volume of Crelle's Journal, has given it as his opinion that the circular species of integrals of the third kind does not admit of the reduction in question; and remarks, that it occurs much more frequently than the other species in the applications of mathematics to natural philosophy.

After having discussed at some length, and by new methods, the properties of elliptic integrals of the third kind, M. Jacobi concludes his work by investigating the nature of two new transcendents which present themselves in immediate connexion with the numerator and denominator of the continued product by which sin am u is expressed. One of them however M. Jacobi had already recognised by a distinctive symbol, in consequence of its intimate connexion with the theory of integrals of the third kind.

Such is the outline of this remarkable work: before it appeared M. Jacobi gave in the third and fourth volumes of

(vide *ante*, p. 255). The specific names are derived from the circumstance that for the former the fundamental formula of addition involves a logarithm, for the latter a circular arc.

Crelle's Journal (III. pp. 192, 303, 403, IV. p. 185) notices, mostly without demonstrations, of the progress of his researches. Almost everything in the first and second of these notices is found in the *Fundamenta.* In the third we find a remarkable algebraical formula for the multiplication of the elliptic integral of the first kind. The fourth and last relates to ulterior investigations, which it was the intention of the author to develope in a second part of his work. It contains an indication of a method of transformation depending on a partial differential equation*; values of the elliptic functions of multiple arguments; a method of transforming integrals of the second and third kinds; a most important simplification of the method of Abel for the division of any integral of the first kind, &c. Of this simplification he had already given some idea in a note in the preceding volume of the same Journal, p. 86.

17. It may not be improper in this place to observe, that in 1818, and thus in the interval between Legendre's first and second systematic works on the theory of elliptic functions, M. Gauss published the tract entitled *Determinatio Attractionis*, &c. The illustrious author begins by remarking that the secular inequalities due to the action of one planet on another are the same as if the mass of the disturbing planet were diffused according to a certain law along its orbit, so that the latter becomes an elliptic ring of variable but infinitesimal thickness. The problem then presents itself of determining the attraction exerted by such a ring on any external point. In the solution of this problem M. Gauss arrives at two definite integrals; they can readily be reduced to elliptic integrals of the first and second kinds. For the evaluation of the integrals to which he reduces those of his problem, M. Gauss gives a method of successive transformation, analogous in some measure to that of Lagrange. But the transformation of which he makes use is a rational one, and is in fact the rational transformation of the second order. The discovery of this transformation appears therefore to be due to M. Gauss. He has remarked, though merely in passing, that his method is applicable to the indefinite as well as to the definite integral.

* Mr Cayley, to whose kindness I have been, while engaged on the present report, greatly indebted, has communicated to me a demonstration of the truth of this equation.

The rational transformation in question leads to a continually increasing series of moduli, or is, to use an expression of M. Jacobi, a transformation 'minoris in majorem.' The law connecting two consecutive moduli is the same as in Lagrange's, which is, as we have seen, an irrational transformation; so that M. Gauss's method does not afford us a new scale of moduli. Nevertheless, as no rational transformation had I believe been noticed when his tract appeared*, his method is, in a historical point of view, of considerable interest.

18. In the second volume of Crelle's Journal, p. 101, we find Abel's first memoir on elliptic functions. It was published in the spring of 1827, and therefore before M. Jacobi's announcement in No. 123 of Schumacher's Journal. But it contains nothing which interferes with M. Jacobi's discovery of the general theory of transformation. Abel's researches on this part of the subject appeared in the third volume of Crelle's Journal, p. 160. This second communication is dated, as we are informed by an editorial note, the 12th of February, 1828, and though it is announced as a continuation of the former memoir, it is yet in effect distinct from it, as its contents are not mentioned in the general summary prefixed to the first communication.

These details may not be without interest, though it is not often that questions of priority deserve the importance sometimes given to them. There is no doubt that Abel's researches were wholly independent of those of M. Jacobi; and though the coincidence of some of their results is therefore interesting, yet the general view which they respectively took of the theory of elliptic functions is essentially different, as different as the style and manner of their writings.

With M. Jacobi the problem of transformation occupied the first place; with Abel that of the division of elliptic integrals. Both introduced a notation inverse to that which had previously been used, and as an immediate consequence recognised the double periodicity of elliptic functions. Expressions of these functions in continued products and series were given by both, but those of Abel were deduced by considering the limiting case of the

* The fundamental formula of his transformation is incidentally mentioned in Legendre's second work (*Traité des Fonct.* i. 61).

multiplication of elliptic integrals, those of M. Jacobi, as we have seen, from the limiting case of their transformation. Hence Abel's fundamental expressions depend on doubly infinite continued products, corresponding to the double periodicity of elliptic functions. On the other hand, M. Jacobi's continued products are all singly infinite.

Other differences might of course be pointed out, but the most remarkable is that which we find in the character and style of their writings. Nothing can be more distinct. In M. Jacobi's we meet perpetually with the traces of patient and philosophical induction; we observe a frequent reference to particular cases and a most just and accurate perception of analogy. Abel's are distinguished by great facility of manner, which seems to result from his power of bringing different classes of mathematical ideas into relation with each other, and by the scientific character of his method. We meet in his works with nothing tentative, with but little even that seems like artifice. He delights in setting out with the most general conception of a problem, and in introducing successively the various conditions and limitations which it may require. The principle which he has laid down in a remarkable passage of an unfinished essay on equations seems always to have guided him—that a question should be so stated that it may be possible to answer it. When so stated it contains, he remarks, the germ of its solution*.

I do not presume to compare the merits of these two mathematicians. The writings of both are admirable, and may serve to show that if ever the modern method of analysis seems to be an *ἐμπειρία* rather than a *τέχνη*, it does so, either because it has not been rightly used, or because it is not duly understood.

To obtain a general view of Abel's writings it may be remarked, that his earliest researches related to the theory of equations. Of the ideas with which he was then conversant he has made two principal applications. The one is to the comparison

* For instance, Is it possible to trisect an angle by the rule and compass? The question thus stated leads us to consider the general character of all problems soluble by the methods of elementary geometry; and following the suggestion thus given, we find that it is to be answered in the negative. But if the last clause be omitted or neglected, we can only proceed, as many persons have done, tentatively, *i.e.* by attempting actually to solve the problem. If we fail, the question remains unanswered; if we succeed, we do answer it, but as it were only by accident.

of transcendents in the manner already described; the other to the solution of the equations presented by the problem of the division of elliptic integrals. The second of these applications is contained in the memoir published in the second volume of Crelle's Journal.

He begins by introducing an inverse notation $\phi(u)$ corresponding to the function denoted in the *Fundamenta Nova* by sin am u, while $f(u)$ and $F(u)$ correspond respectively to cos am u and Δ am u. This notation has the defect of appropriating three symbols which we cannot well spare. On the other hand it is certainly more concise than M. Jacobi's.

He then verifies the fundamental formulæ for the addition of the new functions, and goes on to show that they are doubly periodic*. He next considers the expressions of $\phi n\alpha$, &c. in $\phi\alpha$, &c., and proceeds to prove the important proposition that the equation of the problem of the division of elliptic integrals of the first kind is always algebraically soluble.

In order to illustrate this, which is one of the most remarkable theorems in the whole subject, it may be observed, that as any circular function of a multiple arc can be algebraically expressed in terms of circular functions of the simple arc, so may $\phi n\alpha, fn\alpha, Fn\alpha$ be algebraically expressed by means of $\phi\alpha, f\alpha, F\alpha$.

Conversely, as the determination (to take a particular function) of sin α in terms of sin $n\alpha$ requires the solution of an algebraical equation, so does that of $\phi\alpha$ in terms of $\phi n\alpha$. The equation which presents itself in the former case is, as we know, of the nth or of the $(2n)$th degree as n is odd or even. But the equation for determining $\phi\alpha$ rises to the (n^2)th degree in the former case, and in the latter to the $(2n^2)$th. We may however confine ourselves to the case in which n is a prime number;

* The formulæ in question differ from those already given, only because Abel's form of the elliptic integral is $\int\frac{dx}{\sqrt{(1-c^2x^2)(1+e^2x^2)}}$, which becomes the same as Legendre's on making $e^2=-1$. The double periodicity of the functions is expressed by the formula

$$\phi\theta=\phi\{(-1)^{m+n}\theta+m\omega+n\varpi\sqrt{-1}\},$$

with similar formulæ for f and F. The quantities m and n are integral, and

$$\omega=2\int_0^{\frac{1}{c}}\frac{dx}{\sqrt{(1-c^2x^2)(1+e^2x^2)}},\qquad \varpi=2\int_0^{\frac{1}{e}}\frac{dx}{\sqrt{(1+c^2x^2)(1-e^2x^2)}}.$$

since if it be composite the argument of the circular or elliptic function may first be divided by one of the factors of n, and the result thus got by another, and so on. Thus setting aside the particular case of $n=2$, we shall have to consider, in order to determine $\sin \alpha$ or $\phi\alpha$, an algebraical equation of the nth or (n^2)th degree respectively.

In consequence of the periodicity of $\sin \alpha$, the roots of the equation in $\sin \alpha$ admit of being expressed in a transcendental form; they are all included in the formula $\sin\left(\alpha + \frac{2p\pi}{n}\right)$, in which p is integral, and which therefore admits only n different values.

But elliptic functions are *doubly* periodic, and therefore the roots of the equation in $\phi\alpha$ are expressible by a formula analogous to the one just written, but which involves two indeterminate integers corresponding to the two periodicities of the function, just as p does to the single periodicity 2π. Giving all possible values to these integers, we get n^2 different values for the formula.

The question now is, how are we to pass from the transcendental representation of these roots to their algebraical expression? Or, in other words, how are the relations among the roots deducible from the circumstance of their being all included in the same formula, to be made available in effecting the solution of the algebraical equation?

The answer to this question is to be found in the following principle: that if χu be such a rational function of u that

$$\chi x = \chi y = \ldots = \chi z,$$

$x, y, \ldots z$ being the roots of an algebraical equation, then any of these quantities may be expressed in terms of the coefficients of the equation. This follows at once from the consideration that we shall have

$$\chi x = \frac{1}{\mu}\{\chi x + \chi y + \ldots + \chi z\},$$

μ being the number of the roots $x, y, \ldots z$. For the sum within the bracket being a rational and symmetrical function of the roots, is necessarily expressible in the coefficients of the equation, and the same is therefore of course true of χx, or of any of the other quantities to which it is equal.

If, therefore, by means of the relations which we know to exist among the roots of the equation to be solved we can establish the existence of a system of such functions, χ, χ', χ'', &c., each of which retains the same value of whichever root we suppose it to be a function; and if by combining these functions we can ultimately express x in terms of them, the equation is solved, since each of these functions may be considered a known quantity.

Such is the general idea of Abel's method of solution. The principle on which it depends, namely, the expressibility of any unchangeable function χ, is one which is frequently met with in investigations similar to that of which we are speaking. M. Gauss's solution of the binomial equation is founded upon it.

I have already remarked that an important simplification of Abel's process was given by M. Jacobi. The result which M. Jacobi has stated without demonstration may be proved by means of a theorem established by Abel in the fourth volume of Crelle's Journal, p. 194.

M. Jacobi shows the existence of a system of n^2 functions χ, χ', &c., by combining which we can immediately express the values of the roots. In the last of his *Notices* on elliptic functions we find, as has been said, the explicit determination of all the roots. The formula given for this purpose is, like the former, undemonstrated, and I do not know whether any demonstration of it has as yet been published; but from a note of M. Liouville, in a recent volume of the *Comptes Rendus*, we find that both he and M. Hermite have succeeded in proving it.

But in whatever manner the solution is effected it will always involve certain transcendental quantities, which are introduced in the expressions of the relation subsisting between the different roots. The solution can therefore be looked on as complete, only if we consider these to be known quantities. They are the roots of a particular case of the equation to be solved. They relate to the division of what are called the complete integrals. We may therefore say that the general case is reduced to this particular one. But the latter is not, except under certain circumstances, soluble, though the solution of the equation on which it depends can be reduced to the solution of certain other equations of lower degrees.

But for an infinity of particular values of the modulus, the case in question is soluble by a method closely analogous to that used by M. Gauss for the solution of binomial equations. Thus for all such values the problem of the division of elliptic integrals is completely solved.

The most remarkable of these cases corresponds to the geometrical problem of the division of the perimeter of the lemniscate. Abel discovered that this division can always be effected by means of radicals, and further, that it can be constructed by the rule and compass in the same cases (that is for the same values of the divisor) as the division of the circumference of a circle. Of this discovery we find Abel writing to M. Holmboe, "Ah qu'il est magnifique! tu verras*."

In order to form an idea of the nature of the difficulty which disappears in the case of which we are speaking, let us suppose that we have to solve the algebraical equation which is represented by the transcendental one $\phi\,(3\theta) = 0$, in the same manner as the equation $4x^3 - 3x = 0$ is represented by $\sin\,(3\theta) = 0$.

The roots of $4x^3 - 3x = 0$, are, setting aside zero,

$$\sin\frac{2\pi}{3}, \quad \sin\frac{4\pi}{3}.$$

Those of the former algebraical equation, which, as we know, is of the ninth degree, are, beside zero,

$$\begin{array}{ll} \phi\dfrac{2\omega}{3}, & \phi\dfrac{4\omega}{3}, \\ \phi\dfrac{2\varpi i}{3}, & \phi\dfrac{4\varpi i}{3}, \\ \phi\dfrac{2\,(\omega + \varpi i)}{3}, & \phi\dfrac{4\,(\omega + \varpi i)}{3}, \\ \phi\dfrac{2\,(\omega + 2\varpi i)}{3}, & \phi\dfrac{4\,(\omega + 2\varpi i)}{3}, \end{array}$$

where $i = \sqrt{-1}$.

* It is right to mention that M. Libri has disputed Abel's title to the theory of the division of the lemniscate. I shall, however, not enter on the merits of the controversy which arose on this point between him and M. Liouville. The reader will find it in the seventeenth volume of the *Comptes Rendus*. It appears that M. Gauss had himself recognised the applicability of his method to the equation arising out of the problem of the division of the perimeter of the lemniscate (vide *Recherches Arithmetiques*, § vii. p. 429. I quote from the translation published at Paris, in 1809).

To satisfy ourselves that these are the roots required, we observe that $\phi(m\omega + n\varpi i) = 0$ for all integral values of m and n. Hence the general form of the roots of our equation is $\phi\,\frac{m\omega + n\varpi i}{3}$; but it will be found that if we give any values not included in the above table to m and n, the resulting expression can be reduced to one or other of the forms we have specified in virtue of the formula $\phi(\theta) = \phi\{(-1)^{m+n}\theta + m\omega + n\varpi i\}$. *E. g.* The non-tabulated root $\phi\,\frac{5\omega + 2\varpi i}{3}$ is equal to our sixth root $\phi\,\frac{4(\omega + \varpi i)}{3}$, since the *sum* of their arguments is $3\omega + 2\varpi i$, and the sum of 3 and 2 is an *odd* number.

On considering our table, we observe that it consists of $3 + 1$ horizontal rows, each containing $3 - 1$ terms, and that the arguments of the terms in each row are connected by a simple relation; that of the second being double that of the first. If we were to replace 3 by any odd number p, we should get an equation of the p^2 degree, whose roots, setting aside zero, might similarly be arranged in $p + 1$ rows, each of $p - 1$ terms, the arguments of the terms in each row being as 1, 2, 3, &c.

Moreover, $\sin\frac{4\pi}{3}$ is rationally expressible in $\sin\frac{2\pi}{3}$, and generally $\sin\frac{2p\pi}{2n+1}$ is so in $\sin\frac{2\pi}{2n+1}$, n and p being any integers we please. So too are all the terms in each horizontal row of our table, whether for the particular case we have written down, or for that of any odd number, rationally expressible in the first term.

Hence it may be shown that when the divisor $2n + 1$ is a prime number, an equation whose roots were the terms in any horizontal row could be solved algebraically, by a method essentially the same as that of Gauss, just as we can solve the equation the type of whose roots is $\sin\frac{2p\pi}{2n+1}$. But to construct this equation, *i. e.* to determine its coefficients, requires the solution of an equation of the same degree as the number of horizontal rows, *i. e.* of the degree $2n + 2$. And this equation is in general insoluble. The difficulty we here encounter may be expressed in general language, by saying that although we can pass from one root to another along each horizontal row, yet we cannot pass from row to row.

Our table, however, has the remarkable property, that supposing, as we may always do, $2n+1$ to be a prime number, all the roots are rationally expressible in terms of any two not lying in the same row. This depends on a property of the function ϕ, which it is very easy to demonstrate, and it is intimately connected with the relations which exist among the terms of the same row.

If, then, which is the case for an infinite variety of values of the modulus, we can express any root rationally in terms of another of a different row, say in $\phi \frac{2\omega}{2n+1}$, all the roots become rational in terms of $\phi \frac{2\omega}{2n+1}$. Moreover, it appears that not only are the roots all expressible in one, but they are so in such a manner that the functional dependencies among them fulfil a certain simple condition, which, as Abel shows in a separate memoir (Crelle, IV. p. 131; or Abel's Works, I. p. 114), renders every equation, all whose roots are rationally expressible in terms of one, algebraically soluble.

To take the simplest case, the arc of the lemniscate may be represented by the integral $\int \frac{dx}{\sqrt{1-x^4}}$. If ϕ be the function inverse to this integral, we have the simple relation between roots of different rows, $\phi \frac{2\varpi i}{2n+1} = i\phi \frac{2\omega}{2n+1}$, ω being in this case equal to ϖ.

To apply what has been said to the solution of the general equation for determining $\phi\alpha$ in terms of $\phi(2n+1)\alpha$, it is sufficient to remark that the transcendents introduced in considering the relations among the roots of this equation, are simply $\phi \frac{2\omega}{2n+1}$ and $\phi \frac{2\varpi i}{2n+1}$, or at least may be algebraically expressed in terms of these two quantities.

The remainder of the first memoir contains developments of the functions ϕ, f, and F in doubly and singly infinite continued products and series. They are derived from the expressions of ϕx, &c. in terms of $\phi \frac{\alpha}{n}$, &c., by supposing n to increase *sine limite*, and are therefore analogous to the expression of $\sin\phi$ in terms of ϕ which we have already mentioned.

The second contains the development of what had already been pointed out with respect to the lemniscate, so far as relates to the division of its perimeter by any prime number of the form $4m+1$. In an interesting note which M. Liouville communicated to the Institute in 1844, and which is published in the eighth volume of his Journal, p. 507, he has proved generally that the division of the perimeter of this curve can always be effected whether the divisor be a composite or prime number, real or *complex* (that is, of the form $p+\sqrt{-q}$, p and q being integers). In order to do this, it was only requisite to follow *m.m.*, the reasoning by which Abel has shown that the equation which presents itself in the problem of the division of the circumference of the circle is always resoluble. Thus, as M. Liouville has remarked, his analysis is implicitly contained in Abel's.

This memoir also contains Abel's theorem for the transformation of elliptic integrals of the first kind. It is equivalent to that of M. Jacobi; nor is the demonstration, though presented in quite a different form, altogether unlike M. Jacobi's.

Abel begins by considering the sum of a certain series of ϕ functions whose arguments are in arithmetical progression. He shows that the sum of this series is a rational function of its first term. If we call this sum (multiplied by a certain constant) y, and the first term x, then y is such a function of x as to satisfy the differential equation already mentioned, viz.

$$\frac{dy}{\sqrt{(1-y^2)(1-\lambda^2y^2)}}=\frac{1}{M}\frac{dx}{\sqrt{(1-x^2)(1-k^2x^2)}}, \ldots\ldots \text{(A)},$$

or rather an equation of equivalent form. In fact y is *m.m.* the same function of x that it is in M. Jacobi's theorem. Thus the sum of the series of elliptic functions is itself, when multiplied by a constant, a new elliptic function, having a new modulus, and whose argument bears a constant ratio to that of the first term of the series. It appears also that for the sum of the elliptic functions we may, duly altering the constant factor, substitute their continued product. Thus, beside the algebraical expression of y, there are two transcendental expressions of it, both of which are given by M. Jacobi in the *Fundamenta*

Nova. At the close of the memoir Abel compares his result with the one in Schumacher's Journal, No. 123, and mentions that he had not met with the latter until his own paper was terminated.

19. In the 138th number of this journal, Abel resumed the problem of transformation, and treated it in a more general and direct manner than had yet been done. This memoir appeared in June 1828. M. Jacobi, in a letter to Legendre, has spoken in the highest terms of Abel's demonstration of the formulæ of transformation: he says, "Elle est au-dessus de mes éloges, comme elle est au-dessus de mes travaux." An addition to this memoir, establishing the real transformations by an independent method, appeared in Number 148 of the same journal. These two papers are printed consecutively in the first volume of Abel's Works, pp. 253, 275.

In the first of these two remarkable essays, Abel makes use of the periodicity of the function $\phi\theta$, or, as he here denotes it, $\lambda\theta$, to determine *à priori* what rational function of x, y must be in order that the differential equation

$$\frac{dy}{\sqrt{(1-y^2)(1-k^2y^2)}} = a\frac{dx}{\sqrt{(1-x^2)(1-c^2x^2)}}$$

may be satisfied. [I have altered his notation for the sake of uniformity.] Let ψx be the function sought, then considering $y = \psi x$ as an equation determining x in terms of y, he shows that certain relations necessarily exist among its roots. Let $\lambda\theta$ be one of them and $\lambda\theta'$ another, it will readily be seen that we may put

$$d\theta' = d\theta,$$

since each is equal to

$$\frac{dy}{\sqrt{(1-y^2)(1-k^2y^2)}}.$$

Hence
$$\theta' = \theta + \alpha,$$

α being the constant of integration, or, which is the same thing, being independent of y. Hence $\lambda\theta$ being one root, every other root is necessarily of the form $\lambda(\theta + \alpha)$. Again, we see from hence that

$$y = \psi(\lambda\theta) = \psi\{\lambda(\theta + \alpha)\},$$

which is to be true for all values of θ, and which therefore implies the existence of a series of equations, of which the type is

$$\psi[\lambda\{\theta+(k-1)\alpha\}]=\psi\{\lambda(\theta+k\alpha)\},$$

where k is an integer. Hence $\lambda(\theta+k\alpha)$ is a root, whatever integral value we may give to k. But the equation $y=\psi x$ has but a finite number of roots, and therefore the values of the general expression $\lambda(\theta+k\alpha)$ must recur again and again. This consideration throws light on the nature of the quantity α; it must in all cases be an aliquot part of a period (simple or compound) of the function $\lambda\theta$.

All the values of $\lambda(\theta+k\alpha)$ got by giving different values to k are roots; but the converse is not necessarily true; all the roots are not necessarily included in this expression. But it is not difficult to perceive that all the roots are included in a more general expression, viz. $\lambda(\theta+k_1\alpha_1+k_2\alpha_2\ldots k_n\alpha_n)$, and conversely, that all the values of this expression are roots. The number n is indeterminate: we may have formulæ of the form $y=\psi x$, in which n is unity, others in which it is two, &c.; but in all cases α is an aliquot part of some period of $\lambda\theta$, and k is integral.

It is easy when the roots of $y=\psi x$ are known, to express y in terms of θ. For let

$$\psi x=\frac{fx}{Fx},\ f \text{ and } F \text{ being integral functions. Then}$$

$$yFx-fx=(yp-q)\,[(x-\lambda\theta)\,\{x-\lambda(\theta+\alpha)\}\ldots]$$

is ($yp-q$ being the coefficient of the highest power of x in $yFx-fx$) an identically true equation; whence, to determine y in θ, we have only to assign a particular value to x, or to compare the coefficients of similar powers of it*.

This then determines the form which the function y must necessarily be of: the question which Abel goes on to discuss is this: Under what circumstances will a function of the form thus determined *à priori* be such a function as we require? The character of the reasoning by which this question is treated is similar to that of the method by which Abel had, in his second memoir on elliptic functions, verified the form which, without assigning any reason, he had there assumed for the function y.

The second essay is singularly elegant. If ϕ_k denote the

* I have not noticed an ambiguity of sign at the outset of this reasoning, as given by Abel, as for the purposes of illustration it is immaterial.

function inverse to the integral $\int \frac{du}{\sqrt{(1-u^2)(1-k^2u^2)}}$, and ϕ_c the corresponding function for the modulus c, then, on introducing the inverse notation, the differential equation

$$\frac{dy}{\sqrt{(1-y^2)(1-k^2y^2)}} = a\frac{dx}{\sqrt{(1-x^2)(1-c^2x^2)}}$$

becomes of course $d\theta' = ad\theta$, with $x = \phi_c\theta$ and $y = \phi_k\theta'$. Hence for a given increment α of θ, that of θ' is $a\alpha$.

Let us take the simplest case, and suppose y to be a rational function of x; then, as x or $\phi_c\theta$ remains unchanged when θ increases by a period of the function ϕ_c, y does so too; that is $\phi_k\theta'$ remains unchanged when θ' increases by a times a period of ϕ_c, or in other words, a times a period of ϕ_c is necessarily one of ϕ_k.

Suppose now k and c to be both real and less than unity; then ϕ_k and ϕ_c have each a real period, here denoted by $2\omega_k$ and $2\omega_c$ respectively, and each an imaginary period $\varpi_k i$ and $\varpi_c i$ respectively, ϖ_k and ϖ_c being both real. Let θ receive first the increment $2\omega_c$, and secondly the increment $\varpi_c i$, then, by what has been said,

$$2a\omega_c = 2m\omega_k + n\varpi_k i,$$

$$a\varpi_c i = 2p\omega_k + q\varpi_k i^*,$$

m, n, p, q being certain integers. But can these two equations subsist simultaneously? Not generally, since if we eliminate a and equate possible and impossible parts, we get *two* relations among $\omega_c\varpi_c\omega_k\varpi_k$, which are continuous functions of the *two* quantities k and c. Hence *both* are determinate; and if we wish c to remain indeterminate, we must either make m and q equal to zero, in which case a is impossible, or, making n and p equal to zero, assign a real value to it. When a is real we have

$$a = m\frac{\omega_k}{\omega_c} = q\frac{\varpi_k}{\varpi_c},$$

and hence the remarkable conclusion, that

$$\frac{\omega_k}{\varpi_k} : \frac{\omega_c}{\varpi_c} :: q : m,$$

m and q being integers.

* ϖ here is in M. Jacobi's notation $2K'$, so that $\phi\theta = \phi(\theta + 2m\omega + n\varpi i)$, m and n being any integers.

The commensurability of the transcendental functions $\frac{\omega_k}{\varpi_k}$, $\frac{\omega_c}{\varpi_c}$ is therefore a necessary condition, in order that an integral with modulus c can be transformed into one with modulus k, the *regulator* a being real and c indeterminate. And it may be shown that this condition is not only necessary but sufficient. Similar considerations apply to the case in which a is impossible.

Simple as this view is, it leads to many consequences of great interest. The function q, of which we have already spoken (p. 270), is merely $e^{-\pi\frac{\varpi}{\omega}}$, and as we know for the first real transformation of the nth order, it becomes $e^{-\pi\frac{n\varpi}{\omega}}$. Hence in this case we have $\left(\frac{\varpi}{\omega}\right)_k = n\left(\frac{\varpi}{\omega}\right)_c$ according to the general law. It may be well to remark, that if $k = c$ we have $a = m\frac{\omega_k}{\omega_c} = m$ (an integer). Hence in *multiplying* an integral, the multiplier must be an integer, if y is rational in x, except for particular values of c.

In the paper of which we are speaking Abel has applied precisely similar considerations to the case in which x and y are connected by any algebraical equation.

Passing over one or two shorter papers, one of which has been already referred to at p. 276, we come to a *Précis* of the theory of elliptic functions, published in the fourth volume of Crelle's Journal, p. 236. The work of which it was designed to be an extract was never written, and the *Précis* itself is left unfinished. A general summary was prefixed to it, from which we learn that the work was to be divided into two parts. In the first elliptic integrals are considered irrespectively of the limits of integration, and their moduli may have any values, real or imaginary. Abel proposes the general problem of determining all the cases in which a linear relation may exist among elliptic integrals and logarithmic and algebraical functions in virtue of algebraical relations existing among the variables*.

His first step is to apply his general method for the comparison of transcendents to elliptic integrals, which may be

* In the assumed relation, the amplitude, or rather the sine of the amplitude of each elliptic integral, is to be one of the variables, and *not* a function of one or more of them.

done by what is called Abel's theorem, in at least two different ways: the one, that of which he now makes use; the other, that which we have seen is applied to the case of four functions by Legendre in his third Supplement.

He next determines the most general form of which the integral of an algebraical differential expression of any number of variables is capable, provided it can be expressed linearly by elliptic integrals and logarithmic and algebraical functions. The result at which he arrives admits of many important applications. It is, that the integral in question may be expressed in a form in which the sine of the amplitude of each elliptic integral and the corresponding Δ, and also the algebraical and each logarithmic function are all *rational* functions of the variables and of the differential coefficients of the integral with respect to each.

He proceeds by an interesting train of reasoning to establish the remarkable conclusion, that the general problem which we are considering may ultimately be reduced to that of the transformation of elliptic integrals of the first kind. The problem of this transformation is then discussed, and by a method essentially the same as that of which he had made use in his paper in Schumacher's Journal. The appearance however of the two investigations is dissimilar, because no reference is made to elliptic functions (as distinguished from elliptic integrals) in the first part of the *Précis.* The relations therefore which exist among the roots of $y = \psi x$ are established by considerations independent of the periodicity of elliptic functions; though it is not difficult to perceive that they were *suggested* by the results previously obtained by means of that fundamental property. It is shown, that if the equation $y = \psi x$, where ψx is a rational function, satisfy the differential equation (A), then this equation, considered as determining x in terms of y, is always algebraically soluble. As the multiplication of elliptic integrals may be considered a case of transformation (that, namely, in which the modulus of the transformed integral remains unchanged), this theorem may be looked on as an extension of that which we have spoken of (p. 275) in giving an account of Abel's first memoir on elliptic functions. The two theorems are proved by the same kind of reasoning.

The second part of the memoir was to have related to cases

in which the moduli are real and less than unity; of this however only the summary exists. Abel proposed to introduce three new functions, the first corresponding to that which he had previously designated by $\phi\theta$*. He now denotes it by $\lambda\theta$. The second and third functions are apparently what the second and third kind of elliptic integrals respectively become, when, instead of x, we introduce the new variable θ; x and θ being of course connected by the equation $x = \lambda\theta$. The double periodicity of the function λ and its other fundamental properties having been established, it was his intention to proceed to more profound researches. Some of his principal results are briefly stated. I may mention one, that all the roots of the modular equation may be expressed rationally in terms of two of them†.

One of the last paragraphs of the summary relates to functions very nearly identical with those which M. Jacobi discusses at the close of the *Fundamenta Nova*, and which he has designated by the symbols H and Θ.

The second volume of Abel's collected works consists of papers not published during his life. Two or three of these relate to elliptic functions. The longest contains a new and very general investigation for the reduction of the general transcendent, whose differential is of the form $\frac{P}{\sqrt{R}}$, P being, as usual, rational and R a polynomial of the fourth degree; together with transformations with respect to the parameter of integrals of the third kind.

20. Having now given some account of the revolution which the discoveries of Abel and Jacobi produced in the theory of elliptic functions, I shall mention some of the principal contributions which have been made towards the further development of the subject since the publication of the *Funda-*

* In the *Précis* Abel has adopted the canonical form of the integral of the first kind made use of by Legendre and M. Jacobi; so that the quantity under the radical is $(1 - x^2)(1 - c^2x^2)$. It is worth remarking, that in his first paper in Schumacher's *Nachrichten* this quantity is $(1 - e^2x^2)(1 - c^2x^2)$, while in the second it is the same as in the *Précis*. To this form he appears latterly to have adhered.

† It is not clear whether by roots of the modular equation we are to understand the transformed moduli themselves, or their fourth roots, *i.e.* in M. Jacobi's notation λ or v. Vide *supra*, p. 263.

menta Nova. In Crelle's Journal, IV. p. 371, we find a paper by M. Jacobi, entitled 'De Functionibus Ellipticis Commentatio.' It contains, in the first place, a development of the method of transforming elliptic integrals of the second and third kind, and introduces a new transcendent Ω, which takes the place of Θ, with which it is closely connected. M. Jacobi proves that the numerator and denominator of the value of y, mentioned above, and which have been denoted by U and V, satisfy a single differential equation of the third order. The remainder of the paper relates to the properties of Ω (vide *ante*, note, p. 270). When this function is multiplied by a certain exponential factor it becomes a singly periodic function, and, which is very remarkable, its period is equal to one of the single or composite periods of the elliptic function inverse to the integral of the first kind. By composite period I mean the sum of multiples of the fundamental periods. The exponential factor being properly determined, its product by Ω is equal to Θ multiplied by a constant. In considering this subject, M. Jacobi is led to introduce the idea of conjugate periods. These are periods by the combination of which all the composite periods may be produced. It is obvious that the fundamental periods are conjugate periods; and there are, as may easily be shown, an infinity of others.

In the sixth volume of the same Journal we find a second part of the 'Commentatio.' It contains a remarkable demonstration of the fundamental formulæ of transformation of the odd orders founded on elementary properties of elliptic functions.

In a historical point of view a notice by M. Jacobi in the eighth volume of Crelle (p. 413) of the third volume of Legendre's *Traité des Fonctions Elliptiques* is interesting. It was here, I believe, that M. Jacobi first proposed the name of Abelian integrals for the higher transcendents, which we shall shortly have occasion to consider. After some account of the contents of Legendre's supplements, the first two of which contain the greater part of M. Jacobi's earlier researches, he goes on to generalise a remarkable reduction given by Legendre at the close of his work.

21. I turn to one of the very few contributions which English mathematicians have made to the subject of this report, namely, to a paper by Mr Ivory, which appeared in the *Phil.*

Trans. for 1831. His design is to give in a simple form M. Jacobi's theorem for transformation. The demonstration is essentially the same as that in the *Fundamenta Nova*. But Mr Ivory does not set out with assuming $y=\frac{U}{V}$, U and V being integral functions of x, but with assuming it equal to the continued product of a number of elliptic functions (whose arguments are in arithmetical progression), multiplied by a constant factor. This is one of M. Jacobi's transcendental expressions for y, and the two assumptions are therefore perfectly equivalent in the transformations of odd orders; but in those of even orders, or where the continued product consists of an even number of factors, Mr Ivory's amounts to making y equal to an irrational function of x. Transformations by irrational substitutions, though long the only kind known (since Lagrange's belongs to this class), had not of late been considered in detail. Abel indeed remarked in the beginning of the general investigation contained in Schumacher's *Journal* (No. 138), that the existence of an irrational transformation implied that of a rational one leading to an integral with the same modulus as the other. He was, therefore, in seeking for the most general modular transformation, exempted from considering irrational substitutions; but in a historical point of view it is interesting to see the connection between Lagrange's transformation and those which have been more recently discovered*.

* If $y=(1+b)x\sqrt{\frac{1-x^2}{1-c^2x^2}}$, where $b^2+c^2=1$, then

$$\frac{dy}{\sqrt{(1-y^2)(1-k^2y^2)}}=(1+b)\frac{dx}{\sqrt{(1-x^2)(1-c^2x^2)}},$$

where
$$k=\frac{1-b}{1+b}.$$

This is Lagrange's *direct* transformation. The corresponding rational transformation is

$$y=\frac{1-(1+b)x^2}{1-(1-b)x^2},$$

which satisfies the same differential equation as before.

Again,
$$\frac{dy}{\sqrt{(1-y^2)(1-k^2y^2)}}=\frac{1+c}{2}\frac{dx}{\sqrt{(1-x^2)(1-c^2x^2)}},$$

where
$$k=\frac{2\sqrt{c}}{1+c}$$

is satisfied by
$$2y^2=1+cx^2-\sqrt{(1-x^2)(1-c^2x^2)},$$

The question presents itself, what is the connection between the irrational transformation (that of which Lagrange's is a particular case) and the rational transformation of even orders? Perhaps the simplest answer to it (though every question of the kind is included in the general investigations contained in Abel's *Précis*) is found in a paper by M. Sanio in the fourteenth volume of Crelle's *Journal*, p. 1. The aim of this paper is to develope more fully than Mr Ivory has done the theory of transformations of even orders, and particularly of the irrational transformations, which M. Sanio considers more truly analogous to the rational transformations of odd orders than the rational transformations of even orders; and also to discuss the multiplication of elliptic integrals by even numbers, a subject intimately connected with the other. We have already mentioned the existence of what are called complementary transformations, each of which may be derived from the other by an irrational substitution, by which two new variables are introduced. In the case of transformations of odd orders, the original transformation and the complementary one are both rational, and are both included in the general formula given by M. Jacobi's theorem; but to the rational transformation of any even order corresponds as its complement the irrational transformation of the same order. This remark, which, as far as I am aware, had not before been made, sets the subject in a clear light*.

which may be called Lagrange's *inverse* transformation, k being now the same function of c, that c was before of k. *The corresponding rational transformation is*

$$y = \frac{(1+c)\,x}{1+cx^2},$$

which is M. Gauss's, and is termed in M. Jacobi's nomenclature the rational transformation of the second order. It satisfies the equation

$$\frac{dy}{\sqrt{(1-y^2)(1-k^2y^2)}} = (1+c)\,\frac{dx}{\sqrt{(1-x^2)(1-c^2x^2)}},$$

where, as before, $$k = \frac{2\sqrt{c}}{1+c}.$$

* Lagrange's transformation being

$$y = (1+b)\,x\sqrt{\frac{1-x^2}{1-c^2x^2}}, \text{ let } y = \frac{\sqrt{-1}\,y'}{\sqrt{1-y'^2}} \text{ and } x = \frac{\sqrt{-1}\,x'}{\sqrt{1-x'^2}},$$

then we find that $$y' = \frac{(1+b)\,x'}{1+bx'^2},$$

while the differential equation becomes

$$\frac{dy'}{\sqrt{(1-y'^2)(1-h^2y'^2)}} = (1+b)\,\frac{dx'}{\sqrt{(1-x'^2)(1-b^2x'^2)}},$$

where $$h^2 = 1-k^2.$$

22. In the twelfth volume of Crelle's *Journal* (p. 173), Dr Guetzlaff has investigated the modular equation of transformations of the seventh order: it is, as we know from the general theory, of the eighth degree, and presents itself in a very remarkable form, which closely resembles that in which M. Jacobi, at p. 68 of the *Fundamenta Nova,* has put the modular equation for the third order. Dr Sohncke has given, at p. 178 of the same volume, modular equations of the eleventh, thirteenth and seventeenth orders, none of which apparently can be reduced to so elegant a form as those of the third and seventh. Possibly the transformation of the thirty-first order might admit of a corresponding reduction. The whole subject of modular equations is full of interest. Dr Sohncke has demonstrated his results in a subsequent volume of the *Journal* (XVI. 97).

In the fourteenth volume of Crelle's *Journal* there is a paper by Dr Gudermann on methods of calculating and reducing integrals of the third kind. I have already quoted from this paper the expression of the opinion of its learned author, that it is impossible to express the value of integrals of the circular species in terms of functions of two arguments. If this be so, it is impossible to tabulate such integrals, and therefore our course is to devise series more or less convenient for determining their values when any problem, *e.g.* that of the motion of a rigid body, to which Dr Gudermann especially refers, requires us to do so. The formation of such series is accordingly the aim of this memoir, which contains some remarkably elegant formulæ; one of which connects three integrals of the third kind with three of the second.

In the sixteenth and seventeenth volumes of the same *Journal,* Dr Gudermann has given some series for the development of elliptic integrals; and he has since published in the same *Journal* a systematic treatise on the theory of modular functions and modular integrals, these designations being used to denote the transcendents more generally called elliptic. The point of view from which he considers the subject has been already indicated (vide *supra*, p. 271). In a systematic treatise there is of course a great deal that does not profess to be original, and it is not always easy to discover the portions which are so. Dr Gudermann's earlier researches are embodied and developed in his larger work; and in some of the latter chapters (XXIII. 329, &c.)

we find some interesting remarks on the forms assumed by the general transcendent when the biquadratic polynomial in the denominator has four real roots. Dr Gudermann points out the existence of a species of correlation between pairs of values of the variable.

23. The development of the elliptic function ϕ in the form of a continued product may be applied to establish formulæ of transformation. This mode of investigating such formulæ was made use of by Abel in his second paper in Schumacher's *Journal*, No. 148, which we have already noticed; and a corresponding method is mentioned by M. Jacobi in one of the cursory notices of his researches which he inserted in the early volumes of Crelle's *Journal.* Mr Cayley, in the *Philosophical Magazine* for 1843, has pursued a similar course. Another and very remarkable application of the same kind of development consists in taking it as the definition of the function ϕ, and deducing from hence its other properties. It has been remarked that the continued products of Abel and M. Jacobi are derived from considerations which, although cognate, are yet distinct; those of the latter being singly infinite, while Abel's fundamental developments consist of the product of an infinite number of factors, each of which in its turn consists of an infinite number of simple factors. Thus we can have two very dissimilar definitions of the function ϕ by means of continued products. M. Cauchy, who has investigated the theory of what he has termed reciprocal factorials, that is, of continued products of the form

$$\{(1+x)\ (1+tx)\ \ldots\ldots\}\ \{(1+tx^{-1})\ (1+t^2x^{-1})\ \ldots\ldots\}$$

which is immediately connected with M. Jacobi's developments, has accordingly set out from the singly infinite system of products, and has deduced from hence the fundamental properties of elliptic functions (*Comptes Rendus*, XVII. p. 825).

Mr Cayley, on the other hand, has made use of Abel's doubly infinite products, and has shown that the functions defined by means of them satisfy the fundamental formulæ mentioned in the note at page 266, which, as these equations furnish a sufficient definition of the elliptic functions, is equivalent to showing that the continued products are in reality elliptic functions. He has therefore effected for Abel's developments that which M. Cauchy had done for M. Jacobi's. Mr Cayley's paper appeared in the fourth volume of the *Cambridge Mathematical Journal*, but he has since published a translation of it with modifications in the tenth volume of Liouville's. On the same subject we may mention a paper by M. Eisenstein (Crelle's *Journal*, XXVII. 285).

24. M. Liouville has in several memoirs investigated the conditions under which the integral of an algebraical function can be expressed in an algebraical, or, more generally, in a finite form. This investigation is of the same character as that which occurs in the beginning of Abel's last published memoir on elliptic functions (vide *supra*, p. 286). But while Abel's researches are more general than M. Liouville's, the latter has arrived at a result more fundamental, if such an expression may be used, than any of which Abel has left a demonstration.

He has shown that if y be an algebraical function of x, such that $\int y dx$ may be expressed as an explicit finite function of x, we must have

$$\int y dx = t + A \log u + B \log v + \ldots + C \log w,$$

$A, B, \ldots C$ being constant, and $t, u, v, \ldots w$ algebraical functions of x. The theorem established by Abel in the memoir referred to includes as a particular case the following proposition, that if

$$\int y dx = t + A \log u + B \log v + \ldots + C \log w,$$

then $t, u, v, \ldots w$ may all be reduced to rational functions of x and y.

Combining these two results, it appears that if $\int y dx$ be expressible as an explicit finite function of x, its expression must be of the form

$$t + A \log u + B \log v + \ldots + C \log w,$$

where $t, u, v, \ldots w$ are rational functions of x and y, or rather that its expression must be reducible to this form*.

* An equivalent theorem is stated by Abel in his letter to Legendre for implicit as well as explicit functions (Crelle's *Journal*, VI.).

After establishing these results in the memoir (that on elliptic transcendents of the first and second kinds), which will be found in the twenty-third cahier of the *Journal de l'Ecole Polytechnique*, p. 37, M. Liouville supposes y to be of the form $\frac{P}{\sqrt{R}}$, where P and R are integral polynomials, and hence deduces the general form in which the integral $\int \frac{P}{\sqrt{R}}\,dx$ may necessarily be put, provided it admit of expression as an explicit finite function of x.

He shows from hence that if $\int \frac{P}{\sqrt{R}}\,dx$ cannot be expressed by an algebraical function of x, it cannot be expressed by an explicit finite function of it, and finally demonstrates that an elliptic integral, either of the first or second kind, is not expressible as an explicit finite function of its variable.

In a previous memoir inserted in the preceding cahier, M. Liouville proved the simpler proposition, that elliptic integrals of the first and second kinds are not expressible as explicit algebraical functions of their variable (*Journal de l'Ecole Polytechnique*, t. XIV. p. 137). His attention appears to have been directed to this class of researches by a passage of Laplace's 'Theory of Probabilities,' in which the illustrious author, after indicating the fundamental, and, so to speak, ineffaceable distinctions between different classes of functions, states that he had succeeded in showing that the integral $\int \frac{dx}{\sqrt{1 + \alpha x^2 + \beta x^4}}$ is not expressible as a finite function, explicit or implicit, of x. Laplace however did not publish his demonstration.

In his own *Journal* (v. 34 and 441), M. Liouville has since shown that elliptic integrals of the first and second kinds, considered as functions of the modulus, cannot be expressed in finite terms.

25. In the eighteenth volume of the *Comptes Rendus* (Liouville's *Journal*, IX. 353), we find in a communication from M. Hermite, of which we shall shortly have occasion to speak more fully, a remarkable demonstration of Jacobi's theorem. It is stated for the case of the first real transformation, but might of course be rendered general. This demonstration depends

essentially on the principle already mentioned (p. 276), that any rational function of a root of an algebraical equation which has the same value for every root of the equation is rationally expressible in the coefficients. The equation to which this principle is applied is that to which we have so often referred, viz. $y = \frac{U}{V}$, considered as an equation to determine x in terms of y, and by means of it, M. Hermite shows at once that a certain rational function of x is also a rational function of y, the form of which is subsequently determined.

M. Hermite goes on to prove other theorems relating to elliptic functions.

As elliptic functions are doubly periodic, we may determine certain of their properties by considering to what conditions doubly periodic functions must be subject. This view is mentioned by M. Liouville in a verbal communication to the Institute (*Comptes Rendus*, t. XIX.). He states that he had found that a doubly periodic function which is not an absolute constant and has but one value for each value of its variable must be, for certain values of it, infinite; that from hence the known properties of elliptic functions are easily deduced; and that by means of this principle he had succeeded in proving the expressions of the roots of the equation for the division of an elliptic integral of the first kind, which M. Jacobi had given without demonstration in Crelle's *Journal**. I am not aware that any development of M. Liouville's view has as yet appeared.

In the recent numbers of Crelle's *Journal* there are many papers by M. Eisenstein on different points in the theory of elliptic functions. Among these I may mention one which contains a very ingenious proof of the fundamental formula for the addition of two functions, derived from the differential equation of the second order, which each function must satisfy.

Other contributions to the theory of elliptic functions might be mentioned; some of these, not here noticed, are referred to in the index which will be found at the end of this report. But in general it may be remarked that the form which the subject has

* M. Liouville has mentioned that M. Hermite had demonstrated the formulæ in question in a different manner.

assumed, in consequence of the discoveries of Abel and M. Jacobi, is that which it will probably always retain, however our knowledge of particular parts of it may increase. What has since been effected relates for the most part to matters of detail, of which, however important they may be, it is difficult or impossible to give an intelligible account.

26. It does not fall within the design of this report to consider the various applications which have been made of the theory of elliptic functions; but I shall briefly mention some of the geometrical interpretations, if the expression may so be used, which mathematicians have given to the analytical results of the theory.

The lemniscate has, as is well known, the property that its arcs may be represented by an elliptic integral of the first kind, the modulus of which is $\frac{1}{\sqrt{2}}$. The problem of the division of its perimeter is accordingly a geometrical interpretation of that of the division of the complete integral, and was considered by mathematicians at a time when the theory of elliptic functions was almost wholly undeveloped. Besides Fagnani, whose researches with respect to the lemniscate have been already noticed, we may mention those of Euler, who however did not succeed in obtaining a solution of the problem. Legendre, who seems to have attached considerable importance to geometrical illustrations of his analytical results, assigned the equation of a curve of the sixth order, whose arcs measured from a fixed point represent the sum of any elliptic integral of the first kind and an algebraical expression. He showed also that an arc of the curve might be assigned equal to the elliptic integral, but in order to this both extremities of the arc must be considered variable, so that in effect the integral is represented by the difference of two arcs measured from a fixed point (*Traité des Fonctions Elliptiques*, I. p. 36).

M. Serret, in a note presented to the Institute in 1843 (Liouville's *Journal*, VIII. 145), has proved a beautiful theorem, viz. that the sum and difference of the two unequal arcs, intercepted by lines drawn from the centre of Cassini's ellipse to cut the curve, are each equal to an elliptic integral of the first kind, and that the moduli of the two integrals are complementary. In the lemniscate, which is a case of Cassini's ellipse, one of these arcs disappears, and the moduli of the two integrals are equal, each being the sine of half a right angle. So that M. Serret's theorem is an extension of the known property of the lemniscate.

M. Serret has since considered the subject of the representation of elliptic and hyper-elliptic arcs in a very general manner. His memoir, which was presented to the Institute and ordered to be published in the *Savans Etrangers*, appears in Liouville's *Journal*, x. 257. He had remarked that the rectangular co-ordinates of the lemniscate are rationally expressible in terms of the argument of the elliptic integral which represents the arc, for if we assume

$$x = \sqrt{2}a\,\frac{z+z^3}{1+z^4} \text{ and } y = \sqrt{2}a\,\frac{z-z^3}{1+z^4},$$

we shall have

$$ds = \sqrt{\{dx^2+dy^2\}} = 2a\,\frac{dz}{\sqrt{1+z^4}},$$

and if between the first two of these equations we eliminate z, we arrive at the known equation of the lemniscate*. So that if we state the indeterminate equation

$$dx^2 + dy^2 = Z\,.\,dz^2,$$

(x, y and Z being real and rational functions of z), the lemniscate will afford us one solution of it; and every other solution will correspond to some curve whose arc is expressible by an elliptic or hyper-elliptic integral. Of this indeterminate equation M. Serret discusses a particular case. He succeeds in solving it by a most ingenious method, which is applicable to the general equation, and shows from hence that there are an infinity of curves, the arcs of which represent elliptic integrals of the first kind. M. Serret's researches however have not led him to a geometrical representation by means of an *algebraical* curve of *any* integral of the first kind, though his results are generalized in a note appended to his memoir by M. Liouville. In order that the curve may be algebraical, it is necessary and sufficient,

* On reducing the integral $\int\frac{dz}{\sqrt{1+z^4}}$ to the standard form of elliptic integrals, we find that it is an elliptic integral of the first kind, of which the modulus is the sine of 45^0.

as M. Liouville has remarked, that the square of the modulus of the integral should be rational, and less than unity.

In a subsequent memoir (Liouville's *Journal*, x. 351) he has very much simplified the analytical part of his researches, and in the same *Journal* (x. 421) has proved some remarkable properties of one class of what may be called elliptic curves. In the fourth number of the *Cambridge and Dublin Mathematical Journal* (p. 187), M. Serret has developed this part of the subject, and has also given a general sketch of his previous papers. M. Liouville (*Comptes Rendus*, XXI. 1255, or his *Journal*, x. 456) has given a very elegant investigation of an analytical theorem established by M. Serret.

In the fourteenth volume of Crelle's *Journal* (p. 217), M. Gudermann has considered the rectification of the curve called the spherical ellipse, which is one of a class of curves formed by the intersection of a cone of the second order with a sphere. He has shown that its arcs represent an elliptic integral of the third kind.

In the ninth volume of Liouville's *Journal* (p. 155), Mr W. Roberts proves that a cone of the second order, whose vertex lies on the surface of a sphere, and one of whose external axes passes through the centre, intersects the sphere in a curve whose arcs will, according to circumstances, represent any elliptic integral of the third kind and of the circular species; or any elliptic integral of the same kind and of the logarithmic species, provided the angle of the modulus is less than half a right angle; or (subject to the same condition) any elliptic integral of the first kind; or lastly, by a suitable modification, any elliptic integral of the second kind. The cases here excepted may be avoided by introducing known transformations. The cases in which the arcs represent elliptic integrals of the first kind, Mr Roberts has previously mentioned in the eighth volume of Liouville's *Journal* (p. 263). He has since given in the same *Journal* (x. 297), a general investigation of the subject, in which it is supposed that the vertex of the cone may have any position we please. M. Verhulst has represented the three kinds of elliptic integrals by means of sectorial areas of certain curves, and the function Υ by the volume of a certain solid. It is manifest, however, that it is incomparably easier to do this than to represent these transcendents by means of the arcs of curves.

Besides one or two other papers I may mention a tract by the Abbé Tortolini, on the geometrical representation of elliptic integrals of the second and third kinds. This tract, however, I have not seen.

Lagrange long since proved (vide *Théorie des Fonctions Analytiques*, p. 85), that by means of a spherical triangle a geometrical representation of the addition of elliptic integrals of the first kind may easily be obtained, and that hence by a series of such triangles we are enabled to represent the multiplication as well as the addition of these integrals.

M. Jacobi has given (Crelle's *Journal*, III. p. 376, or vide Liouville's *Journal*, x. p. 435) a geometrical construction for the addition and multiplication of elliptic integrals of the first kind. It is founded on the properties of an irregular polygon inscribed in a circle, and the sides of which touch one or more other circles. It is to be remarked that Legendre, in giving an account of this construction in one of the supplements to his last work, has only considered its application to multiplication and not to addition, and has been followed in this respect by M. Verhulst, whose treatise on elliptic functions has been already mentioned. In consequence of this, M. Chasles was led to believe that until the publication of his own researches, no construction for addition excepting that of Lagrange was known. But he has recently (*Comptes Rendus*, January 1846) pointed out the error into which he had fallen.

27. In the *Transactions of the Royal Irish Academy* (IX. p. 151), Dr Brinkley gave a geometrical demonstration of Fagnani's theorem with respect to elliptic arcs, and in the sixteenth volume of the same *Transactions* (p. 76), we find Landen's theorem proved geometrically by Professor MacCullagh.

M. Chasles has considered the subject of the comparison of elliptic arcs by geometrical methods, and with great success. His fundamental proposition may be said to be, that if from any two points of an ellipse we draw two pairs of tangents to any confocal ellipse, the difference of the two arcs of the latter respectively intercepted by each pair of tangents is rectifiable. Or, what in effect is the same thing, if we fasten a string at two points in the circumference of an ellipse, and suppose a ring to move along the string, keeping it stretched, and winding it on and off the arc which lies between its two extremities, the ring will trace out a portion of an ellipse confocal to the former. If for the first ellipse we substitute an hyperbola confocal with the second, the *sum* of the arcs will be constant. From hence a series of theorems is deduced, remarkable not only for their elegance, but also for the facility with which they are obtained. They furnish constructions for the addition and multiplication of elliptic integrals. The whole of this investigation, of which an account is given in the *Comptes Rendus* (Vol. XVII. p. 838, and Vol. XIX. p. 1239), shows, like others of M. Chasles, how much is lost in treating geometrical questions by an exclusive adherence to what may be called the method of co-ordination. Invaluable as this method is, it yet often introduces considerations foreign to the problem to which it is applied*.

III.

28. The first outline of a detailed theory of the higher transcendents was given by Legendre in the third supplement to his *Traité des Fonctions Elliptiques.* He proposes to classify the transcendents comprised in the general formula

$$\int \frac{f(x)\, dx}{(x-\alpha)\sqrt{\phi x}}$$

according to the degree of the polynomial ϕx, the first class being that in which the index of this degree is three or four; the second that in which it is five or six, and so on. The first class therefore consists of elliptic integrals; all the others may be designated as *ultra-elliptic.* This epithet, however, which was proposed by Legendre, has not been so generally used as *hyper-elliptic*, which was, I believe, first used by M. Jacobi. M. Jacobi, however, has proposed to call the higher transcendents Abelian integrals.

The principle of Legendre's classification is to be found in the mininum number of integrals to which the sum of any

* M. Chasles has also considered the subject of spherical conics, as well as that of the lines of curvature and shortest lines on an ellipsoid. The latter has recently engaged the attention of several distinguished mathematicians—MM. Jacobi, Joachimsthal, Liouville, Mac Cullagh and Roberts may be particularlv mentioned.

number of them can be reduced. As we know, this number is unity in the case of elliptic integrals, and by Abel's theorem we find that it is two in the first class of the higher transcendents, three in the next, and so on.

Following the analogy of elliptic integrals, Legendre proposed to recognise three canonical forms in each class of hyper-elliptic integrals, and thus to divide it into three orders. The sum of any number of functions of the first kind will, when the required conditions are satisfied, be equal to a constant; that of any number of the second and third kinds respectively will, under similar conditions, be equal to an algebraical or logarithmic function.

Much the greater part of the remainder of the supplement consists of a discussion of the particular transcendents

$$\int \frac{dx}{\sqrt{1-x^5}} \text{ and } \int \frac{dx}{\sqrt{1+x^5}}.$$

It contains a multitude of numerical calculations, and if the writer's age be considered (he was then almost eighty), is a very remarkable production. By means of the numerical calculations he recognised, as it were empirically, the values to be assigned in different cases to the above-mentioned constant: what these values ought to be, he did not attempt to determine *à priori.*

At the close of the supplement we find a remarkable reduction of an integral, apparently of a higher order to elliptic integrals. The method employed has been generalised by M. Jacobi, in a notice of Legendre's *Supplements,* inserted in the eighth volume of Crelle's *Journal* (p. 413).

29. In the ninth volume of Crelle's *Journal* (p. 394), we find a most important paper by M. Jacobi (*Considerationes Generales, &c.*) which may be said to have determined the direction in which the researches of analysts in the theory of algebraical integrals were to proceed.

The writer proposes two questions, both suggested by the cases of trigonometrical and elliptic functions. First, as in these cases we consider certain functions to which circular and elliptic integrals are respectively inverse, and which are such that functions of the sum of two arguments are algebraically expressible in terms of functions of the simple arguments, what are the corresponding functions to which the hyper-elliptic or Abelian integrals are inverse, and how by means of them can Abel's theorem be stated?

Secondly, as in the same cases we obtain algebraical integrals of differential equations, whose variables are separated, but which nevertheless can only be directly integrated by means of transcendents*, what are the differential equations of which Abel's theorem gives us algebraical integrals? These two questions are, it is obvious, intimately connected.

M. Jacobi first takes the particular case in which the polynomial under the radical is of the fifth or sixth degree. If we call this polynomial X, it follows from Abel's theorem, that if

$$\phi x = \int \frac{dx}{\sqrt{X}},$$

$$\text{and } \phi_1 x = \int \frac{x dx}{\sqrt{X}},$$

we shall have the equations

$$\phi a + \phi b = \phi x + \phi y + \phi x^1 + \phi y^1,$$
$$\phi_1 a + \phi_1 b = \phi_1 x + \phi_1 y + \phi_1 x^1 + \phi_1 y^1,$$

where a and b are given as algebraical functions of the independent quantities x, y, x^1, y^1.

Let
$$\phi x + \phi y = u, \qquad \phi x^1 + \phi y^1 = u^1,$$
$$\phi_1 x + \phi_1 y = v, \qquad \phi_1 x^1 + \phi_1 y^1 = v^1.$$

Then x and y are both given as functions of u and v. We may therefore put

$$x = \lambda (uv), \qquad y = \lambda (uv);$$

and similarly,

$$x^1 = \lambda (u^1 v^1), \qquad y^1 = \lambda_1 (u^1 v^1);$$

and as
$$\phi a + \phi b = u + u^1$$
$$\phi_1 a + \phi_1 b = v + v^1,$$
we shall have
$$a = \lambda (u + u^1,\ v + v^1)$$
$$b = \lambda (u + u^1,\ v + v^1).$$

* *E.g.* $\dfrac{dx}{\sqrt{1-x^2}} + \dfrac{dy}{\sqrt{1-y^2}} = 0$, of which the algebraical integral is

$$x\sqrt{1-y^2} + y\sqrt{1-x^2} = C.$$

Each term of this differential equation is a differential of a transcendent function $\sin^{-1}x$ or $\sin^{-1}y$.

Hence the functions $\lambda(u+u^1, v+v^1)$ and $\lambda_1(u+u^1, v+v^1)$ are expressible as algebraical functions of

$$\lambda(uv), \quad \lambda_1(uv), \quad \lambda(u^1v^1), \quad \lambda_1(u^1v^1).$$

These then are functions to which the integrals are in a certain sense inverse, and which have the same fundamental property as circular and elliptic functions.

In the general case of Abel's theorem, we introduce (when the degree of the polynomial is $2m$ or $2m-1$), $m-1$ functions analogous to λ, each being a function of $m-1$ variables. These functions will, it may easily be shown, have the fundamental property just pointed out for the case in which m is equal to three.

Again, the differential equations of which Abel's theorem gives us algebraical integrals, are, if the degree of the polynomial X be five or six, the following:

$$\frac{dx}{\sqrt{X}}+\frac{dy}{\sqrt{Y}}+\frac{dz}{\sqrt{Z}}=0,$$

$$\frac{x\,dx}{\sqrt{X}}+\frac{y\,dy}{\sqrt{Y}}+\frac{z\,dz}{\sqrt{Z}}=0;$$

and generally, if the degree of the polynomial be $2m$ or $2m-1$, there are $m-1$ such equations, the numerators of the last containing the $(m-2)$th power of the variables.

M. Jacobi concludes by suggesting as a problem the direct integration of these differential equations, so as to obtain a proof of Abel's theorem corresponding to that which Lagrange gave of Euler's (vide *ante*, p. 241).

30. Another important paper by M. Jacobi is that which is entitled *De Functionibus duarum Variabilium quadrupliciter periodicis*, etc. (Crelle, XIII. p. 55). It is here shown that a periodic function of one variable cannot have two distinct real periods. In the case of a circular function, though we have for all values of x

$$\begin{aligned}\sin x &= \sin(x+2m\pi)\\ &= \sin(x+2n\pi),\end{aligned}$$

m and n being any integers, yet $2m\pi$ and $2n\pi$ do not constitute two *distinct* periods, since each is merely a multiple of 2π, which is the fundamental period of the function. But if we had for all the values of x

$$f(x) = f\{x + \alpha\} = f\{x + \beta\},$$

we should also have

$$f(x) = f(x + m\alpha + n\beta),$$

where m and n may be any integers, positive or negative. Hence $m\alpha + n\beta$ may, provided α and β are incommensurable, which is implied in their being distinct periods, be made less any assignable quantity, so that we may put

$$fx = f(x + \epsilon),$$

where ϵ is indefinitely small, and this manifestly is an inadmissible result. Accordingly we see that one at least of the periods of elliptic functions is necessarily imaginary.

Again, similar reasoning shows that in a triply periodic function, that is in one in which we have

$$f(x) = f\{x + m(\alpha + \beta\sqrt{-1}) + m'(\alpha' + \beta'\sqrt{-1}) + m''(\alpha'' + \beta''\sqrt{-1})\}$$

for every value of x, m, m', m'' being any integers, and in which the three periods $\alpha + \beta\sqrt{-1}$, &c. are *distinct*, we can make

$$f(x) = f(x + \epsilon)$$

by assigning suitable values to m, m', m''; ϵ being as before less than any assignable quantity. Hence as this result is inadmissible, it follows that there is no such thing as a triply periodic function. Whenever therefore a function appears to have three periods they are in reality not distinct, and so *à fortiori* when it appears to have more than three. But now we come to a difficulty. For M. Jacobi proceeds to show that if we consider a function of one variable inverse to the Abelian integral

$$\int \frac{(\alpha + \beta x)\, dx}{\sqrt{X}},$$

X being of the sixth degree in x, this function has four distinct and irreducible periods. His conclusion is that we cannot consider the amplitude of this integral as an analytical function of the integral itself. In the present state of our knowledge, this conclusion, though seemingly forced on us by the impossibility of recognising the existence of a quadruply periodic function of one variable, is not, I think, at all satisfactory. The functional dependence, the existence of which we are obliged to deny, may be expressed by a differential equation of the second order; and therefore it would seem that the commonly received opinion that every differential equation of two variables has a primitive, or expresses a functional relation between its variables, must be abandoned, unless some other mode of escaping from the difficulty is discovered. It is probable that some simple consideration, rather of a metaphysical than an analytical character, may hereafter enable us to form a consistent and satisfactory view of the question, and this I believe I may say is the opinion of M. Jacobi himself. The same difficulty meets us in all the Abelian integrals: as in the case of those of Legendre's first class, namely where X is of the fifth or sixth degree, so also generally, the inverse function has more than its due number of periodicities.

Abel, in a short paper in the second volume of his works, p. 51, has in effect proved the multiple periodicity of the functions which are inverse to the integrals to which his theorem relates. The difficulty to which this gives rise did not strike him, or was perhaps reserved for another occasion.

M. Jacobi next proves that his inverse functions of two variables are quadruply periodic, but that quadruple periodicity for functions of two variables is nowise inadmissible.

A difficulty however seems to present itself, which is suggested by M. Eisenstein in Crelle's *Journal*, viz. that if for each value of the amplitude the integral ϕx or $\int \frac{dx}{\sqrt{X}}$ (vide *supra*, p. 302), has an infinity of magnitudes real and imaginary, and the same is the case for ϕy, it is by no means easy to attach a definite sense to the equation $u = \phi x + \phi y$, or to see how the value of u is determined by it*.

31. Two divisions of the theory of the higher transcendents here suggest themselves, which are apparently less intimately connected than the corresponding divisions in the theory of

* The difficulty here mentioned may perhaps be met by saying that the value of ϕx determined by the integral $\int_0^x \frac{dx}{\sqrt{X}}$ is necessarily determinate, and so likewise is that of u. That considerations connected with the conception of a function inverse to ϕx make the latter quantity appear indeterminate is undoubtedly a difficulty; but it is, so to speak, a difficulty collateral to M. Jacobi's theory, and therefore need not prevent our accepting it.

elliptic functions, viz. the reduction and transformation of the integrals themselves, and the theory of the inverse functions.

But before considering these I shall give some account of what has been done in fulfilment of the suggestion made by M. Jacobi at the close of the *Considerationes Generales*. Mathematicians have succeeded in effecting the integration of the system of differential equations to the consideration of which we are led by Abel's theorem, and which is commonly designated by German mathematicians as the "Jacobische system;" its existence and its integrability having been first pointed out by M. Jacobi.

In Crelle's *Journal* (XXIII. 354), M. Richelot, after modifying the form in which Lagrange's celebrated integration of the differential equation of elliptic integrals is generally presented, extended a similar method to the system of two differential equations which occurs when we consider the Abelian transcendents of the first class. He thus obtains one algebraical integral of the system. In the case of Lagrange's equation one integral is all we want; but in that which M. Richelot here discusses we require two. Now if in the former case we replace each of the variables by its reciprocal, we obtain a new differential equation of the same form as the original one, and integrable therefore in the same manner; and if in its integral we again replace each new variable by its reciprocal, that is by the original variable, we thus, as it is not difficult to see, get the integral of the original equation in a different form. That the two forms are in effect coincident may be verified *à posteriori*. But the same substitutions being made in M. Richelot's equations, which are of course those we have already mentioned at p. 302, the first of them becomes similar in form to the second, and *vice versâ* the second to the first. Thus the system remains similar to itself; and if in the algebraical integral we obtain of it we again replace the new variables by their reciprocals, we fall on a new algebraical integral of the original system; this integral being, which is remarkable, independent of that previously got. Thus the system of two equations is completely integrated. Extending his method to the general system of any number of equations, M. Richelot obtains for each two integrals, but of course these are not all that we want. At the conclusion of his memoir M. Richelot derives from Abel's theorem the algebraical integrals of the "Jacobische system."

Though in this memoir M. Richelot only obtained by direct integration two of the $m-1$ algebraical integrals of the "Jacobische system," yet he put the problem of its complete integration into a convenient and symmetrical form. As there are m variables and $m-1$ relations among them, we may suppose each to be a function of an independent variable t. Lagrange, as we know, in integrating the equation

$$\frac{dx}{\sqrt{X}}+\frac{dy}{\sqrt{Y}}=0,$$

introduced such an independent variable by the assumption

$$\frac{dx}{dt}=\sqrt{X},$$

which of course implied that $\frac{dy}{dt}=-\sqrt{Y}$. This assumption is unsymmetrical, and it is therefore difficult to see how to generalise it. But if we assume $\frac{dx}{dt}=\frac{\sqrt{X}}{x-y}$, we shall of course have $\frac{dy}{dt}=\frac{\sqrt{Y}}{y-x}$, and therefore t is symmetrically related to x and y. Let $Fu=0$ be an equation whose roots are x and y, then, as we know, when $u=x$, $F'u=x-y$, and, when $u=y$, $F'u=y-x$, so that using an abbreviated notation

$$\frac{dx}{dt}=\frac{\sqrt{X}}{F'(x)} \text{ and } \frac{dy}{dt}=\frac{\sqrt{Y}}{F'(y)}.$$

Nothing is easier than to generalise this result. For instance, the "Jacobische system" of two equations is

$$\frac{dx}{\sqrt{X}}+\frac{dy}{\sqrt{Y}}+\frac{dz}{\sqrt{Z}}=0,$$

$$\frac{xdx}{\sqrt{X}}+\frac{ydy}{\sqrt{Y}}+\frac{zdz}{\sqrt{Z}}=0.$$

Now if $Fu=0$ have x, y, z for its roots, the two preceding equations may, in virtue of a very well-known theorem, be replaced by the three following,

$$\frac{dx}{dt}=\frac{\sqrt{X}}{F'x},\ \frac{dy}{dt}=\frac{\sqrt{Y}}{F'y},\ \frac{dz}{dt}=\frac{\sqrt{Z}}{F'z},$$

which introduce an independent variable t, symmetrically related to x, y and z; and so in all cases.

M. Richelot* then takes a symmetrical function of $x, y, \ldots z$, viz. their sum, and by means of the last written equations arrives at an integrable differential equation of the second order, the principal variable being the said sum and the independent variable being t. From the first integral of this equation it is easy to eliminate the differentials, and we have thus an algebraical relation in $x, y, \ldots z$, from which, in the manner already mentioned, M. Richelot deduces another. We now see that if we could find any other symmetrical function which would lead to an integrable equation we should get a finite relation among the variables.

In the next volume of Crelle's *Journal* M. Jacobi took the following function as his principal variable,

$$\sqrt{\{(\mu - x)(\mu - y) \ldots (\mu - z)\}},$$

μ being a root of $X = 0$ or $fx = 0$ if we suppose $X = fx$. Calling this function v, we get a simple differential equation in v and t, and a corresponding integral of the system. Now $fx = 0$ has $2m$ or $2m - 1$ roots, and we only want $m - 1$ integrals. The integrals therefore which we get by making μ the first, second, &c. root of $fx = 0$ are not all independent.

* As M. Richelot's method of demonstrating Euler's theorem is more symmetrical and far more easily remembered than Lagrange's, it ought, I think, to be introduced into all elementary works on elliptic functions. The equation to be integrated being

$$\frac{dx}{\sqrt{\alpha + \beta x + \gamma x^2 + \delta x^3 + \epsilon x^4}} + \frac{dy}{\sqrt{\alpha + \beta y + \gamma y^2 + \delta y^3 + \epsilon y^4}} = 0,$$

assume
$$\frac{dx}{dt} = \frac{\sqrt{\alpha + \beta x + \gamma x^2 + \delta x^3 + \epsilon x^4}}{y - x}.$$

Then
$$\frac{dy}{dt} = \frac{\sqrt{\alpha + \beta y + \gamma y^2 + \delta y^3 + \epsilon y^4}}{x - y}.$$

Let $p = x + y$. Then after a few obvious reductions

$$\frac{d^2p}{dt^2} = \frac{1}{2}\delta + \epsilon p.$$

Hence
$$\frac{dp^2}{dt^2} = \delta p + \epsilon p^2,$$

or
$$\{\sqrt{\alpha + \beta x + \gamma x^2 + \delta x^3 + \epsilon x^4} - \sqrt{\alpha + \beta y + \gamma y^2 + \delta y^3 + \epsilon y^4}\}^2$$
$$= \delta(y - x)(y^2 - x^2) + \epsilon(y^2 - x^2)^2,$$

the algebraical integral sought. It may easily be expressed in other forms.

In the twenty-fifth volume of Crelle's *Journal* M. Richelot resumed the subject of his former paper, and discussed it in a very interesting memoir. This fundamental or principal result may be said to be a generalisation of M. Jacobi's. It is that in the function

$$\sqrt{\{(\mu - x)(\mu - y) \dots (\mu - z)\}}$$

μ may have any value whatever. The resulting differential equation, though rather more complicated than when, with M. Jacobi, we suppose μ a root of $fx = 0$, is still very readily integrable. We have thus an indefinite number of algebraical integrals, since the quantity μ is arbitrary, but of course not more than $m - 1$ of them are independent.

In the same volume of Crelle's *Journal*, p. 178, there is a curious paper by Dr Hædenkamp, in which the algebraical integrals of Jacobi's system are for the case of a polynomial of the fifth degree under the radical deduced from geometrical considerations. It is shown that in a system of curvilinear coordinates (those of which MM. Lamé and Liouville have made so much use), the equations of the system are the differential equations given by the Calculus of Variations for the shortest line between two points. Consequently the finite equations of a straight line are the integrals sought. This very ingenious consideration is afterwards generalised.

32. In the twelfth volume of Crelle's *Journal*, p. 181, M. Richelot has considered the Abelian integral of the first class. The principal result at which he arrives is, that the only rational transformation by means of which such an integral may be changed into one of similar form is linear in both the variables which it involves. By means of this substitution, he transforms, under certain conditions, the integral in question into a form analogous to the standard forms of elliptic integrals. The subject of the division of hyper-elliptic integrals of each class into three genera is also considered, and the same principle of classification as Legendre's is made use of. The paper concludes by pointing out an error which Legendre committed in the application of his principle. Legendre had thought that the formula of summation given by Abel's theorem for integrals of the form $\int \frac{x^e dx}{\sqrt{\phi x}}$ could not involve a logarithmic function.

Thus these integrals would belong to the first or second kind, according to the value of the index e, and of λ the degree of ϕx. But in reality, though the integrals in question are of the first kind (that is, they admit of summation without introducing either an algebraical or logarithmic function) if e be less than a certain limit, yet if it be not so their formula of summation will in general involve both algebraical and logarithmic functions. Either may, under certain conditions as to the form of ϕx, disappear, but while ϕx is merely known as the polynomial of the λth degree, we cannot decide whether the integral is to be referred to the second or third kind.

I may mention here a very elegant result due to M. Jacobi. It appears in the thirtieth volume of Crelle's *Journal*, p. 121, and is a generalisation of the fundamental formula for the addition of elliptic arcs. With a slight modification it may be thus stated. If ϕx involve only even powers of x, the highest being x^{4m}, then the sum of the integrals $\int_0^x \frac{x^{2m}dx}{\sqrt{\phi(x)}}$ is equal to the product of their arguments, that is of the different quantities denoted by the symbol x. In this case then the logarithmic function disappears, and the integral belongs to the second kind.

In the twenty-ninth volume of Crelle's *Journal* there is a paper by M. Richelot on a question connected with hyper-elliptic integrals. The reader will find in it a good many fully-developed results, which may be considered as particular cases of Abel's theorem. They illustrate the learned author's criticism of Legendre's classification of hyper-elliptic integrals, though they are not adduced for that purpose.

The function M (vide *ante*, p. 244) is a function of the arbitrary quantities $a, b, \ldots c$, which, as has been remarked, may themselves be considered functions of the arguments $x_1, x_2, \ldots x_\mu$. To determine M as a function of the last-written quantities is a necessary ulterior step in almost any special application of Abel's theorem, and this M. Richelot has done in several interesting cases, establishing at the same time the relations which exist among the quantities in question. His investigations, however, have an ulterior purpose, and are not to be considered merely as corollaries from Abel's theorem.

Another paper of M. Richelot, on the subject of the Abelian integrals, is found in the sixteenth volume of Crelle's *Journal*,

p. 221. The aim of it is to furnish the means of actually calculating the value of the Abelian integral of the first class by a method of successive transformation, that is, by a method analogous to that used for elliptic integrals. M. Richelot's process depends essentially on an irrational substitution, by means of which we can replace the proposed integrals by two others which differ only with respect to their limits. In the development of this idea the author confines himself to the first kind of the Abelian integrals of the first class, though the same method may $m.m.$ be more generally applied*. From the formula which expresses the proposed integral as the aggregate of two others is deduced another, in which it is expressed by means of four integrals, the inferior limits of all being zero. The first and second of these integrals differ only in their amplitude, and the same is true of the third and fourth. There are two principal transformations, either of which may be repeated as often as we please; and though it might seem that the number of integrals would in the successive transformations increase in a geometrical progression, yet by the application of Abel's theorem we can always reduce them to the same number. But the development of this part of the subject M. Richelot has reserved for another occasion†.

At the close of his memoir, M. Richelot has given some numerical examples of his method for the case of a complete hyper-elliptic integral. The third example he had previously given in a brief notice of his researches, published in No. 311 of Schumacher's *Journal*.

33. For many years after the death of Legendre the subject of the comparison of transcendents was studied principally by German and Scandinavian writers‡: a young French mathematician, M. Hermite, has recently made important discoveries in this theory; but as the principal part of what he has done

* The integral to be transformed is $\dfrac{(M+Nz)\,dz}{\sqrt{\{z(1-z)(1-\kappa^2 z)(1-\lambda^2 z)(1-\mu^2 z)\}}}$, M and N, &c. being constant.

† His transformations ultimately reduce the hyper-elliptic integral to elliptic integrals; the latter may be considered known quantities, "vel per paucas adjectas transformationes directe computentur."

‡ The papers of M. Liouville, already noticed, may be said to be an exception to this remark.

is as yet not published, a very imperfect outline is all that can be given.

In the seventeenth volume of the *Comptes Rendus*, we find the report of a commission, consisting of MM. Lamé and Liouville, on a memoir presented to the Institute by M. Hermite. This report is reprinted in the eighth volume of Liouville's *Journal*, p. 502. A remark which incidentally occurs in it, namely, that Abel was the first to give the general theory of the division of elliptic integrals, led to a very warm discussion between MM. Liouville and Libri, on the subject of the claims which, as I have already remarked, the latter had made with reference to this theory.

It appears from the report, that M. Hermite has succeeded in solving the problem of the division of hyper-elliptic integrals. The division of elliptic integrals depends on the solution of an algebraical equation; that of the hyper-elliptic integrals (as the functions inverse to them involve, as we have seen, more than one variable), on the solution of a system of simultaneous algebraical equations. This solution can, M. Hermite has shown, be effected by means of radicals assuming, as in the analogous case of elliptic functions, the division of the complete integrals. M. Hermite's method depends for the most part on the periodicity of the functions considered. A transcendental expression of the roots of the equation of the problem having been obtained, their algebraical values are deduced from it.

These researches, in themselves of great interest, are yet more interesting, when we consider how completely they justify the views of M. Jacobi as to the manner in which Abel's theorem ought to be interpreted, by showing that his theory of the higher transcendents is no barren or artificial generalisation.

At page 505 of the volume of Liouville's *Journal* already mentioned, we find an extract from a letter of M. Jacobi to M. Hermite, in which, after congratulating him on the important discovery he had made, he points out that the transcendental functions $\lambda(uv)$, $\lambda_1(uv)$ (vide *ante*, p. 302) are algebraical functions of transcendental functions which involve but one variable.

M. Hermite's subsequent researches have embraced a much more general theory than that of the Abelian integrals, namely, that of the integrals of any algebraical function whatever. Thus his views bear the same relation to Abel's general theory, developed in the *Savans Etrangers*, that those of M. Jacobi in the *Considerationes Generales* do to Abel's theorem.

All that has yet been published with respect to them is contained in the *Comptes Rendus*, XVIII. p. 1133, in the form of an extract of a letter from M. Hermite to M. Liouville. This extract is reprinted in Liouville's *Journal*, IX. p. 353. It was communicated to the Institute in June 1844.

Following the course of M. Jacobi's inquiries, M. Hermite proposed to determine what are the differential equations of which Abel's investigations give the complete algebraical integrals. When this is done it suggests the nature of the inverse functions which are to be introduced. The number of these functions will of course vary in different cases, just as in M. Jacobi's less general theory. Let us suppose this number to be denoted by γ, then each function will involve γ variables. And if each of these variables be replaced by the sum of two new variables, then all the functions are given as the roots of an equation of the γth degree, whose coefficients are rational in terms of the corresponding functions of each of the new variables and of certain known algebraical functions. From hence is derived the theory of the periodicity of these functions.

After some other remarks on the theory of the higher transcendents, M. Hermite states that the method of division of which he made use in the problem of the division of Abelian integrals extends also to the new transcendents now considered, but that in the theory of transformation he had not as yet been successful. The greater part of the remainder of this remarkable communication relates to elliptic functions, and has been already noticed. The remark just mentioned as having been made by M. Jacobi for the functions which are inverse to the Abelian integrals, extends, M. Hermite observes, to the functions which he considers.

In conclusion, M. Hermite remarks that the method of differentiation with respect to the modulus of which Legendre made so much use in the theory of elliptic functions, may be applied to all functions of the form

$$\int f(xy)\, dx,$$

where y is given by the equation

$$y^n - X = 0.$$

In concluding this report, it may be remarked that the subject of it is still incomplete, and that there is yet much to be done which we may hope it will not be found impossible to do. It is however difficult to predict the direction in which progress will hereafter be made. Yet I think we may reasonably suppose that the question of multiple periodicity, from the paradoxical aspect in which it has presented itself, and from its connexion with the general principles of the science of symbols, will sooner or later attract the attention of all philosophical analysts. M. Liouville's idea of considering the conditions to which a doubly periodic function must as such be subject, can scarcely be developed or extended to the higher transcendents without leading to results of great generality and interest.

The detailed discussion of different classes of algebraical integrals, their transformations and reductions, form an endless subject of inquiry. But in this, as in other cases, the increasing extent of our knowledge will of itself tend to diminish the interest attached to the full development of particular portions of it; and with reference to analytical problems arising out of questions of physical science, the theory of the higher transcendents will it is probable never become of so much importance as the theory of elliptic functions. We have occasion to make use of circular much more frequently than of elliptic functions, and similarly we shall, it may be presumed, have less frequently to introduce the higher transcendents than elliptic functions. Numerical calculations of the values of the higher transcendents are therefore less important than similar calculations in the case of elliptic functions*.

The following index is intended to contain references to all the papers in the first thirty-one volumes of Crelle's *Journal*, and in the first ten volumes of Liouville's *Journal*, more or less connected with the subject of this report, together with a considerable number of others.

* The Academy of Sciences has proposed as the subject of the great mathematical prize for 1846 the following question:—"Perfectionner dans quelque point essentiel la théorie des fonctions abéliennes ou plus généralement des transcendantes qui résultent de la considération des intégrales de quantités algébriques." The memoirs are to be sent in before the 1st of October.

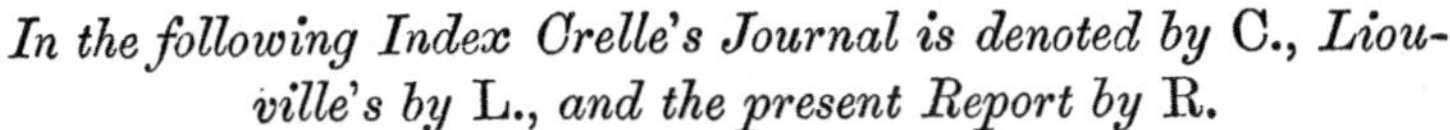
In the following Index Crelle's Journal is denoted by C., *Liouville's by* L., *and the present Report by* R.

ABEL. Ueber die Integration der differential Formel $\frac{\rho dx}{\sqrt{R}}$ wenn R und ρ ganze Functionen sind.—C. I. 185. This paper contains formulæ of reduction. It is mentioned by M. Liouville, 'Journal de l'École Polytechnique,' 23d cah. p. 38. It appears in French in Abel's collected works, Tom. I. 33.

—— Recherches sur les Fonctions Elliptiques.—C. II. 101, and III. 160 [1827]. V. Abel's works, I. 141 ; also R. p. 273.

—— Remarques sur quelques Propriétés Générales d'une certaine sorte de Fonctions Transcendantes.—C. III. 313. This paper contains the theorem commonly known as 'Abel's Theorem.' V. Abel's works, I. 288 ; also R. p. 247.

—— Sur le Nombre de Transformations Différentes qu'on peut faire subir à une Fonction Elliptique par la Substitution d'une Fonction donnée du premier dessé.—C. III. 394. V. Abel's works, I. 309.

—— Théorème Général sur la Transformation des Fonctions Elliptiques de la seconde et de la troisième espèce.—C. III. 402. The theorem is stated without demonstration. V. Abel's works, I. 317.

—— Note sur quelques Formules Elliptiques.—C. IV. 85. This paper contains developments of elliptic functions, &c. V. Abel's works, I. 299.

—— Théorèmes sur les Fonctions Elliptiques.—C. IV. 194. They relate to the demonstration of the theorem stated by M. Jacobi in the third volume of Crelle's Journal, p. 86, by means of which Abel's method for the division of elliptic integrals is greatly simplified. V. Abel's works, I. 318 ; also R. p. 277.

—— Démonstration d'une Propriété Générale d'une certaine Classe de Fonctions Transcendantes.—C. IV. 200. This short paper contains the fundamental idea of the memoir presented to the Institute in 1826. V. Abel's works, I. 324 ; also R. p. 246.

—— Précis d'une Théorie des Fonctions Elliptiques.—C. IV. 236 and 309. This *précis* was left unfinished. V. Abel's works, I. 326 ; also R. p. 286.

—— Extracts from Letters to M. Crelle, one of which relates to the comparison of Transcendents.—C. V. 336. V. Abel's works, II. 253.

—— A Letter to M. Legendre.—C. VI. 73. Works, II. 256. It

contains a theorem proved by M. Ramus in the twenty-fourth volume of Crelle's Journal, p. 78; and another, proved by M. Liouville in the twenty-third cahier of the 'Journal de l'École Polytechnique.' V. R. p. 249 and p. 294.

—— Mémoire sur une certaine classe de Fonctions Transcendantes. Presented to the Institute, Oct. 30, 1826, published in 1841 in the 'Mémoires des Savans Étrangers,' VII. 176. It is the only memoir of Abel's not contained in the collected edition of his works published in 1839; the editor, M. Holmboe, not having been able to procure a copy of it. V. R. p. 246.

—— Solution d'un Problème Général concernant la Transformation des Fonctions Elliptiques.—Schumacher's Astronomische Nachrichten, No. 138, VI. 365. V. Abel's works, I. 253; also R. p. 282.

—— Addition au Mémoire Précédent.—Schumacher's Astronomische Nach., No. 147, VII. 33. V. Abel's works, I. 275; also R. *ubi supra.*

The following papers were published for the first time in Abel's collected works. The references are to the second volume:—

—— Propriétés remarquables de la Fonction $y=\phi x$, etc., p. 51. The multiple periodicity of a function inverse to a hyper-elliptic integral is here mentioned. V. R. p. 305.

—— Sur une Propriété remarquable d'une Classe très étendue de Fonctions Transcendantes, p. 54. This paper contains a generalisation of a theorem relating to elliptic functions.

—— Extension de la Théorie Précédente, p. 58.

—— Sur la Comparaison des Fonctions Transcendantes, p. 66. This paper contains a somewhat fuller development of his general theory than that which is inserted in the fourth volume of Crelle's Journal, p. 200. V. R. p. 247.

—— Théorie des Transcendantes Elliptiques, p. 93. V. R. p. 287.

—— Démonstration de quelques Formules Elliptiques, p. 210.

BROCH. Sur quelques Propriétés d'une certaine classe des Fonctions Transcendantes.—C. XX. 178. An extension of Abel's theorem. V. R. p. 248.

—— Mémoire sur les Fonctions de la forme

$$\int x^{s-\gamma p-1} F(x^p)\,(R(x^p))^{\pm\frac{s}{rp}}\,dx, \text{ etc.—C. XXIII. 145 and 201.}$$

This memoir, of which the first part may be considered a generalisation of the preceding, is accompanied by a report of MM. Liouville and Cauchy. V. R. p. 248.

BRONWIN. On Elliptic Functions.—Camb. Mathematical Journal, III. 123. Mr Bronwin puts the transcendental formula of transformation in a very neat form.

—— On M. Jacobi's Theory of Elliptic Functions.—Lond., Ed. and Dub. Phil. Mag. XXII. 258. V. R. p. 267.

—— Reply to Mr Cayley's Remarks.—L., E. & D. Phil. Mag. XXIII. 89. V. R. *ubi supra.*

CATALAN. Sur la Réduction d'une Classe d'Intégrales Multiples.—L. IV. 323.

—— Sur les Transformations des Variables dans les Intégrales Multiples. Mémoires Couronnés par l'Académie Royale de Bruxelles, XIV. 2de partie, p. 1. The third part contains a transformation of a multiple integral leading to properties of hyperelliptic integrals analogous to known properties of elliptic integrals.

CAUCHY. Comptes Rendus, XVII. 825.—V. R. p. 292.

CAYLEY. Mémoire sur les Fonctions doublement Périodiques.—L. X. 385. An enlargement of his paper on the inverse elliptic functions, published in the fourth volume of the Cambridge Mathematical Journal. V. R. p. 292.

—— Remarks on the Rev. B. Bronwin's paper.—L., E. and D. Phil. Mag. XXII. 358.

—— Investigation of the Transformation of certain Elliptic Functions.— L., E. and D. Phil. Mag. XXV. 352. V. R. p. 292.

—— On the Inverse Elliptic Functions.—Camb. Math. Journal, IV. 257. V. R. p. 292.

CHASLES. Comptes Rendus de l'Institut, XVII. 838, and XIX. 1239. M. Chasles in these two communications presents to the Institute notices of his geometrical researches illustrative of the theory of elliptic functions. V. R. p. 300.

CLAUSEN. Schumacher's Nachrichten, XIX. 178. On a particular Integral mentioned by Legendre.

—— Schumacher's Nachrichten, XIX. 181. It is shown that the arcs of one of the curves, known as the Spirica of Perseus, may be rectified by means of an elliptic integral.

EISENSTEIN. Théorèmes sur les Formes Cubiques.—C. XXVII. 75. At the end of this paper we find some developments of elliptic functions in continued fractions. This subject is continued in the following paper of M. Eisenstein's.

—— Transformations remarquables de quelque Séries.—C. XXVII. 193, and XXVIII. 36. See also Theorema, C. XXIX. 96.

—— Bemerkungen zu den elliptischen und Abelschen Transcendenten.—C. xxvii. 185. M. Jacobi has criticised this paper (of which a translation appears in Liouville's Journal; x. 445) in the thirtieth volume of Crelle's Journal.

—— Elementare Ableitung einer merkwürdiger Relation zwischen zwei unendlichen Producten.—C. xxvii. 285. V. R. p. 293.

—— Beitrage zur Theorie der Elliptischen Functionen.—C. xxx. 185, 211. This paper contains a demonstration of the fundamental formula of elliptic functions. V. R. p. 295.

Gudermann. Integralia Elliptica Tertiæ Speciei Reducendi Methodus Simplicior, &c.—C. xiv. 159, 185. V. R. p. 291.

—— Einige Bemerkungen über Elliptische Functionen.—C. xvi. 78.

—— Series novæ quarum ope Integralia Elliptica Primæ et Secundæ Speciei computantur, &c.—C. xvi. 366, and xvii. 382.

—— Theorie der Modular Functionen und der Modular Integrale.—C. xviii. 1, 142, 220, 303; xix. 46, 119, 244; xx. 62, 103; xxi. 240; xxiii. 301; xxv. 281. A systematic treatise on elliptic functions. V. R. p. 291.

Gützlaff. Æquatio Modularis pro Transformatione Functionum Ellipticarum Septimi Ordinis.—C. xii. 173. V. R. p. 291.

Haedenkamp. De Transformatione Integralis

$$\iint \frac{d\phi \, d\psi}{\sqrt{(\sin^2\nu - \sin^2\phi \cos^2\psi)}}.$$

—C. xx. 97. It is shown to be the product of two elliptic integrals.

—— Über Transformation vielfacher Integrale.—C. xxii. 184. Analogous to the researches of M. Catalan in the 'Mémoires de Bruxelles,' which appears to have been previously published.

—— Über Abelsche Integrale.—C. xxv. 178. V. R. p. 309.

Hermite. Sur la Théorie des Transcendantes à Différentielles Algébriques.—L. ix. 353. Extracted from the 'Comptes Rendus' [June 1844]. This note, which contains scarcely more than an indication of M. Hermite's results, may be said to mark the furthest advance yet made in the theory of the comparison of transcendents. V. R. p. 312.

Hill. Exemplum usus Functionum Iteratarum, &c.—C. xi. 193. This paper contains some interesting applications of the calculus of functions to the comparison of transcendents. V. R. p. 250.

Jacobi. Addition au Mémoire de M. Abel sur les Fonctions Elliptiques.—C. iii. 86. A short note, containing an important simplification of Abel's method of solving the equation of the problem of division. V. R. p. 277.

—— Note sur la Décomposition d'un Nombre donné en quatre

quarrés.—C. III. 191. The demonstration referred to is founded on elliptic functions.

—— Note sur les Fonctions Elliptiques.—C. III. 192.

—— Suite des Notices sur les Fonctions Elliptiques.—C. III. 303.

—— Suite des Notices, etc.—C. III. 403.

—— Suite des Notices, etc.—C. IV. 185. These notes contain theorems stated for the most part without demonstration. V. R. p. 271.

—— Ueber die Anwendung der elliptischen Transcendenten auf ein bekanntes Problem der Elementar-geometrie, u. s. w.—C. III. 376. This paper contains a geometrical construction for the addition and multiplication of elliptic integrals of the first kind. A translation of the most important part appears in Liouville's Journal, x. 435. V. R. p. 272.

—— De Functionibus Ellipticis Commentatio.—C. IV. 371. Transformations of integrals of the second and third kinds, &c. V. R. p. 288.

—— De Functionibus Ellipticis Commentatio altera.—C. IV. 397. We find here an elementary demonstration of M. Jacobi's theorem. V. R. p. 288.

—— Note sur une nouvelle application de l'Analyse des Fonctions Elliptiques à l'Algèbre.—C. VII. 41. It relates to the development in continued fractions of a function of the fourth degree.

—— Notiz zu Théorie des Fonctions Elliptiques de Legendre, Troisième Supplément.—C. VIII. 413. V. R. p. 288.

—— De Theoremate Abeliano—C. IX. 99. V. R. p. 249.

—— Considerationes Generales de Transcendentibus Abelianis.—C. IX. 394 [1832]. This memoir lays the foundation of the theory of the higher transcendents. V. R. p. 301.

—— De Functionibus Duarum Variabilium quadrupliciter Periodicis, &c.—C. XIII. 55. M. Jacobi here proves the impossibility of a function of one variable being triply periodic. V. R. p. 303.

—— De usu Theoriæ Integralium Ellipticorum et Integralium Abelianorum in Analysi Diophanteâ.—C. XIII. 353. It is here pointed out that a problem of indeterminate analysis, discussed by Euler in the posthumous memoirs recently published by the Academy of St Petersburg, is in effect that of the multiplication and addition of elliptic integrals. Suggestions are made as to the corresponding application that might be made of the Abelian integrals.

—— Formulæ novæ in Theoriâ Transcendentium Fundamentales.—C. XV. 199. Elegant elementary formulæ.

JACOBI. Note von der Geodätischen Linie auf einem Ellipsoid, u. s. w. —C. XIX. 309. M. Jacobi has here announced the important discovery that the equation to the shortest line on an ellipsoid is expressible by means of Abelian integrals of the first class. As this is perhaps the first application made of Abelian integrals since their recognition as elements of analysis, I have thought it well to mention it in this place. A translation of the note is found in Liouville's Journal, VI. 267.

—— Demonstratio nova Theorematis Abeliani.—C. XXIV. 28. V. R. p. 308.

—— Zur Theorie der elliptischen Functionen.—C. XXVI. 93. This paper contains series for the calculation of elliptic functions, and a table of the function *q*.

—— Ueber die Additions-theoreme der Abelschen Integrale zweiter und dritter Gattung.—C. XXX. 121. We find here some remarkable formulæ. V. R. p. 310.

—— Note sur les Fonctions Abéliennes.—C. XXX. 183. This note relates principally to the fact announced in M. Jacobi's letter to M. Hermite. V. L. VIII. 505.

—— Ueber einige die elliptischen Functionen betreffenden Formeln.—C. XXX. 269.

—— Extrait d'une Lettre à M. Hermite.—L. VIII. 505. V. R. p. 312.

—— Extraits de deux Lettres de M. Jacobi, &c.—Schumacher's Nachrichten, VI. 33 [Sept. 1827]. They contain the first announcement of his theorem.

—— Demonstratio Theorematis ad Theoriam Functionum Ellipticarum Spectantis.—Schumacher, VI. 133. The first published demonstration of his theorem. See also Legendre at p. 201 of the same volume. V. R. pp. 259 and 260.

JÜRGENSEN. Sur la Sommation des Transcendantes à Différentielles Algébriques.—C. XIX. 113.

—— Remarques Générales sur les Transcendantes à Différentielles Algébriques.—C. XXIII. 126. V. R. p. 248.

IVORY. On the Theory of the Elliptic Transcendents.—Phil. Trans., 1831, p. 349. V. R. p. 288.

LIBRI. Sur la Théorie des Nombres.—Mémoires des Savans Étrangers, V. 1.

—— Sur la Résolution des Equations Algébriques, &c.—C. X. 167. These memoirs are referred to in the controversy between MM. Liouville and Libri.

LIOUVILLE. Sur les Intégrales de Valeur Algébrique.—Journal

de l'École Polytechnique, cah. XXII. 124 and 149. These two memoirs are printed also in the fifth volume of the 'Mémoires des Savans Etrangers,' pp. 76, 105. Poisson's report on them is inserted in the tenth volume of Crelle's Journal, v. *infra*.

LIOUVILLE. Sur les TranscendantesElliptiques de Première et de Seconde Espèce.—Journ. de l'Ecole Polytech., cah. XXIII. 37. V. R. p. 294.

—— Note sur la Détermination des Intégrales dont la Valeur est Algébrique.—C. x. 347. This note is appended to Poisson's report.

—— Sur l'Intégration d'une Classe de Fonctions Transcendantes.—C. XIII. 93. On the same general subject as the preceding memoirs.

—— Sur la Classification des Transcendantes.—L. II. 56, and III. 523. These papers contain an exposition of the principles on which this classification is to be effected.

—— Sur les Transcendantes Elliptiques de Première et de Seconde Espèce considérées comme Fonctions de leurs Modules.—L. v. 34 and 441. It is proved that these transcendents so considered cannot be reduced to algebraical functions.

—— Rapport fait à l'Académie des Sciences, &c.—L. VIII. 502. Report on M. Hermite's memoir. V. R. p. 312.

—— Sur la Division du Périmètre de la Lemniscate.—L. VIII. 507. V. R. p. 281.

—— Rapport sur le Mémoire de M. Serret sur la Représentation Géometrique des Fonctions Elliptiques et Ultra-elliptiques.—L. x. 290. A note is appended to this report generalising M. Serret's theory. V. R. p. 297.

—— Sur un Mémoire de M. Serret, &c.—L. x. 456. V. R. p. 73.

LOBATTO. Sur l'Intégration de la Différentielle

$$\frac{dx}{\sqrt{x^4 + \alpha x^3 + \beta x^2 + \gamma x + \delta}}.$$

—C. x. 280.

LUCHTERHANDT. De Transformatione Expressionis

$$\frac{dy}{\sqrt{[\pm (y-\alpha)(y-\beta)(y-\delta)]}}, \text{ \&c.}$$

—C. XVII. 248.

MACCULLAGH. Transactions of the Royal Irish Academy, XVI. 76. An elegant geometrical proof of Landen's theorem.

MINDING. Théorème relatif à une certaine Fonction Transcendante.—C. IX. 295. The function in question was shown by M. Richelot to be reducible to elliptic integrals.

MINDING. Sur les Intégrales de la forme

$$\int \frac{dx\, P \sqrt{p}}{c - x}, \text{ \&c.}$$

—C. x. 195. An addition to this memoir is found at p. 292 of the same volume.

—— Recherches sur la Sommation d'un certain nombre de Fonctions Transcendantes, &c.—C. XI. 373. These researches relate to an extension of Abel's theorem.

—— Propositiones quædam de Integralibus Functionum Algebraicarum unius variabilis e principiis Abelianis derivatæ.—C. XXIII. 255. This memoir is mentioned by M. Hermite.

POISSON. Rapport sur deux Mémoires de M. J. Liouville, &c.—C. x. 342. V. *supra, Liouville.*

—— Théorèmes relatifs aux Intégrales des Fonctions Algébriques.—C. XII. 89. V. R. p. 249.

RAABE. Bemerkungen zum Principe der doppelten Substitution, u. s. w.—C. XV. 191.

RAMUS. De Integralibus Differentialium Algebraicarum.—C. XXIV. 69. V. R. p. 249.

RICHELOT. Note sur le Théorème, &c.—C. IX. 407. V. *supra, Minding.*

—— De Integralibus Abelianis Primi Ordinis Commentatio Prima.—C. XII. 181. V. R. p. 309.

—— De Transformatione Integralium Abelianorum Primi Ordinis Commentatio.—C. XVI. 221 and 285. V. R. p. 310.

—— Ueber die Integration eines merkwürdigen Systems Differential-gleichungen.—C. XXIII. 354. These equations are those known as the "Jacobische System." V. R. p. 306.

—— Einige neue Integral-gleichungen des Jacobischen Systems Differential-gleichungen.—C. XXV. 97. The results contained in this paper are much more general than those of the preceding one. V. R. p. 309.

—— Nova Theoremata de Functionum Abelianorum cujusque ordinis Valoribus, &c.—C. XXIX. 281. V. R. p. 310.

—— Ueber die auf wiederholten Transformationen beruhende Berechnung der ultra-elliptischen Transcendenten.—Schumacher Astr. Nach. XIII. 361. [July, 1836]. V. R. p. 311.

ROBERTS. Sur une Représentation Géométrique des Fonctions Elliptiques de Première Espèce.—L. VIII. 263.

—— Sur une Représentation Géométrique des Trois Fonctions Elliptiques.—L. IX. 155. Mr Roberts's papers relate to curves formed by the intersection of a cone of the second order with a sphere. The following paper contains a more general exposition of his views.

ROBERTS. Mémoire sur quelques Propriétés Géométriques relatives aux Fonctions Elliptiques.— L. x. 297. V. R. p. 298.

ROSENHAIN. Exercitationes Analyticæ in Theorema Abelianum de Integralibus Functionum Algebraicarum.—C. XXVIII. 249, and XXIX. 1. V. R. p. 249.

SANIO. De Functionum Ellipticarum Multiplicatione et Transformatione quæ ad numerum parem pertinet Commentatio.—C. XIV. 1. V. R. p. 290.

SERRET. Note sur les Fonctions Elliptiques de Première Espèce.—L. VIII. 145. V. R. p. 296.

—— Propriétés Géométriques relatives à la Théorie des Fonctions Elliptiques.—L. VIII. 495.

—— Note à l'occasion du Mémoire de M. William Roberts, &c.—L. IX. 160.

—— Mémoire sur la Représentation Géométrique des Fonctions Elliptiques et Ultra-Elliptiques. Addition au mémoire précédent.—L. x. 257 and 286. It was on this memoir that M. Liouville made so favourable a report to the Institute. V. R. p. 297.

—— Développemens sur une Classe d'Equations relatives a la Représentation des Fonctions Elliptiques.—L. x. 351.

—— Note sur les Courbes Elliptiques de la Première Classe.—L. x. 421.

—— Sur la Représentation des Fonctions Elliptiques de Première Espèce.—Camb. and Dublin Math. Journ. I. p. 187.

SOHNCKE. Æquationes Modulares pro Transformatione Functionum Ellipticarum et undecimi et decimi tertii et decimi septimi ordinis.—C. XII. 178. M. Sohncke here gives the results which he investigates by a general method in the following paper.

—— Æquationes Modulares, &c.—C. XVI. 97. V. R. p. 291.

TALBOT. Researches in the Integral Calculus.—Phil. Trans. 1836, p. 177; 1837, p. 1. V. R. p. 249.

SOLUTION OF A DYNAMICAL PROBLEM.

A PERFECTLY rough sphere is placed upon a perfectly rough horizontal plane which is made to rotate with a uniform angular velocity about a vertical axis: to determine the path described by the sphere in space*.

A sphere, resting on a perfectly rough horizontal plane, receives a tangential impulse when the plane is made to move in its own plane. This impulse gives a velocity to the centre of the sphere and produces an angular velocity about a horizontal axis. The centre of the sphere moves parallel to the impulse, the axis of rotation is perpendicular to it; therefore the point of contact moves parallel to the impulse and therefore to the direction of motion of the centre. Therefore, as there is no sliding, the centre moves in the same direction as that of the motion of the plane supposed rectilineal. Moreover it is easily seen that the velocity of the centre is to that of the point of contact, or, which is the same thing, to that of the plane, as $1 : 1+\frac{a^2}{k^2}$, a being the radius of the sphere, k its least radius of gyration. While the direction and velocity of the plane's motion remain unaltered, no farther action occurs; when a change takes place, a new tangential impulse is given to the sphere, producing a new velocity of the centre parallel to its own direction, and a new velocity of rotation about an axis at right angles to it. The new velocity of rotation bearing to the old the same ratio as the new velocity of the centre to the old, the result is a compound velocity of the centre bearing the same ratio as before to the velocity of the point of contact, and as before parallel to it, and therefore still parallel to the direction

* Walton's *Problems in Theoretical Mechanics*, p. 540.

of motion of the plane; and so on; whether the motion of the plane varies continuously or discontinuously, in direction, in velocity, or in both. In the case proposed the motion of the plane (by which throughout I mean the element thereof in contact with the sphere) is always normal to a line drawn to a fixed point. Therefore the motion of the centre is so too, therefore the centre describes a circle whose centre is perpendicularly over the said fixed point. Q.E.D.

ON THE TAUTOCHRONISM OF THE CYCLOID*.

CONCEIVE two points, not acted on by gravity, to move on the circumference of a stationary circle towards its lowest point: the plane of the circle we will suppose to be vertical. Let their motion be such that the ratio between their distances from the lowest point may be invariable. Then their velocities towards that point must be in that invariable ratio. Their heights above it are in the duplicate ratio: their initial heights above it were also in the duplicate ratio: so likewise therefore are their vertical descents towards it. In other words, the squares of their velocities towards the lowest point are as their vertical descents.

Conceive one of the points to be, when the other starts, at the highest point of the circle, and to move with a constant velocity $(ga)^{\frac{1}{2}}$, a being the radius of the circle. Then, when it has descended through a vertical space z from its initial position, the square of its velocity towards the lowest point of the circle is equal to $\frac{1}{2}gz$. On this supposition with respect to one point, it appears, from what has been said before, that the square of the velocity of the other point towards the lowest point of the circle is similarly equal to $\frac{1}{2}gz'$, z' being the quantity corresponding to z, viz. its vertical descent below its initial position.

Now suppose the circle to move horizontally in its own plane with a velocity equal at every instant to the velocity, along the arc, of one of the points, the direction of the motion of the circle being towards the right or the left accordingly as the point is to the right or the left of the vertical diameter. If the former point be chosen, then the velocity of the circle will be constant, if the

* Walton's *Problems in Elementary Mechanics*, p. 245.

latter point, it will be variable. In either case, the path of the point selected obviously becomes a cycloid, and it is easily seen that the velocity of the point towards the lowest point of the circle is destroyed by the motion of the circle itself, while the velocity at right angles to this direction is doubled: consequently the whole velocity of the point will have, for its square, $2gz$ or $2gz'$, and we have thus a perfect representation of cycloidal motion under the action of gravity. But it is obvious from the fundamental hypothesis that the two points will reach the lowest point of the circle at the same time: that is to say, the descent to the lowest point of the cycloid is tautochronous.

The analogy to the descent to the lowest point of a circle along its chords is in the essential point complete, but here the motion is not along the chord but along the arc, and is complicated with the motion of the circle itself. In both cases it is easily seen that a medium, the resistance of which varies as the velocity, does not affect the tautochronism.

ON NAPIER'S RULES*.

To the Editor of the "Quarterly Journal of Mathematics."

Some time ago, my friend Mr R. L. Ellis sent me the following remarks upon Napier's rules.

NAPIER'S rules for the solution of right-angled spherical triangles are generally presented merely as a *memoria technica;* and when so presented do not exhibit the principle upon which they depend. To investigate and exhibit that principle is the purpose of this paper.

LEMMA.

If the three sides of a tetrahedron are right-angled triangles, no right angle being at the apex, then the base is also a right-angled triangle.

Let OAB be a right-angled triangle in A, and similarly OAC. Let the third side OCB be right-angled in C.

Then will the base ACB be also right-angled in C.

* Since Napier's method of investigating his rules was independently discovered by Mr Ellis, the Editor is of opinion that this Essay, written by the Dean of Ely, will not be without interest to the readers of this volume.

$$OA^2 + AC^2 + CB^2 = OC^2 + CB^2,$$
$$= OB^2,$$
$$= OA^2 + AB^2;$$
$$\therefore\ AC^2 + CB^2 = AB^2,$$

or the angle at C is a right angle. Q.E.D.

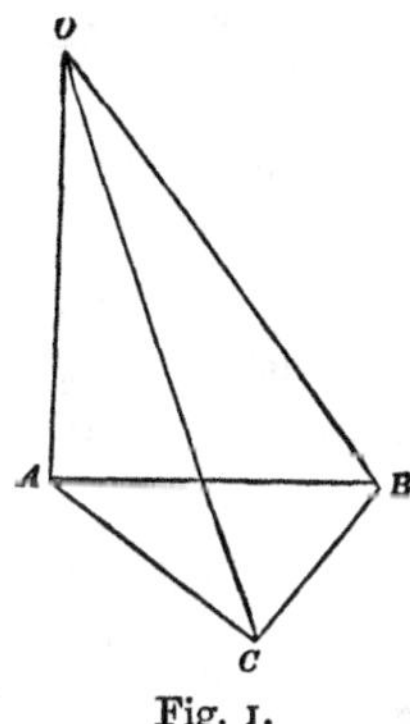

Fig. 1.

Cor. It follows from this that the dihedral angle AC is a right angle.

Considering the figure $OABC$, we observe that it is in some sort symmetrical with regard to the line OB. CB being at right angles to the plane OAC, and OA at right angles to the plane CAB; and as the dihedral angle OC is a right angle, so also is the dihedral angle BA.

Now the three lines OA, OB, OC evidently represent any right-angled spherical triangle; and similarly the three lines BO, BA, BC represent another, between which and the former a certain relation exists. One angle, namely, the dihedral angle BO, is common; the side ABC is the complement of the dihedral angle OA; the hypothenuse OBC is the complement of the side BOC; the side OBA is the complement of the hypothenuse AOB; and, lastly, the angle BC is the complement of the side AOC.

Hence this conclusion. If a_1, a_2, a_3, a_4, a_5 represent the parts of a right-angled spherical triangle taken in order, and beginning at the hypothenuse, then $\frac{\pi}{2} - a_3$, $\frac{\pi}{2} - a_4$, $\frac{\pi}{2} - a_5$, $\frac{\pi}{2} - a_1$, and a_2 are the parts also taken in order and beginning at the hypothenuse of another right-angled spherical triangle. If therefore to characterize the former triangle, we introduce a new set of quantities p, such that $a_1+p_1=a_2+p_2=a_5+p_5=\frac{\pi}{2}$, the original triangle being characterized by p_1, p_2, p_3, p_4, p_5 the secondary triangle is similarly characterized by p_3, p_4, p_5, p_1, p_2; and as the secondary triangle gives rise to a third, and so on, we thus see that every right-angled spherical triangle is one of a system of five such triangles.

It is obvious, that the transformation just employed depends upon the circumstance that the complement of the complement of an angle is the angle itself, and would succeed equally if the parts of the secondary triangle, which are the complements of those of the first, had been any function (f) of them, provided that $f^2=1$.

Returning to the figure we observe, that if, instead of taking a point B in OB we had taken one, as A, in OA, we should by a similar construction have got another of the four spherical triangles, which with the original one make up the system of which we have been speaking. Further, if in BC we assume any point as a first centre related to B as B to O, and make a similar assumption of a fifth centre in AC, the system will be complete. But as a figure so drawn would be complicated, it is better to adopt a different plan. Since BA is at right angles to OA, and lies in the plane OAB, it is easy to represent the corelate spherical triangles, to which the system of lines of which we have been speaking gives rise.

Let BAC (fig. 2) be the original triangle right-angled in A.

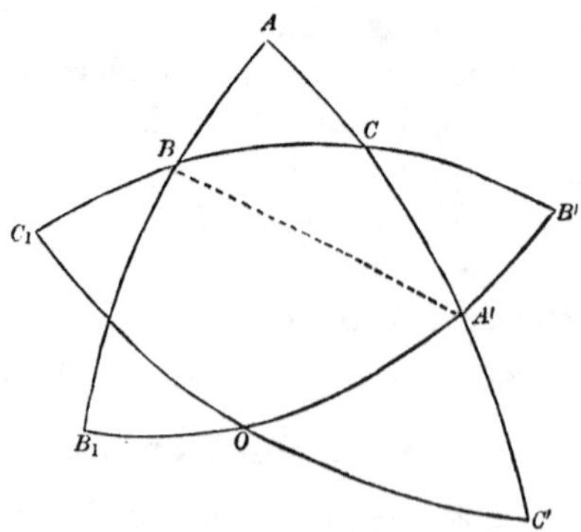

Fig. 2.

Produce BC to B', and AC to A', making $AA' = BB' = \frac{\pi}{2}$; and join $A'B'$.

Then $A'B'C'$ is the secondary triangle; and that it is a right-angled triangle may be proved, independently of what has been already said, by joining BA'. For since AA' is a quadrant, and A a right angle, BA' is as well as BB' a quadrant, and therefore B is the pole of $A'B'$, and therefore B' a right angle. The hypothenuse of the secondary triangle $A'C$ is the complement of AC, a side of the first: the angle C is common to the two triangles: CB' is the complement of the hypothenuse BC: $B'A' =$ angle $B'BA$ and is therefore the complement of the angle ABC: lastly, the angle $B'A'C$ has its complement measured by the arc AB.

Again producing $B'A'$ to O, $B'O$ being a quadrant, and similarly CA' to C', CC' being also a quadrant, we obtain a third triangle $OC'A'$; and finally completing the figure (the obvious details of demonstration are omitted) we get a reentering pentagon, all the angles of which are right angles, each being the right angle of one of the system of five right-angled triangles.

From this point of view it is plain that Napier's rules may just as naturally be considered as the statement of similar properties of five associated triangles, as in the usual mode, namely, as the statement of dissimilar properties in one triangle; and thus we obtain the rationale of their uniformity. Every relation between a middle and opposites, is the relation between two sides and the hypothenuse: every relation between a middle and adjacents is the relation between two angles and the hypothenuse. I only give these relations, of course, as instances, for any other would do as well.

A convenient mode of establishing these relations is afforded by the Lemma.

If we project OB in fig. 1, into OA, or if we project OB into OC and then OC into OA, the coincidence of the two results gives rise to the equation

$$\cos c = \cos a \cos b,$$

Again, $$OA \tan c = AB,$$

and $$AB \cos A = AC = OA \tan b;$$

$$\therefore \tan b = \tan c \cos A,$$

or,
$$\cos A = \tan b \cot c,$$

which gives the relation between any middle part and the two adjacent ones.

But in whatever way this relation and the preceding one are established, the point to which I wish to call attention is this, that by considering the system of five associated triangles we immediately generalize any particular result without having to demonstrate it in the separate cases.

It is clear that an equiangular spherical pentagon would in every case give rise to the system of five equiangular spherical triangles mutually corelated in a manner analogous to the right-angled triangles of which we have been speaking. But in the general case the relation would be too complicated to be useful.

Having perused the preceding I was anxious to know in what manner the subject had presented itself to Napier's own mind, and having referred to a copy of the *Mirifici Logarithmorum Canonis Descriptio* in the Cambridge University Library, I found to my astonishment that the mode of treatment invented by Mr Ellis was in reality a re-invention of Napier's own conception of the subject, which owing to some cause or another has dropped out from all Cambridge books upon Spherical Trigonometry*.

The following is translated from Napier. After giving a general explanation of the use of circular parts, he proceeds thus:

"This uniformity of the circular parts becomes very evident in the case of right-angled triangles formed on the surface of a sphere by five great circles, of which the first cuts the second, the second the third, the third the fourth, the fourth the fifth, and the fifth the first at right angles; the other intersections taking place at oblique angles.

"Thus, for example, let the meridian DB (fig. 3) cut the horizon BE in the point B. Let the horizon BE cut the circle EC of which the sun S is the pole in E. Let EC cut the sun's declina-

* This defect has been recently corrected in Mr Todhunter's *Treatise on Spherical Trigonometry*.

tion circle *CF* in *C*. Let *CF* cut the equator *FD* in *F*. And, lastly, let *FD* cut the meridian *DB* in *D*. Then all these intersections take place orthogonally in the points *B*, *E*, *C*, *F*, *D*;

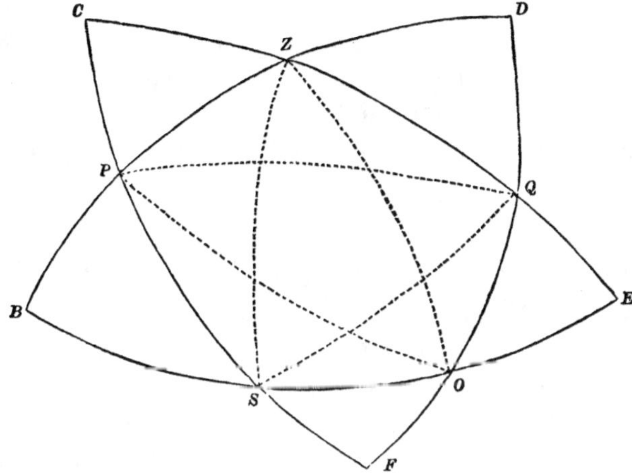

Fig. 3.

the other intersections taking place obliquely in the points *Z*, *P*, *S*, *O*, *Q*. And by this means five right-angled triangles will be formed, namely, *PBS*, *SFO*, *OEQ*, *QDZ* and *ZCP*, of which though the parts are different the *circular* parts are the same.

"The same uniformity of circular parts for quadrantal triangles may be made to appear by joining in the preceding figure *PQ*, *QS*, *SZ*, *ZO*, *OP*; by which means five quadrantal triangles will be formed having different parts with the same *circular* parts.

"The rules for the solution of the circular parts may be proved in each case separately, but besides this proof the general truth of the rules may be seen from what precedes. For the homologous constitution of the circular parts argues the similarity of the relations connecting them; so that any proposition which can be enunciated concerning the relation of any middle part to the adjacents or opposites may be at once concluded to be true of every other part regarded as the middle part."

Thus it appears that Napier himself did not regard his rules as a mere *memoria technica*, but saw them in their mutual relation, and in fact conceived them in the very best manner possible. Hence, it is very strange that we should find such statements as the following:—

Woodhouse. "There is no separate and independent proof of these rules; but the rules will be manifestly just, if it can be shewn that they comprehend every one of the ten results, (1), (2), (3), &c."

Airy, *Encyclopædia Metropolitana*. "These rules are proved to be true only by shewing that they comprehend all the equations which we have just formed."

The same view has been adopted (I believe) in all our Cambridge books.

Soon after receiving Mr Ellis' paper, given above, I received from him the following:—

Addition to the foregoing.

In any spherical rectangular polygon all the sides but three may have any assigned length, the length of the other three being functions of them. In the case of the pentagon there are thus two elements arbitrary, the same number as in a right-angled spherical triangle, between which and the pentagon exists a close analogy, which may be developed from the following lemmas.

LEMMA 1.

Of three consecutive sides of a right-angled spherical polygon, the extremes cut off quadrants of one another.

LEMMA 2.

The angle between the said extremes at their point of intersection (measured by the deflection of the direction of motion of a point which travels along the boundary of the polygon) is equal to the supplement of the next; in other words, the complement of the angle added to the complement of the side is equal to zero.

Consider any side of a pentagon and the two sides which are opposite to it and which meet one another at a right angle: they form with that side a right-angled spherical triangle. We will for distinction call that side (1), so that the other sides will similarly be called (3) and (4). Then as (4), (5) and (1) are consecutive, the angle between (4) and (1) is equal to the supplement of (5), and similarly the angle between (3) and (1) is the supplement of (2). Also the lengths of the sides of the triangle are by Lemma 1 the negative complements of (3) and (4), that is, they are those sides diminished by a quadrant. As the points of intersection of (3) and (4) with (1) are respectively distant by quadrants from the two extremities of (4), it follows that the segment they cut off between them is the supplement of (4). Hence the sides of the triangle and the complements of its angles and hypothenuse are severally equal to the negative complements of the sides of the pentagon, taken in the same order as the five parts of the triangle. The same demonstration will of course apply to the triangle formed by any other side of the pentagon and its two opposites, the only difference being in the starting-point of the cycle. Hence whatever relation connects any one part of a right-angled spherical triangle *as such* with its two opposites connects every other part with its two opposites. Q.E.D.

Scholium.

The general principle that the sines of the sides are as those of the opposite angles of course gives a relation between a part and its opposites. But I should prefer to begin from that between hypothenuse and the sides, proving that, as in p. 331, from a right-angled tetrahedron, which might be made the foundation of spherical, as the right-angled triangle is of plane trigonometry.

The relation connecting adjacent parts will be most simply got by considering three consecutive relations of those which connect opposites, multiplying the extremes and dividing by the means.

The general theory of spherical polygons must certainly lead to some curious general results, but of course they are out of my reach.

The preceding paper was not sent to the *Quarterly Journal*, because it was pointed out to me by a friend that a similar resuscitation of Napier's own conception of his rules and a similar remark upon the phenomenon of the disappearance of this conception from modern English books had been already made in the Course of Mathematics used in the Royal Military College —Dr Hutton's *Course of Mathematics*, edited by the late T. S. Davies*.

* Tutor to Mr Ellis; see *Biographical Memoir*.

ON THE RETARDATION OF SUNRISE *.

SUPPOSE the sun on the eastern horizon: his actual apparent motion until next sunrise may obviously be replaced by a fictitious one composed of an unscrew rotation round Π, the pole of

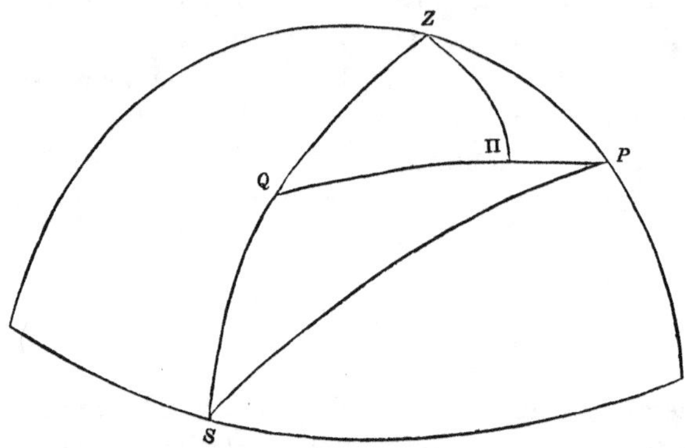

the ecliptic, combined with a screw rotation round P, the pole of the earth, the magnitude of the latter rotation measuring the retardation. Combine these rotations supposed small into one: the resultant axis must lie in some point Q, where $P\Pi$ produced cuts ZS, (Z the zenith, S the sun). The angle between the component rotations is constant, being $P\Pi$ or ω. Therefore the ratio of inequality between the given rotation round Π, measured by the sun's diurnal motion in his orbit, and rotation, round P, which measures the retardation, is then greatest when the resultant axis Q lies as near as possible to Π, or, in other words, when PQ is a minimum. Now, S being the pole of the arc $Z\Pi$, it will easily be seen that the angle $QZ\Pi$ is a right angle and the angle QZP is an obtuse angle, while near the equinox the angle ZQP must always be acute. Therefore PQ is least when Q lies in PZ, and then Π does so too. In other words, the retardation is least when Π culminates at sunrise, that is, at the equinoxes.

The common expression for retardation may easily be deduced from the expression furnished by what has been said, viz. $\dfrac{m \sin \Pi Q}{\sin PQ}$.

* Now first published.

A SOLUTION OF PROBLEM IX. OF THE FIRST BOOK OF NEWTON'S PRINCIPIA*.

THE fundamental principle of the proof I gave Dr Goodwin† is, that as the time varies as the increment of area, or as the square of the radius into the increment of angle, the force varies as the increment of angle, so that the acceleration in an instant dt is equal to $\frac{\mu}{h} d\theta$. The following application of this principle

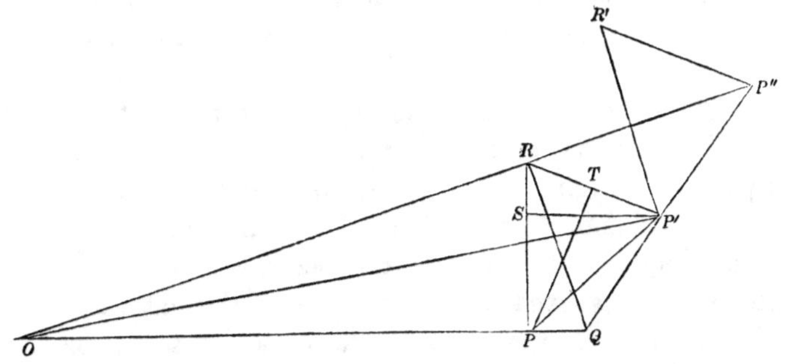

is perhaps more simple. Let O be the centre of force, P a position of the moving body. Take PR at right angles to PO and equal to $\frac{\mu}{h} dt$. Join RP', P' being the position of the body at the next instant. Similarly draw $P'R'$ at right angles to $P'O$ and equal to PR. Join $R'P''$, P'' being a third position of the body. Draw RQ parallel to $R'P'$. Therefore the angle at R being equal to the increment of the angle vector, QP is

* Dictated to the Editor of this volume and now first published.

† Goodwin's *Course of Mathematics*.

equal to the acceleration into dt produced by the central force during the instant dt. Likewise PP' is the space described with the initial velocity in the time dt. Compounding this with the acceleration we have $P'Q$ equal in magnitude and coincident in direction with $P'P''$. Therefore in the triangles RQP', $RP'P''$, we have, neglecting quantities of the third order, RQ equal and parallel to $R'P'$, QP' equal and in same direction with $P'P''$, and therefore the third side RP' equal and parallel to $R'P''$. That is to say, the point so moves as that its velocity may always be resolved into two elements, both of them constant; one, i.e. PR or $P'R'$, normal to the radius vector, the other, RP' or $R'P''$, parallel to a fixed straight line.

2. The velocity along the radius vector arises wholly from the component RP', and varies as $P'S$. The velocity normal to RP' arises wholly from the component RP and varies as PT. Now these lines $P'S$, PT, lie in similar triangles and are as the hypotenuses, and therefore in a constant ratio. But as the velocity towards the centre of force is proportional to the velocity in a fixed direction, it is clear the point moves in a conic section, because the fundamental property is, the distance of any point from the focus is proportional to that from the directrix.

SCHOLIUM. It is obvious that if a boat rows at a given rate through still water so that a line at right angles to the keel always passes through a given point, the boat moves in a circle, and the above demonstration shews that if the water is flowing in a given direction with a constant velocity, the boat will move as if attracted to the fixed point according to the natural law of force, that is, it will describe a conic section with major axis transverse to the stream.

ON ROMAN AQUEDUCTS*.

THE aqueducts by means of which Rome was supplied with water are in many points of view interesting, not only from the magnificence of their construction but also from the problems which they suggest respecting the scientific knowledge possessed by those who originally formed or subsequently managed them. The accounts which have come down to us with respect to the water supply of Rome are unsatisfactory. The treatise of Frontinus is the most detailed, and, on that account perhaps, the most disappointing. We find in it a great deal of information with respect to the sizes of the pipes by means of which the distribution of water took place, but that is nearly all; and, as Prony has remarked, the fulness of details on this matter makes the total silence on another not less important appear more strange than it otherwise would do; I mean the depth at which the pipes were placed below the level of the surface of the water in the reservoirs. To this remark we may add, that there is no estimate of the amount of water which in a given time was actually received through a pipe of given size. Prony is of opinion that a general rule must have prevailed, and that pipes of all sizes must have been inserted at the same depth, because we find it assumed that the discharge was proportional to the area of the section. The inference being natural, the same writer is of opinion that the notions of Frontinus, who of course knew all that was known in his time, were exceedingly unscientific, and it is difficult to see how this can be denied considering the account he gives of the attempts he made to gauge the different aqueducts. I cannot attempt to enter

* Now first published.

upon the general subject, but there are one or two points connected with it which admit of a kind of elucidation we have hitherto perhaps not received.

I.

It was doubtful even in the time of Frontinus why a pipe called quinaria, which, as is well known, is the fundamental modulus of the whole system, had received that designation. Setting aside the explanation, in itself improbable and resting on no authority, that Agrippa had introduced a modulus equivalent to five of the original units, and therefore called quinaria, two others remain; the one that of Vitruvius, which Frontinus mentions in connection with him, the other that which Frontinus himself preferred. According to Vitruvius leaden pipes receive their names from the number of fingers (digiti) which the sheet they were formed of was in width. Thus a centenaria was formed of a sheet of lead 100 inches in width, and so in all other cases. Frontinus observes that one opinion touching the origin of the word quinaria is, that it was introduced by the plumbarii, and Vitruvius because it was a pipe made of lead 5 digiti in width; and we may remark that it is a way in which a plumber would be likely to designate a particular pipe, inasmuch as it depends on the way of making it, and not on a measurement to be taken after it is made. Frontinus goes on to say that this is uncertain, meaning apparently the size of the pipe so formed, for that in rolling up the lead the outside surface was stretched and the inner one contracted. A modern writer who has devoted much attention to Roman antiquities, without being, so far as I have seen, happy in his conjectures in respect to them, namely, Dureau de la Malle, mistranslates the text, and makes Frontinus say that one part of the lead overlaps the other; the inference appears to be not only that he mistook the meaning of his author, but that he did not know how pipes were made. At present of course pipes are made like wires, by extension, and I do not know whether the old plan is ever used; but comparatively speaking this is a modern improvement, and in not very old books one may find described the process of cutting the lead into strips, folding it up so as to bring the edges together and then laying solder over the line of junction. Frontinus's own account is that the name probably indicated the circumstance that the diameter, being as it appears a clear internal diameter, was equal to $\frac{5}{4}$ of a digit: and undoubtedly in his time the names of a number of other pipes were formed in analogy with that of quinaria in a way which indicates that the magnitude of the bore was regarded as the foundation of the nomenclature. Thus senaria was a pipe of six quarters bore, and so on; but it by no means follows that the received opinion as to the origin of the word is true, and it seems impossible to set aside the authority of Vitruvius, who speaks without doubt, and must have been speaking of a matter with which he was familiar: nor is there any thing improbable in the supposition that in the interval between his time and that of Frontinus a new nomenclature may have grown up linked with the old one at one point only, namely, the word quinaria. Dureau de la Malle brings an objection against the supposition that the width of the lead is the cause of the name, that the circumference of 5 digits does not correspond to a diameter of one and a quarter, adding sagely, what is by no means to any practical purpose, that the problem of the rectification of the circle has hitherto been found insoluble. But this objection, rightly considered, leads to a curious confirmation of the statement of Vitruvius, which I shall now attempt to explain.

The diameter spoken of is the bore or lumen, whereas the diameter corresponding to a circumference of 5 inches must be that of a circle lying somewhere in the thickness of the lead between its two surfaces, one of which, as Frontinus remarks, is stretched and the other diminished, either by condensation or *puckering* in the process of formation. If we assume, as we may do without sensible error, that the lead is stretched to one half of its thickness and no more, we have to calculate the diameter corresponding to a circumference of five, and to diminish this by the thickness of the lead, that is, by twice the half thickness, in order to get the bore. The question then is, what data have we for determining the thickness of the lead from which these pipes are made, and the answer to it is given by a passage in Vitruvius, and which has been copied by Pliny and Palladius.

Vitruvius, in speaking of aqueducts, says that leaden pipes ought to be made in pieces of 10 feet long, their weight to vary with the width of the lead of which they are made, from 1200 lbs.

in the case of a centenaria to 60 in that of a quinaria. It will be observed that these weights and the intermediate ones mentioned in the same passage are proportioned to the corresponding widths, which indicates that the lead employed was always of the same thickness. We see from these statements that a strip of lead 10 feet long and 16 digits wide would weigh 192 lbs.; 192 being to 16 as 12 to 1. Now it is not necessary to convert this number of Roman pounds into English pounds, and similarly the digits into inches, in order to obtain the result we seek. We may make use of a method which, if not perfectly accurate, has at least the advantage of not depending on any experimental comparison of ancient and modern standards. For we know that the Romans reckoned 80 lbs. as the weight of a cubic foot of water. The specific gravity of lead is about 11·4; but as lead is seldom pure, and when alloyed is alloyed with substances lighter than itself, we will take that of the pipes at 11. A cubic foot of lead will therefore weigh 880 lbs., and by what we have just seen the weight of a prism whose base is a square foot and height the thickness sought will be (16 digits being of course equal to a foot) 19·2 lbs. Dividing the latter number by the former, and multiplying the quotient by 16, we obtain the thickness of the lead in digits: it is approximately 0·35. Subtracting this from the diameter corresponding to a circumference of 5 (that is, from 1·59, the ratio of the circumference of the diameter being taken at 3·14), there remains for the diameter of the bore 1·24, or within a 100th of a digit of the 5 quarter digits assigned by Frontinus to the diameter of the quinaria, which is a nearer coincidence than we were entitled to expect. So far therefore from the statement of Vitruvius being contradicted by that of Frontinus, they appear to be in perfect harmony, the quinaria being at once the pipe made of a strip of lead 5 digits in width, and that whose bore was $\frac{5}{4}$ of a digit. Not so, however, the other moduli of the system. The vicenaria of Frontinus is not that made of a strip of lead 20 digits in width, but that whose bore is equal to 20 quarter digits. The discrepancy between the two things is not inconsiderable, as may be seen by a little calculation. The truth is, that the nomenclature used with reference to the water supply of Rome, appears to have been very unsettled. We find in the treatise of Frontinus traces more or less developed of four different systems; that of Vitruvius is a fifth; and there are certain exceptional cases mentioned by Frontinus in speaking of the frauds of the aquarii, which it is not very easy to connect with any of the five. In the midst of all this complexity it is well to have one point fixed, namely, the historical significance of the name quinaria applied to what appears in Frontinus as the fundamental modulus to which in practice all others were referred. This I think has been effected by the considerations just suggested, and it is curious to remark that neither Frontinus nor any one else was likely to have fallen on them. He was probably aware of the real state of the case, but if the difficulty suggested by Dureau de la Malle had occurred to him, he could only have got rid of it by actual measurement, and not by a calculation founded on the specific gravity of the material employed. In conclusion, it may be well to observe that there are many cases in which the uncertainty arising from the experimental comparison of ancient and modern standards may be avoided: for instance, if it is stated that a modius of a particular kind of grain weighed 26 lbs. (I refer of course to what Pliny says of the far of Clusium), we may determine at once what an imperial bushel of it would weigh by very simple considerations. The congius of water like the gallon weighed 10 lbs., and therefore if a congius and a gallon are filled with the same substance, the former will weigh, whatever the substance may be, as many Roman pounds as the latter weighs pounds avoirdupois. Now the modius is 3 congii, and the bushel 8 gallons, and consequently the grain of which Pliny speaks would weigh 69 lbs. and a third the bushel. This is more than our best English wheat, but some specimens of Australian wheat have been known to weigh as much as 70 lbs., and it is said that in Spain this weight is not unusual. Another point of interest suggested by the statement of Vitruvius is the comparative thickness of Roman and English sheet lead. By the process of milling, that is, of rolling a plate of lead between heavy rollers whose distance is gradually diminished as the sheet becomes thinner, the weight corresponding to a given surface may be reduced much below the minimum possible in the old process of casting. Thus milled lead sometimes weighs only 4 lbs. in the square foot. But it is more natural to compare the Roman lead which was doubtless produced by casting with that which is now made in a similar manner. Of the thickest kind of the latter the weight is said to be 11 lbs. to the square foot, and we have seen that Vitruvius' estimate gives a weight of 19·2 lbs. to the square foot. Reducing this to English weights and measures it would appear, speaking in round numbers, that an English square foot of Roman lead would weigh about 15 lbs. avoirdupois. It is therefore considerably thicker than that which is usually made in England. But it is to be understood that the Romans may have had a thinner kind than that which they were in the habit of using in the formation of pipes. It has already been remarked that pipes of moderate bore are now made by extension. The thickest kind of pipes of an inch bore are made in lengths of 15 feet, and weigh 56 lbs. Fifteen English feet of fistula quinaria would weigh about 68 lbs. avoirdupois. I have thought these details worthy of notice, because they illustrate the economy which is the result of improved modes of manufacture, and of our power of calculating the strength of materials.

II.

The most interesting consideration which the treatise of Frontinus suggests relates to the state of knowledge current at the time it was composed of what used to be called Hydraulics. The details into which he enters are so precise (at least on some points), and the provisions of the Senatus Consulta so strict, that one finds it difficult to believe that he and his contemporaries were ignorant of any of the circumstances which it is necessary to attend to in distributing a supply of water (necessary, that is, unless considerations of justice are altogether to be neglected); but then the difficulty arises, why, when so much is said, so many details, not at all more likely to be generally known than those mentioned, should have been omitted.

Prony has remarked that it has been the universal practice to estimate the supply of water by the quantity which in a given time and under a given pressure is discharged through a pipe of given bore from a reservoir whose surface is kept at a fixed level. But although he conceives that this remark applies to the Roman system, yet neither the text of Frontinus, nor any inference which can be deduced from it, appears to justify us in supposing that they had any idea of the necessity of making the water in the reservoir from which the distribution took place remain at a constant height, or, to use Prony's own expression, of making it stagnate. It is quite true, that Fabretti asserts that the measurements were made in the piscinæ, where the water was comparatively speaking at rest, but although measurements in the piscinæ were undoubtedly made, yet it is quite clear that Frontinus had no scruple as to the propriety of measuring the section of a running stream, calculating its area in quinariæ by a process of simple reduction, and then assuming the result as an expression of the number of quinariæ which the stream was capable of supplying. This is so evident, that Prony remarks that Frontinus could have had but vague ideas as to the efflux of fluids, and that a method which took no account of the velocity of the stream must have led to strange results. He adds, and rightly, that Frontinus appears to have avoided making his measurements in places where the velocity was small. Nor is this inconsistent with Fabretti's remark as to measurements in piscinæ, for we do not in reality know that the water was in any sense reduced to rest in the places in which these measurements were made.

The same conclusion, namely, that one of the conditions which Prony speaks of as essential, was not attended to at all, appears from the passage in which Frontinus describes the position which the *calix* ought to occupy with respect to the stream of water in which it is placed. The passage is obscure, but a stream is plainly spoken of. A third consideration would lead to the same result. I refer to the description which Vitruvius gives of the three basins immediately supplied from the castellum. From two of these there appears to have been a waste pipe into the third;—a circumstance which it is hard to reconcile with the supposition that the surface was kept at a constant level by means of pipes employed in its ulterior distribution.

In this respect then at least the system of the Romans appears to omit one of the conditions which in modern times have been found necessary, and there is at least another point in which the discrepancy between their practice and ours deserves to be noticed. It is well known, that if water escapes from an orifice in the side of a vessel, the quantity discharged will, other things remaining alike, vary with the length of the adjutage, as it is called, through which it passes. If the side of a reservoir be very thin the discharge is less than if it be thick, or if, which comes to the same thing, a short horizontal tube is adjusted to it. Now, at first sight, the Romans seem to have understood this, for Frontinus states that a calix of not less than 12 digits in length ought to be applied to the side of the reservoir. This calix, he says, should be made of brass in order that its aperture may not be tampered with; but between the calix and the adjutages of which we have been speaking there is the essential distinction that the water escapes freely from the extremity of the one, while from the other, I mean the calix, it passes at once into the fistula. The Roman system of distribution is essentially a system of distribution through pipes inserted into the reservoir from which the erogation was made. To prevent fraud the commencement of the pipe was made of brass, and moreover, it was enacted, that for 50 feet the diameter of the pipe should be the same as that of the calix to which it was joined. It is one of Frontinus's complaints that the aquarii would sometimes insert pipes without using a calix, and thereby gave an opportunity, to use a homely phrase, of playing tricks with the aperture of the orifice. There is nothing in this which resembles an adjutage or short pipe through which the water is allowed to flow freely; and it is, moreover, especially remarkable, that although the Senatus Consultum would not allow the grantee of the pipe to vary its size within 50 feet of the calix, it is silent with respect to the direction in which it was carried, though nothing can be clearer, that if the pipes slope downwards the supply of water would be greater than if it were turned in an upward direction, and of two persons who lived on different levels one of them must have had an advantage over the other. Of this, Frontinus appears to have been aware, as he states that according to the relative position of the acceptorium and the castellum, the erogation was to be burthened or relieved. But according to what rule this correction was to be made, he has not told us.

The question of the nature of the calix is particularly worth considering, because Prony's hypothesis for determining the quantity of water brought by the aqueducts to Rome rests upon an assumption that the calix was in the usual sense of the word an adjutage.

It so happens that in modern Rome the unit of distribution is a pipe of an uncia in diameter, the pressure on the depth of its centre and its length being each $\frac{5}{4}$ of a palm.

Now, remarks Prony, although Frontinus has given us no information as to the pressure on the orifice, that is, as to the depth of its centre below the surface of the water in the reservoir, yet the Romans must have had some rule, and as in modern Rome the rule is to make the depth in question equal to the length of the adjutage, let us assume that this relation existed also in his time. The next step is to convert Frontinus's rule, that the calix must not be less than 12 digits in length, into a statement that an adjutage must be employed of that precise length, and then the inference follows at once that the pipes were placed 12 digits below the surface. But, however ingenious this is, there are several objections to it besides the one already noticed.

In the first place, the argument from tradition is worth little or nothing in the case of a matter which nowise depends on popular usage, but is settled at the will and pleasure of persons in authority, who are free to adopt whatever may seem to them to be an improvement. In the second place, Italy has for many reasons long been remarkable for the degree of attention paid to the subject of hydraulics. The reasons for this are connected with the physical geography of that country. Not to dwell upon them, it is sufficient to remark, that in a country in which so much attention has been paid to the subject there is little reason to suppose that any point of modern practice is founded on old tradition; and, as we have seen, the most eminent Italian writers confess that they cannot discover that the Romans had any rule as to the pressure on the orifice. Moreover, no one knew better than Prony how vague were Frontinus's views or hydraulics.

Quid te exempta juvat spinis de pluribus una?

What is gained by getting over one difficulty by an ingenious assumption, while others, and especially the question as to the different directions in which water is distributed, remain untouched?

On this hypothesis Prony estimates the quantity of water

supplied by a quinaria at about 56 cubic metres in 24 hours, that is to say, at more than 1200 gallons, an estimate which appears excessive, notwithstanding all we know of the magnificence of Rome. It is about three times as much as that given by Dureau de la Malle, who simply assumes, without assigning any reason for doing so, that the quinaria was subject to the same pressure as the Pouce de Fontainier in the French system of distribution, that is, to a pressure of 7 lines on the centre. The old French system of distribution (and I am not aware what improvements have of late been introduced) was comparatively speaking rude and inartificial, and it is on that account more likely to represent the Roman practice than the scientific method to which Prony would compare the latter. No attention was paid to the presence or absence of an adjutage, and allowance being made for the thickness of the lead, the upper part of the outside surface of the pipe would be just at the surface, a mode of placing it which is perhaps indicated by Frontinus's phrases *ad libram* and *ad lineam*. The objection at once occurs that pipes of different bores would have their centres at different depths below the surface. Prony has remarked that the Romans must always have subjected the centre of the orifice to the same pressure (and therefore must have had a different rule as to depth), because they estimated the discharge of pipes of different sizes by simply comparing the area of their sections with that of the quinaria. The inference would be valid if we had any reason to believe that these estimates were either founded on or verified by observation. In the present state of our knowledge it involves a *petitio principii*, and it is remarkable that Prony should have attached any weight to it, as he admits that Frontinus estimated the product of a stream of water as if it bore a constant ratio to its section.

Several considerations may be suggested which make it easier for us to believe that neither Frontinus nor any of his contemporaries were sufficiently acquainted either theoretically or practically with the principles of hydraulics to be able to avoid enormous errors. Why, it may be said, was so much care bestowed on a proper determination of the size of the pipes, if other elements of the question of supply were neglected? Surely they must have learned to give up making calculations on which a little observation would have shown that no reliance was to be placed. In reply to these remarks, which of course are not without weight, it is to be observed that the questions with which Frontinus undertook to deal had not pressed themselves on the attention of any one at Rome until a comparatively short time before. Several hundred years had no doubt elapsed since the first aqueduct was constructed, but until concessions were made to individuals, and further, until such concessions had acquired sufficient importance to attract the attention of the government, there was no occasion to attempt to determine how much water any aqueduct brought, or how it was distributed—all went to public uses: private persons were only allowed to appropriate what was afterwards called *aqua caduca*, that is, water which would otherwise have run to waste; and of this the amount was probably insignificant.

Not until the time of Agrippa, we are told by Frontinus, did it become usual to make concessions of water, and he adds, that it was not until after Agrippa's death that any attempt was made to reduce these concessions to a system. Moreover, he has particularly spoken of the substitution of a larger pipe in cases in which a concession had been made of several quinariæ, as of a comparatively recent practice, and it is obvious that the difficulties of the question could not be felt to their full extent, as long as no other pipes than quinariæ were employed as moduli. We thus see that not much more than a hundred years, if so much, can have elapsed during which any attention need have been paid to the subject. At the end of this period, when Nerva appointed Frontinus to the office of Curator Aquarum, everything was, according to his own account, in the wildest confusion. He blames the aquarii unsparingly, and no doubt they had sins enough of their own to answer for; but that there were other vices in the system than the frauds with which he charges them, appears sufficiently from his own statements. All his calculations are completely at variance with the results which his predecessors had obtained, and which were recorded in the *Commentarii Principum.* He has no explanation to give of these discrepancies.

He seems to have been struck by finding that the amount of water as estimated in quinariæ and known to be distributed was greater than the estimate of the amount supplied. With a strong impression, therefore, that his predecessors had under-estimated the supply, he attempted to re-measure it, and formed a new estimate, greatly exceeding theirs. The supply being now in excess, he explained the discrepancy by imputing frauds to the aquarii. Of many they were probably guilty, but if any scientific knowledge of the subject had existed in his time or that of his predecessors, there could scarcely have been the great difference between his estimate of the supply and theirs, unless the supply varied greatly at different seasons of the year, and that he and they omitted taking this variation into account.

If I were to guess how the distribution of water was really made, my conjecture would be something of this kind. The distribution to private persons having at first been quite a secondary object, their pipes were inserted so as not to receive any water when the water in the reservoir fell below a certain level, which was known as the *libra* or *linea*. If, therefore, the water rose above this height, the grantees had the benefit of the increased supply, whatever it may have been. When the pipe first became full, the level of the water was higher than the libra by the diameter of the tube, and at this state the phrase *implere mensuram* may perhaps have been especially applied. The determination of the libra was probably made on the principle that when the water fell below it, the supply was observed to be no more than adequate to the public purposes of the aqueduct. Vitruvius speaks of a reservoir especially devoted to private purposes, but it does not appear that this arrangement was actually followed; on the contrary, we know by the practice being prohibited, that at one time private pipes were placed, but also by the running streams.

Two or three things would be explained by this hypothesis. 1st, The absence of any statement of the depth *below* the surface of the water at which the pipe was inserted. This question would of course not arise, if the rule were to place the pipe *above* what was considered low-water level. 2ndly, The absence of any statement in Vitruvius as to how the water in the Castella was to be kept at a constant level. He certainly speaks of the reservoirs overflowing—that is, of two of them overflowing—but the third, which in his arrangement was that into which private pipes were to be inserted, was fed by the other two, and from the nature of things must have been fed unequally in different seasons: nor, as I have said, do we know that his arrangement was really adopted; but if the waters in the reservoirs rose and fell, the depths of the pipes below the surface must have varied, for the position of the pipe was plainly fixed. We know so little about matters of detail, that much stress cannot be laid upon what I have mentioned, but the difficulty of shifting the pipes, and the absence of any information as to how the water in the reservoir was kept at a constant level, are worth considering. 3rdly, This would account for the absence of any statement as to the quantity of water delivered by a quinaria. The question, how much does such a pipe deliver in twenty-four hours, must have occurred to the most unthinking person, if any steps had been taken to make the quantity constant. 4thly, Frontinus speaks of a case in which the erogation had ceased, on the ground that the modulus was exhausted—a passage which would be easily understood, if we suppose that from a deficient supply the water had sunk below what was accounted low-water level.

5thly. The fraud imputed to the aquarii of making the large pipes too large would thus appear to have at least had its origin in a reasonable principle of compensation. For if the lowest point of six vicenariæ, and that of one pipe of 120 square digits were all on the same level, the six small pipes, though nominally equivalent to the one large one, would, by having their centres lower, deliver a great deal more water, and it would be easy to calculate the circumstances under which they would be equivalent to the larger pipe which the aquarii substituted for one of 120 digits.—Lastly. The old French water system would then be an improvement on the Roman, and not a retrocession from it. The former seems to run so far back in the middle ages that we may fairly suppose it traditionally connected with the Roman. The transition is easy, and may not improbably have been made at Rome itself. You have only to take the pipe out of the reservoir so as to allow the efflux of water to be free, and to let the water just rise so as fairly to cover the whole aperture. This with making the standard modulus one inch would in effect be the old French system, which may easily have suggested itself in answer to a question—what would be the discharge through a given opening when its mensura was just filled and no more, and the disturbing influences (due to the direction of the pipe, &c.) removed?

If what has been said is the correct view, it would follow that any attempt to estimate the water supply of Rome from the data given by Frontinus is mere guess-work. Paris appears to have been ill-supplied with water in 1814, and the supply not to have exceeded that which a writer quoted by Prony speaks of as sufficient, namely, an inch of water to each thousand of the population. Prony himself thinks this to be as much as is absolutely required, though more may be desirable. It would appear, if I remember Frontinus's numbers correctly, that in his time (that is, before three more aqueducts had been added by Trajan) the water supply of Rome would, on Prony's theory, have sufficed at the Paris rate for a population of 42 millions. Dureau de la Malle, as I have said, gives a far more moderate estimate; but there is really no authority for either. My impression is that the supply of London is at least double what was *thought* sufficient at Paris. The highest estimate I have seen of the French water inch is Prony's, who makes it nearly 19500 litres per diem, which would be under 4½ gallons to each head of the population. The Abbé Bossut's estimate would reduce the amount under 4 gallons. That Rome was far better supplied than London is clear, but at Rome there were no breweries, no steam-engines, and probably far less manufactures than in London; and though there were many more baths there was much less washing of clothes. A good many fountains, &c. could be filled with the present London supply if these sources of expenditure were taken away.

ON THE FORM OF BEES' CELLS*.

MARALDI seems to have been the first person who determined the form of the rhombs of the pyramidal ending of the bee's cell: his observations on bees were published in the *Memoirs of the French Academy* for 1712. They are very good and clear: one ambiguous expression, however, has given rise to a misconception which is still current and is repeated by every writer, or nearly so, on the subject. Maraldi states that the angles of the rhomb are 110° and 70°, and that the angles of the trapeziums which form the sides of the body of the cell, are also 110° and 70°. I mean, of course, the angles next the apex of the cell, those at the open extremity of the tube are right angles; the point to be observed is this, Maraldi when he gives the results of his measurements makes no attempt to do more than state the value of the angles to the nearest degree, an amount of accuracy beyond which it is scarcely possible, under the circumstances, to go, and which is amply sufficient, considering the irregularities which undoubtedly exist in the structure of individual cells; but a little further on in his paper, he remarks, that as the angles of the rhombs are equal to those of the trapeziums, they must be respectively 109° 28' and 70° 32'. Thus, he remarks, the angles chosen by the bees have this advantage which conduces to the elegance and symmetry of the structure, namely, that only two angles are used throughout. He does not state how he is led to this conclusion, but an attentive reader will see that after having asserted that the result of measurement was 110 degrees, Maraldi did not proceed in the next page to contradict himself by asserting that the result of his measurements was 109° 28'. There can be no question but that, being struck by the equality (so far as his measurements could ascertain it) of the angles in the rhombs and in the trapeziums, he assumed that in the normal or typical cell, this

* Now first published.

equality was absolute, and hence the determination of the angles became, as I need hardly point out, a matter of spherical trigonometry, namely to determine the sides of the equilateral spherical triangle whose angles are each equal to 120°, that is, to the angle between the two adjacent sides of a hexagon. This angle has for its cosine $\frac{1}{3}$, and is, I believe, to the nearest second 109° 28' 16".

Unfortunately Reaumur chose to look upon this second determination of Maraldi's as being, as well as the first, a direct result of measurement, whereas it is in reality theoretical. He speaks of it as Maraldi's more precise measurement, and this error has been repeated in spite of its absurdity, to the present day: nobody appears to have thought of the impossibility of measuring such a thing as the end of a bee's cell to the nearest minute. One can only suppose that Maraldi made so many observations, varying between 109° and 110°, that the arithmetical mean came out with this excessive amount of accuracy, a supposition in itself highly improbable, and at variance with Maraldi's distinct statement. The subsequent history of the matter is this,—Reaumur employed Kœnig to determine the form the rhombs ought to have in order to give the greatest volume to the cell with the least expenditure of wax. Kœnig, for a reason which will be mentioned by and by, gave as his solution the following values to the angles of the rhomb, 109° 26' and 70° 34'. He was agreeably surprised, says Reaumur, to find that his result agreed within two minutes with Maraldi's measurements, whereas in reality they only agreed with Maraldi's theoretical determination. With this determination they ought to have been absolutely coincident: for it is easy to show that when the surface of the cell is made a minimum, the angles of the rhombs are equal to those of the trapeziums. It was soon afterwards shown that Kœnig's results were wrong, and thus we have been pleasantly told, that the bees proved to be right and the mathematician wrong, that there was no mistake on the part of the bees, and so on. Maclaurin was, I believe, the first to correct Kœnig: he has not always had credit for this priority. Thus Dr Carpenter in his *Physiology*, informs us that Lord Brougham, not satisfied with Kœnig's determination, took into account certain small quantities previously neglected, and showed that the coincidence between theory and observation was absolute.

Lord Brougham's own remark is that the deviation between Kœnig and Maraldi had always been ascribed either to an error of measurement or to an inaccuracy in the construction of the cells. It is certainly not so ascribed by Maclaurin, who remarks it could only have arisen from Kœnig's not having carried his computation far enough. Throughout Lord Brougham's elaborate discourse on the matter, one circumstance is omitted, namely, Maraldi's statement that the angles, according to his measurements, were 110° and 70°.

Either the habit of an advocate's mind,—for Lord Brougham may be regarded as counsel for the bees,—or his not having read Maraldi's paper, must have been the cause of this omission. Nor has he mentioned the advantage which Maraldi conceived to result from the angles being what they are: namely, that only two plane angles occur throughout the structure. He quotes at second-hand the opinion of Boscovich that the angles could not be measured with the supposed degree of precision.

I have not seen what Boscovich says, but have no doubt that Lord Brougham has misunderstood him: for he makes Boscovich's opinion to be, that Maraldi merely deduced his result by assuming that the dihedral angles are all equal to 120°, that is, he makes Boscovich accuse Maraldi of dishonesty, whereas, in reality, he probably only repeated what Maraldi in effect says: that he deduced the precise angles from assuming that the angles in the trapezium were precisely equal to those of the rhomboid, his measurements giving the same value to both, namely, as I have already said, 110° and 70°.

The matter has been so long confused that it is worth while to quote Maraldi's own words.

'Chaque base d'Alvéole est formée par trois rhombes presque toujours égaux et semblables, qui suivant les mesures que nous avons prises, ont les deux angles obtus chacun de 110 degrés et par conséquent les deux aigus chacun de 70 degrés.'

* * * * * *

'Ces six mêmes côtés des trois rhombes sont autant de bases sur lesquelles les Abeilles élevent des plans qui forment les six côtés de chaque Alvéole. Chacun de ces côtés est un trapèze qui a un angle aigu de 70 degrés, l'autre obtus de 110 degrés, et les deux angles du trapèze qui sont du côté de l'ouverture, sont droits. Il faut remarquer ici que l'angle aigu du trapèze

23—2

est égal à l'angle aigu du rhombe de la base, et l'angle obtus du même rhombe égal à l'angle obtus du trapèze.'

* * * * * *

'Outre ces avantages qui viennent du côté de la figure de la base, il y en a encore qui dépendent de la quantité des angles des rhombes; c'est de leur grandeur que dépend celle des angles des trapèzes, qui forment les six côtés de l'Alvéole; or on trouve que les angles aigus des rhombes, étant de 70 degrés 32 minutes et les obtus de 109 degrés 28 minutes, ceux des trapèzes qui leur sont contigus doivent être aussi de la même grandeur.'

The '*étant*' in the last sentence is plainly the prothesis of a hypothetical proposition, *if* they are &c. *then* so and so.

The best part of Lord Brougham's essay is his argument against the theory contained in the Article on Bees in the *Penny Cyclopædia*, though rightly or wrongly, he has deprived himself of the fatal objection to this theory furnished by Barclay's observation of the doubleness of the walls.

A system of associated cells has many curious properties. The following may perhaps not have been noticed. Round each corner which is not an apex of any cell, the apices are arranged in groups of 6 at 6 of the corners of a cube of which the first-named point is the centre, and each apex belongs to three such groups.

The peculiar difficulty as to the instinct shown by bees is this, that one does not see how they perceive when the true form of their cell is attained. In common cases of instinct, though the impulse is mysterious, one sees how the animal knows that its end has been obtained: not so in this case. The following is my guess. Beside the complex eyes of bees, they have three single eyes placed lower down, and probably serving for the vision of near objects. Assume that the axes of these eyes diverge so as to be respectively normal to three ideal planes forming a solid angle, each dihedral angle of which is of 120 degrees. Geometry shows that *every* solid angle of the bee's cell is precisely similar to this type, so that a bee looking at it with his three single eyes, might have direct vision with each eye, of one of the three planes of the solid angle. This direct vision may correspond to a particular sensation, so that a bee is not satisfied till it is attained. If we had three eyes, the axes mutually at right angles, we should, I think, be well able to judge whether the walls and ceiling of a room were truly at right angles to each other. And so in the case proposed. To this guess two objections have been made, the first, that most if not all hymenoptera have similar eyes, but do not make similar cells; the second, that there is very little light inside a hive: but neither appears to me conclusive.

I remember a little inaccuracy in the way I explained my notion as to the manner in which bees are guided in making their cells. I said that all the angles of the cell were of a certain type; I should have said all the trihedral angles. There are of course three others, each bounded by four sides, but these also have the fundamental property that all the dihedral angles are of 120 degrees, so that the bee could always obtain direct vision of the faces of each dihedral angle.

Take two equal cubes, divide one into six pyramids, the base of each being the face of the cube, and the apex at the centre of the cube, fit each of the six pyramids by its base on a face of the other cube, divide the solid thus got by a plane through the centre of the central cube, normal to a diagonal. Each half is the typical form of the bee's cell, except that the prismatic portion of the latter is a little longer than according to this construction it would be. There is of course no theory as to the length of this portion of the cell, and I believe no accurate observations.

The above is, I think, the simplest way of conceiving the form of the cell, and as far as I know it is not given in any book: it would certainly be the easiest way of modelling a cell.

ON THE THEORY OF VEGETABLE SPIRALS*.†

THE circumstances under which the following remarks have been composed would be a sufficient excuse for any defects, if they were not rather a reason why no such task should have been attempted. But the error, if it be one, will not be repeated, and, like Socrates, Ἐγκαλυψάμενος ἐρῶ‡.

The following remarks on the Theory of Vegetable Spirals, appear to possess some interest, though it is very possible that it is only my imperfect acquaintance with the history of the subject which makes me think they contain anything not already noticed. On the chance however that they are new I shall endeavour to put them down, though my state of health obliges me to do so in a hurried and imperfect manner.

The fundamental principle of what follows is the resolution of the symmetrical spiral into portions actually unsymmetrical yet capable of becoming regular without essential alteration. Symmetry is therefore regarded as something superadded to the essential principle of the spiral, and we are thus led to trace the existence of this principle in cases in which it exists apart from any appearance of symmetry.

A spiral may be divided in two ways: by two *azimuths* or by a horizontal circle. We may either obliterate all the leaves

* Now first published.

† Though the nature of the subject and my own ignorance make the whole of these remarks unsatisfactory, no part of them seems to me more so than that which relates to the way in which simpler forms are included in the more complex. I have not expressed my own idea, imperfect as that is. The following statement would be nearer it: "Every form includes all simpler forms, modifying them alternately in opposite directions, *i. e.* by expansion and contraction."

‡ Plat. *Phædr.*

which grow, for instance, on the north side of a tree, or all those belonging to a given spiral which rise above a circle drawn round the axis of growth. The two modes of resolution give, as we shall see, similar results; but we shall set out from the first. If we were to take off all the leaves on the north side of a tree, those left could not be arranged in spiral order at all, except in particular cases: but there is always one mode in which spirals may be divided by means of two azimuths, so that the essential principle of the spiral remains untouched. This will be clear from the figure which, like all the other figures,

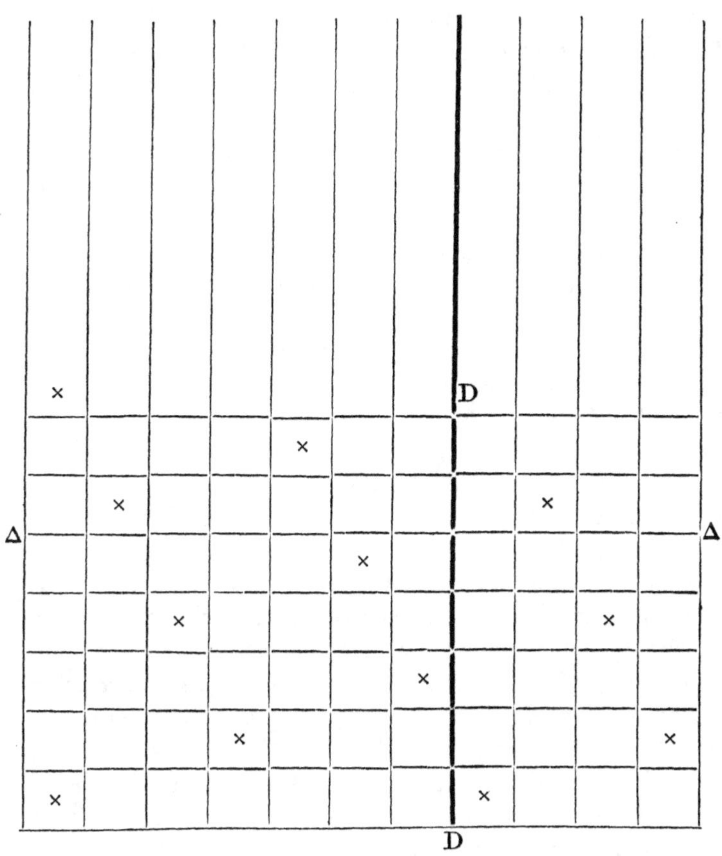

Fig. 1.

represents a spiral in a state of development, that is, as if unwound from the cylindrical axis on which it grew, so that the right-hand and left-hand sides of the rectangle are in reality one and the same line. The spiral in the figure has for its angle of divergence the fraction $\frac{7}{11}$, that is, it consists of 11 leaves and goes 7 times round the axis; counting always from left to right the successive leaves are found at intervals of 7 cells. Now if we blot out all that lies to the right of the line *DD* and consider the first 7 cells by themselves as having been unwound from a smaller cylinder, the leaves will be found to stand at equal intervals from each other, namely, at intervals of 3 cells. There is indeed a dislocation as to height, but height is throughout the Theory of Vegetable Spirals an unessential, or at least very variable element. Similarly what lies to the right of the line *DD* forms a system of 4 leaves, similarly arranged at intervals of 3 cells. Thus the spiral $\frac{7}{11}$ is resolved into two whose angles of divergence are respectively $\frac{3}{7}$ and $\frac{3}{4}$. The former is of course the same as a spiral whose angle of divergence is $\frac{4}{7}$ only running in the opposite direction; and as it is usual to give the name of Fundamental Spiral to the flattest or most sloping spiral which can be drawn through the leaves of a given system, the resolution we have effected may be described by saying that the original system has been divided into two, whose fundamental lines run in opposite directions, thus suggesting the notion of two opposing or antagonistic growths, making up, by the mutual influence of their antagonism, one symmetrical whole. Now I would propose, as a sort of postulate, the assumption that every system which forms a part of a system actually existing in nature is capable of independent existence, that it only wants, if the expression may be used, to be allowed to get possession of the whole axis of growth in order to shape itself into a symmetrical spiral. The consequences of this assumption are sufficiently interesting to make it worth while to trace them in detail. They afford a remarkable instance of the confirmation by facts of an *à priori* principle: the principle in question being, so to speak, an interpretation of Schelling's celebrated maxim*, that the whole is in every part and every part in the whole; a maxim which, though it may have led Oken wrong (I allude of course to his speculations as to the significance of the bones of the skull), yet enabled him to lead others right.

2. Let us in the second place make an assumption naturally

* This maxim is in effect equivalent to the fundamental principle of Leibnitz's philosophy, that each monad represents the universe.

connected with what has been already said, namely, that the portions into which any real system can be resolved are, as in the example just given, opposed to one another in the direction of growth. This assumption, combined with the postulate previously stated, enables us to determine *à priori* what systems are real or possible, and what not.

If the angle of divergence is $\frac{N}{D}$ in the original spiral, those in the resolved portions have the same numerator, namely, $2N-D$, the denominators being respectively N and $D-N$, and the two fundamental spirals run in opposite directions, provided the original fraction is not greater than $\frac{2}{3}$ or less than $\frac{3}{5}$. The proof of these propositions I will not stop to give; a little consideration will enable any one to supply the omission. Our assumption therefore excludes all fractions lying beyond the limits just stated, and in fact we need not consider the second limit at all, as the first is sufficient to guide us to the conclusions we require.

In order to determine the successive spirals into which the original one may be divided and subdivided, we proceed as if we were seeking the greatest common measure of the numerator and denominator of the angle of divergence, which of course have no common measure but unity. As long as the successive fractions are less than $\frac{2}{3}$, the successive quotients are of course unity; but whenever the fraction transcends that limit, the next step of the process will introduce 2 or some higher number as a quotient; and conversely no such quotient can appear until this limit has been transgressed. Hence this conclusion, every real system has both its numerator and denominator consecutive terms in a certain recurring series, namely, 1, 2, 3, 5, 8, 13, 21, 34, 55, 89, &c. of which the law is that each term is the sum of the two preceding ones. For these are the only fractions which, according to the method of continued fractions, can present themselves when all the quotients are unity. Thus $\frac{3}{5}$, $\frac{5}{8}$, and so on, are the angles of real systems, and there can be no system not similarly included in the preceding series. Every other system will, when our dichotomizing process is carried far enough, present us with two sub-systems, which have lost the character of antagonism, and have their fundamental spirals in the same direction. We have thus arrived *à priori* at the conclusion established inductively by Schimper and Braun, namely, that these are the only numbers which present themselves in vegetable spirals. I believe at least I am justified in saying so, for though Professor Henslow*, in giving an account of Braun's researches, does not say that no other numbers occur than those of the series, yet all his examples belong to it, and the principle is broadly stated as a general law by Unger in his Botanical Letters†. To this numerical law we can now append an equivalent statement, which not being numerical brings us a good deal nearer to a theory of the actual genesis of vegetable spirals, for we may substitute the following as an equivalent law, namely, that every vegetable spiral is resoluble into antagonistic portions, actually unsymmetrical but potentially symmetrical. In this form the law serves to give morphological significance to many obvious phenomena; as for instance, the three larger petals of the pansy stand in opposition to the two smaller, the balance of symmetry not having been fully established. Again, the five petals of papilionaceous flowers present us with two petals opposed to three, the three latter containing a sub-system of two petals opposed to one. Thus perhaps too we are led to recognise a certain duplicity in the structure of cruciferous flowers; but I will not venture to enter on any but obvious examples.

It is interesting to observe how consciously or unconsciously we are always influenced by the forms of nature. That man is her minister and interpreter is true, not only of his knowledge and his power, but in that which combines both, his works of art. That volutes and tracery reproduce the forms of the vegetable world is obvious, but we may be reminded of the numerical relations of which we have been speaking when we recollect how often three lights in a western window are opposed to five in an eastern: the higher symmetry in the holier place and related to the lower as the flower to the leaf.

Sed fugit interea, fugit irreparabile tempus;
Singula dum capti circumvectamur amore.

3. The result at which we have arrived may perhaps be more clearly expressed by saying that every natural spiral can be resolved into two parts, of which the smaller stands in the

* In Lardner's *Cabinet Cyclopædia.*

† Naumann appears to have recognised a spiral of 377 elements, thus reaching the thirteenth term of the series.

same relation to the greater, that the greater does to the whole; thus when a system of 8 leaves is divided into 5 and 3, as the first contains the second once, so does the second the third; moreover the second is opposed in direction to the first, and so similarly is the third to the second. Or we might say, though it is dangerous to use mathematical terms except with mathematical precision, that every natural spiral can be divided in extreme and mean ratio.

Thus much respecting division by azimuths. We now come to consider horizontal division, that is, division in which all above and below a given horizontal circle is successively supposed to be obliterated, so as to leave in the first instance only the lower part of the spiral and in the second only the higher. It may be easily shown—but as before I shall not stop to give the mathematical demonstration—that if N and D are respectively the numerator and denominator of the angle of divergence of any spiral, and N' and D' respectively are the numerator and denominator of the last converging fraction to $\frac{N}{D}$; then D' and $D-D'$ are the number of leaves in the two portions, N' and $N-N'$ respectively being the corresponding number of times that the spiral goes round the axis. Thus recurring to the former example, the spiral whose angle of divergence is $\frac{7}{11}$ is resoluble into two whose fractions respectively are $\frac{5}{8}$ and $\frac{2}{3}$. In figure 1 the line ΔΔ indicates this division, there being 3 leaves above and 8 below. Of course the division might have been made so as to leave 3 rows below and 8 at top. But this is scarcely worth remarking; it may be more needful to point out that in examining the figure we are not, as before, to be guided by the number of cells, but must look at each set of leaves as we would at a spiral existing in nature. It will be observed that there is a deviation from the perpendicular in both parts of the figure; a retardation, so to speak, in the one and an acceleration in the other, and that these, on the whole, balance. As in the former mode of division there was a certain amount of vertical displacement, so here there is of horizontal.

Let us now observe the *prærogativa* which, with reference to this mode of division, belongs to the natural systems. It is this, that whether divided vertically or horizontally the result is the same. Not so with the case of the figure; 11 was before divided into 7 and 4, now into 8 and 3. But when the quotients in the process of finding the greatest common measure are all unity, then the successive fractions formed by the remainders are themselves the series of converging fractions. Thus $\frac{5}{8}$ is the last converging fraction to $\frac{8}{13}$, and so is $\frac{3}{5}$ to $\frac{5}{8}$. In consequence of this the character remarked at the beginning of this section with respect to vertical division presents itself again with respect to horizontal, namely, that natural systems can be divided into parts, of which the first is to the second as the second to the whole; whereas, to recur for a moment to the figure, $\frac{3}{8}$ is not the last converging fraction to $\frac{5}{8}$, although $\frac{5}{8}$ is so to $\frac{7}{11}$.

There is something very interesting in this recurrence of similar relations among the parts of natural systems. One of the Bernoullis engraved upon his tomb the equiangular spiral as an image of Immortality, giving it the motto 'Eadem mutata resurgo:' a vegetable spiral might similarly be chosen by Braun or Schimper.

But what gives this mode of division a peculiar interest is that it brings us nearer to the actual genesis of the spiral. For the lower leaves are formed before the upper, and as the process of dichotomy may in this as in the former case be carried on indefinitely, we thus come to the remarkable conclusion that every vegetable spiral *has been** every simpler spiral before it becomes what it is. It is impossible not to be reminded of the similar doctrine which has been held with respect to animal or rather with respect to organic life in general, namely, that the higher forms are not merely typically connected with the lower, but actually developed from them. It would not be wise to carry our inferences too far, but with respect to vegetable forms, so far as these consist of arrangements of leaves, the matter admits of demonstration. It is no more than anybody may see who will take the trouble to count the scales on a fir cone.

Though I have said that we must not carry our inferences too far, it would, I think, be impossible to stop at the limits which in the present state of our knowledge lie within the range of precise demonstration. The production of leaves mark, to use the German phrase, a series of similar moments of develop-

* In saying this, we of course neglect the small acceleration or retardation already mentioned.

ment in the life of the plant; but what is true of such series must, we can hardly doubt, be true of all. We can hardly look at figure (2), in which the spiral $\frac{8}{13}$ is represented, without seeing

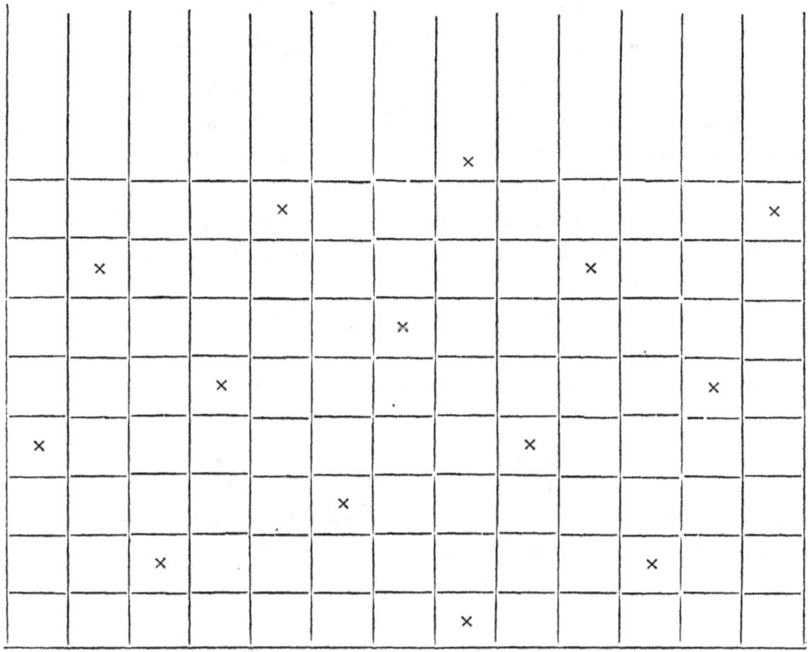

Fig. 2.

at its base the primitive phyton with its two leaves and radicle: nor having got thus far not to remember that the laws of vegetable spirals are simply the result of their capability of division, and that the ideal division of the spiral is only a type of the actual division of the cell. The Monad, said the Neo-platonists, produced the Duad, and the Duad all things.

4. Maxims of this kind may well be called 'oracula mentis;' they envelope the truth rather than express it, and like oracles do not teach us how we are to interpret them. They consequently resemble oracles in this also, that for the most part those who are guided by them are led astray. For it is difficult to resist the sort of charm which philosophical dicta possess when having reached the boundary land between the finite and the infinite, they pass into the region of poetry*. In thinking, for

* See W. von Humboldt on the "Bhagavad Gita."

instance, of the relation of species to the individual, of the whole to its constituent parts, of the unity of type and the endless diversity of forms, we feel that our doubts and difficulties cannot be better expressed than in the words of the old Orphic poet:

πῶς δέ μοι ἕν τι τὰ πάντ' ἔσται, καὶ χωρὶς ἕκαστον;

Is a cell a unit, or is a leaf, or is a whole plant, or is the idea of unity a relative one, realised in different degrees in lower and higher organisms? However unsatisfactory this view certainly is, so much must be admitted, that the constituent parts of lower organisms have a greater tendency to commence an independent existence than those of higher: the bond of unity is weaker.

If, following the train of ideas suggested in the last section, we attempt to express the difference between monocotyledonous and dicotyledonous plants, we may perhaps say that the former assume the condition of independent phyta before they have raised themselves above the system expressed by the number two, or if we look to the leaf alone above the first system, whereas the latter stand one step above them in complexity. And it is interesting to observe that they keep at the same distance throughout, at least, in the greater number of cases three being the prevailing number in the flowers of the former class and five in those of the latter. There is no real spiral structure in what for the sake of regularity we call the spiral $\frac{1}{2}$, because there is nothing to determine which way we are to turn; growth is in this case an oscillation in a single plane, and the formation of a structure possessing three dimensions must be the result of the superposition of growths in parallel planes. There is no single spiral growth upwards and outwards, binding together indissolubly the growth in both horizontal dimensions; the primary direction of growth is therefore directly upwards, and any growth upwards and outwards corresponds to another with which it may be said to coalesce, directed outwards and downwards. Similarly there is no spiral growth downwards and inwards, no tap-root, and therefore in effect no true root at all; what supplies its place is the downward and outward growth tending to recur at every part of the structure. I do not say that any of this amounts to a demonstration, but it certainly coincides in a remarkable manner with what we know, as a matter of fact, of the growth of monocotyledons, not only with respect to the absence of a tap-root and to the tendency to put out adnate roots, but also as relates to Mohl's discovery of the crossing of the vascular bundles or fibres; a discovery which might, I think, have been anticipated by an attentive consideration of morphological principles; or which, at any rate, now that it has been made, seems to confirm them. Figure* (3) is intended to illustrate what has now been said, and we shall see hereafter that the same view is confirmed by a consideration of the leaves of this class of plants. That they stand lower in point of organization than dicotyledons is admitted by every one, and seems to be indicated on the principle we have already noticed by their tendency to put out roots above the ground as well as by many other circumstances, and we seem to be justified in inferring from this that the spirals which are numerically more complex are really types of higher organizations.

It may be well, in order to obviate a misconception of my meaning, to explain what has been said as to upward and outward growth in dicotyledons. No doubt the stem or trunk of any tree belonging to this class is a portion of a conical surface whose apex is above the ground and not below it, and the same is true of the surface of contact of cambium and wood, that is, of the surface where fresh wood is, at any given moment, in process of being formed, so that in one sense the growth is upwards and inwards. But the corresponding conical surface a year ago lay within and below the other, so that the transition from the one to the other is outwards as well as upwards, and that it takes place in a spiral as if the surface turned on its axis, and at the same time enlarged, may reasonably be presumed in accordance with the views which I have been endeavouring, though with much diffidence, to support. Of course it is not an objection to these views to say that the woody fibre does not consist of spirals, actual growth being one thing and the developement of forms, or what may be called ideal growth, another.

5. In endeavouring to apprehend the process of transition from one kind of spiral to another, it is difficult in the present

* The figures (2) and (3) referred to in the text, were not to be found among the papers placed in the Editor's hands: fig. (2) on page 365 has been drawn by the Editor, who is however unable to infer with any confidence the form of the lost figure (3).

state of our knowledge, or perhaps impossible, to arrive at an explanation not based on morphological grounds or expressed in morphological language. Yet it is certain that we ought not to acquiesce in a merely morphological explanation of any phenomenon, though it is going too far to say, that such explanations are merely transitional or provisional, to be accepted only until something else can be got. For teleological considerations, whereof morphological are one kind*, neither exclude nor are excluded by those which relate to actual genesis. The formal and the final cause alike exist and coincide, if I may presume to touch upon the subject, never to be mentioned but most humbly and most reverently, in the Unity of the Divine Wisdom, from which both proceed, and the explanations derived from either are, or rather would be, if our knowledge were complete, equally and alike true, though in different ways. But to return to the matter in hand. It seems reasonable to suppose that when vegetable growth is proceeding according to any given system, say for instance, that whose characteristic fraction is $\frac{8}{13}$, the tendency to rhythmical recurrence on which the whole depends still exists among the parts of which the whole is made up, especially as these two parts are as we have seen nearer in point of position to constituting a symmetrical spiral than any other two portions into which the whole could be divided. The first or last five, or the first or last eight are within $\frac{1}{13}$ of the circumference of going round the axis an integral number of times. Now as we have seen, it is the larger portion of a spiral which possesses a character of antagonism to the whole. If the interval between adjacent points exceeds the limit towards which this interval continually tends†, then that between adjacent

* It is surely only from want of attention to the History of Philosophy that the two are put in opposition to one another, whereas in the sense in which Aristotle would have used the term, the doctrine of those who speak of the Unity and Modification of Types, and that of those who refer every thing to the advantage of the individual or species, are alike based on the Final Cause. Moreover there seems no sufficient reason for setting the two doctrines in opposition, still less for rejecting either. Truth comes to us in fragments, and so we must be content to accept it, not letting ourselves be deterred by any *antinomy* from embracing all of it that we can apprehend.

† Namely four right angles multiplied by the positive root of the equation $x^2+x=1$. I use the word *tend* here in the mathematical sense of tendency to a limit. It is, I believe, Bravais' doctrine that there is in reality but one angle of

points of the spiral next below it falls short of this limit, and *vice versâ*. In consequence of this, under the influence of circumstances, which we cannot appreciate, the last eight points may repeat themselves in a symmetrical position, and maintain this form of symmetry permanently. On the contrary, if after having thus repeated themselves they are again subjected to the influence of the thirteen preceding points, a sort of compromise is effected, of which the result is a spiral of twenty-one points. The two cases resemble a contest between the constitution of a whole kingdom and that of a particular province; the province may revolt and succeed in establishing its own law universally, or a mixture of the two may become the common law of the land. Thirdly, things may remain as they are and the provincial law be altogether abrogated. Thus the opposition of the greater part to the whole may under varying circumstances produce an increase or a decrease in the complexity of the system, and the same cause, namely the rhythmical tendency which seems to belong to all growth, may be at one time a cause of change, and at another of permanence. If we try to go beyond this, "ad ulteriora tendentes ad proximiora recidemus," our explanations must become vaguer and yet more unsatisfactory. Something might be said of the contest between the principles of recurrence and opposition, but it would too much resemble the *lis* and *amicitia* of Empedocles[1].

6. The attempt I am about to make to trace the general characters of vegetable arrangement in the structure of leaves is based upon so simple a remark that I can hardly suppose it has escaped observation. As however I have not met with it, it seemed worth while to set it down here though I cannot trace its consequences in detail. The primary division of the nerves of leaves is into anguli-form and curvi-form nerves, that is, those which have a straight portion at their base and those which exhibit curvature throughout. Now just as the straight

divergence, namely, this limiting one; and that if botanists have thought they recognised distinct systems, it has been because they have stopped sometimes at one degree of approximation to it and sometimes at another; but this doctrine appears to be opposed to ascertained facts, the real unity of nature does not in this way exclude variety.

[1] When the spiral passes into a whorl, the tendency to rhythmical recurrence seems generally to cease and to be replaced by a tendency to symmetrical antithesis or alternation, as if, speaking mathematically, positions of equilibrium previously stable had become unstable, and *vice versâ*.

growth of the stem is the result of a *quaquaversal* tendency to curvature, so more generally I think it may be asserted, that straight growth in any part of a plant is constrained growth, the result of a tendency to curve in opposite directions. What characterises, therefore, the venation of such a leaf as that of the lily is in this sense the absence of constraint. To this remark an hypothesis is to be added which seems to be justified by the phenomena presented by compound leaves, namely, that the nerves are originally boundary lines, showing where the leaf in its growth paused in its development, just as a range of pebbles marks the old boundary of sea and land after the sea has receded. Thus the lily leaf shows us a *single* growth in various stages of progress. If we now look at the second division of monocotyledonous leaves, those which have a mid-rib and angular formed nerves, we see the successive stages of a *double* growth lying on the right hand and left hand respectively of the mid-rib, which is to be regarded as the straight resultant of the opposite curvatures of the boundaries of the two portions; the remaining part of the boundary remaining free and therefore curved. When therefore we say, that with few exceptions, monocotyledonous leaves follow one or other of the types just described, we in effect say that they are either single or double growths, that they belong to the first or second forms of vegetable development, and rise no higher. Let us now look at the commonest form, namely the penni-nerved of dicotyledonous leaves; in these not only the mid-rib but the diverging nerves are, at their base at least, straight, exhibiting evidence therefore of constraint, or opposing growth on their two sides. The space between the apex and the first pair of lateral ribs seems to me to be a leaf in itself, that between the first pair and the second to be similarly two independent portions osculating with one another and with the first. By their osculation with one another, they form an additional portion of mid-rib, by their osculations with the central part they form the first pair of lateral ribs, and if we follow the course of either of these ribs we find that it gives out secondary branches, which freed from constraint, or comparatively so, become curved. In the space they leave, however, we find a secondary central portion, so that the meeting of three elements is exhibited at the sides of the leaf as well as in its central growth: thus the reticulation is throughout polygonal, and not properly speaking, curved, though from the shortness of the sides of the polygons it may assume a curvilinear appearance. At the sides, however, there often runs a sort of fringing nerve between which and the edge there is nothing but parenchyma, and this nerve being free from constraint appears truly curvilinear. In a simple leaf the lateral growths remain subordinate to the central; the contest between them being shewn by the jagged edges, the cicatrices veteris vestigia pugnæ. But if the lateral growths prevail, the central one ceases, the osculations no longer taking place in the manner described, and thus the leaf becomes compound, the fissures now reaching to the axis.

From what has been said, it appears that a penni-nerved leaf belongs to that system of vegetable growth of which the characteristic number is 3, divided as such a system must be, into 1 against 2, that the 1, so to speak, assimilates 2 and 3, so as to form a second unit or 1′; and this entering similar relations with similar lateral portions 2′ 3′ forms a third unit, a wave of form being thus propagated from the apex towards the base, and increasing in size as it goes along. To enter into an examination of the cases in which a simple leaf belongs to a higher system than the third or of the structure of compound leaves would detain us too long. It is enough for my purpose to have pointed out the two principles by which I believe all the details are governed, and by the help of which they may be traced out. The distinction between a leaf and a floral whorl which makes the numerical relations so much less easy to recognise in the former case than in the latter, is, that in a floral whorl the axis of such symmetry as exists coincides with that of growth, whereas in the case of an ordinary leaf, the two are at right angles. It is interesting to observe that the distance between monocotyledonous and dicotyledonous plants which shows itself in the commonest cases in their respective phyta and flowers, is also seen in the intermediate growth of the leaf, the commonest number which presents itself in the dicotyledonous leaf being that which occurs most frequently in the monocotyledonous flower. We are, I think, entitled to consider the leaf as intermediate between the phyton and the flower, since reckoning the sheath as one, it consists of three elements, as the phyton of two and the flower of four: the respective element of all three fulfilling Goethe's law of alternate contraction and expansion.

The ordinary form of the edge of a dicotyledonous leaf, supposing the jags filled up, is sufficiently well represented by the projection on a vertical plane of an equiangular spiral, chased on an inverted cone (I mean of course the outline of the half leaf, commencing at the base and going thence to the apex). The curve in question cuts the axis at a greater angle at the base than at the apex, and has its maximum ordinate nearer to the former than to the latter. Something may be said on theoretical grounds in favour of taking this as the typical leaf curve in cases in which the petiole and lamina lie in the same plane, since when they are at right angles, or nearly so, there is an approximation towards a circular form, as if the plane of projection had now become parallel to the base of the cone, and the slope of the spiral were diminished. In truth this principle of projection might be applied to explain many variations of form in the flower as well as in the leaf.

7. The preceding remarks can scarcely be free from mistakes, and yet they may contain a portion of important truth, namely, the principle of tracing numerical relations in the unsymmetrical as well as in the symmetrical aspects of vegetable growth. Just as spirals are less symmetrical than whorls, and yet enable us to understand the latter better than we otherwise could have done, so likewise unsymmetrical portions of spirals and of whorls may throw new light upon both[1].

[1] For an account of recent speculations on the theory of vegetable spirals, see Braun, *Betrachungen über die Erscheinung der Verjüngung in der Natur*, p. 125.

SOME THOUGHTS ON COMPARATIVE METROLOGY*.

In many respects it resembles comparative philology; one is, the necessity of avoiding general hypotheses. Corresponding to the attempts to deduce all languages from some one assumed to be primitive, or from certain elementary sounds, such as Alexander Murray's nine words, are such hypotheses as that of Gosselin, who, following Bailly, thought that at some early age in the history of the world, the length of the meridian had been accurately determined, and that all the guesses which have come down to us as to the magnitude of the earth were all equally accurate statements of the result of this primitive geodesy, the difference being solved by assuming the existence of a number of different stadia. Touching Bailly's astronomical speculation, it is hard to deny the truth of De Lambre's judgment, who describes it as "le plus cru des songes," or by some equivalent phrase. Another general hypothesis seems little better, I mean that of Boeckh, namely, that the weights and measures of antiquity all came from Babylon, and were connected with the use in astronomy of the clepsydra. Thus weight is the primitive element, then capacity is derived from weight, and lastly linear measure from capacity; the unit of the last being the edge of a cube whose volume was the unit of the second; the unit of the first being the weight of that volume of water. Starting from hypotheses so unproved as these, and modifying them as occasion may require, we may derive anything from anything and explain everything.

2. In another point of view metrology resembles philology, namely, that similarity is a very bad test of real connection, and that much better evidence is furnished by formal relations than

* Now first published.

by material resemblance. The tradition of magnitudes is always inaccurate even when there is not, as in the case of coinage, any advantage to be gained by inaccuracy. The French *livre*, for instance, though shrunk to a shadow of its former self, preserved to the last the due number of *sous* and *deniers*. The tradition of magnitudes is, so to speak, physical, that of relations, mental.

3. William Von Humbolt wrote an ingenious essay (first published in the *Berlin Transactions*) on the influence which is exerted on a language by its having been reduced to writing. It is easy in a general way to see that writing not only gives greater fixity to language but also greater influence to the educated class by whom alone it is familiarly employed. Corresponding to this is the reduction of weights and measures to tables, &c., which are made a part of ordinary education, and here we find an analogy which both the sciences I have been speaking of bear to jurisprudence. In the famous essay on the vocation of the age to codification, Savigny has pointed out that in the earlier period of the history of a nation law is a living and popular thing, "volitat per ora virum," and does not fall into the hands of a particular class, at least not to any great extent, until a later time. It ascribes this mainly to its increasing complexity, but it arises also from the division of labour inseparable from the progress of civilization, and likewise from other causes. In English literature it is curious to see how much more frequent legal allusions are in old than in more recent writers. Some have fancied that Shakespeare must have had a legal education, but the same thing might have been thought with regard to Chaucer. Hugo has pointed out that many phrases of quotation or reference in the *Digest* relate not to the maxims of earlier writers but to what may be called legal proverbs. To these correspond in the history of Metrology such maxims as "a pound is a pint," or the Arabs' saying that a mile is as far as one can tell a man from a woman. Of much the same nature are the names given to measures of land from the quantity of seed which they were supposed to require. Half legal, half metrological are such things as Mr Justice Buller's well-known rule, and the curious determinations in the *Sachsen Spiegel*, whereby a man was hindered from making his

house his fortress. He was to dig no deeper trench than he could throw the earth out of with his spade, to raise the bottom of the door no higher than his knee, to enclose himself by no wall higher than a man on horseback could look over. Cases of the limitation of jurisdiction by throwing a stone or shooting an arrow from a given spot and noting where it falls, are mentioned by Grimm, and there is even now a ceremony of the same kind at Cork, the mayor annually throwing a javelin into the sea. The meaning of his doing so was doubtless in its origin some determination of a boundary.

Of popular pieces of metrology I may mention that I have heard that a certain popular hymn-book passes current in Hanover as a pound weight. For all these things picturesque and inaccurate as they are, advancing civilization substitutes abstract numerical determinations, but the natural tendency to connect standards of different kinds with one another, and all with natural objects, remains. For the foot or the fathom we employ the length of the meridian or of the pendulum. It may be quite true that as things now are it is better to trust to standards accurately made and carefully preserved, but the desire to recognize something ideal in practical life and to connect ourselves with something less perishable than our own handy-work, is not easily to be dismissed. How interesting, says the report of the French Metrical Commission, for the father of a family to know what proportion of the surface of the earth is the field which supports his family! The remark is a little in the tone of the golden age of Rousseau, but nevertheless it has its foundation in the realities of human nature.

4. What the historian of Metrology must do is, dismissing general hypotheses, to observe the nature of the relations which the popular mind seeks and establishes between different standards, making large allowance both for popular inaccuracy and for the love of simplicity. He must also observe the nature of the traditions by which different standards are handed down, and the greater or less respect in which they are held: the curse, for instance, on him who removes a land-mark, and the unscrupulous way in which arbitrary governments have always dealt with coinage. Then again with regard to what may be called

positive metrology, we must take into account the degree of scientific culture and of practical accuracy which may be supposed to have existed in different ages and countries. What follows is merely some remarks on detached points.

5. I do not now recollect Boeckh's explanation, founded on the general view already noticed of the relation between the Greek and the Roman foot, but a different one to which I think he makes no reference, seems to me far more probable. The Romans starting from the assumption that the pace is five feet (which is as near the truth as any relation expressed so simply could be) made a unit of itinerary measure of a thousand paces length. Observe by the way how like this is to the derivation of the sestertium from the sestertius. Now, it is a general principle that units tend to divide themselves into halves, quarters, &c., although they are for the most part multiplied by tens or twelves, the decimal system seeming much more natural in the upward than in the downward scale. Thus the Romans would speak of distances as being a quarter or half-a-quarter of a mile; and this last fraction was sufficiently near the stadium to be identified with it. On the other hand, the Greeks, starting from another natural assumption, made the orgyia six feet, and a hundred of the former became their unit of distance. The stadium would seem less an itinerary than, if the word may so be used, an athletic measure, for it is certainly more natural to measure distances by paces than by fathoms. So it was, however, that 600 Greek feet were identified with the eighth of a Roman mile—that is—with 625 Roman feet, which gives the common relation of 25 to 24. I do not mean that the average Greek foot was not longer than the Roman. We know that it was, and the difference probably corresponded to a difference in the two races of men. But what we have to explain is the origin of the precise relation in question. Again, it is not to be supposed that after it was recognized, the measure-makers of Greece and Rome thought themselves bound to conform to it. Writers on metrology perpetually fall into a kind of realism, and speak as if they believed that there was for instance an ideal Roman foot, an archetype to which the Romans perpetually endeavoured to conform themselves, and which it is our business to discover. Whereas there

was of course in reality no general archetype. One measure-maker copied another with more or less accuracy, and all we can do with respect to absolute magnitudes, is to determine as well as we can their average value. The *intention* only is ideal, and this cannot have reference to the tradition of standards of the same denomination, but to the relations among different standards. I add a few words of etymology. There can be no doubt that orgyia is properly oregyia, and is derived from orego. Thus the word means *the stretch*, which is not, properly speaking, the meaning of either, of fathom or of the French brasse, though all three words in effect mean the same thing, namely, the greatest distance at which the finger-tips of the two hands can be placed from each other. Brasse comes from embrasser, and suggests the idea of a man putting his arms round a pillar or a tree. Whether any appreciable difference exists between the girth of a tree which a man could just embrace and what we call a fathom I do not know, nor am I likely to try the experiment. Fathom has the same origin as brasse, as we may see without referring to Anglo-Saxon in the modern Danish, in which favn, a fathom, is obviously connected with the verb favne, to embrace. The latter, by the way, we have in English, in to fawn, though we only use it figuratively, or with reference to dogs who cannot properly be said to embrace those whom they caress. The word orgyia, besides confirming the common etymology of agyia from ago suggests another which I do not remember to have seen, namely, aithyia from aitho. Nothing can be more natural than to call a sea-bird swift on the wing, and whose feathers are often sprinkled with sea-water when it dashes down after its prey, the gleamer or flasher. I have noticed, probably after a shower, absolute flashes of light from pigeons wheeling in the sunshine, and the same thing must happen more frequently in the case of sea-birds.

6. Nothing of the kind is more permanent than the fixation of land-marks. Many causes which affect the tradition of other measures produce no effect on that of agrarian. It is probable that the majority of our fields had their present limits, or nearly so, at the time Domesday Book was compiled, and yet the estimation of their magnitude may have varied. The tradi-

tion of agrarian measure may be of two kinds. First, direct, as when the magnitude of certain pieces of ground being known and admitted, others are measured in accordance with them; or, second, indirect, as when what tradition preserves, is the relation of square to the units of linear measure, the latter being, for obvious reasons, the more common case. As an instance of it I will take the relation of the acre to the jugerum. Of these seven jugera, which long formed the Roman unit of landed property, we may reasonably suppose that one was devoted to house, offices, and garden. Here it is worth noting that Pliny remarks that *hortus* originally meant not merely a garden, but what in his time was called a villa, that is, I presume, a country-house with its homestead. Something of the same kind still perhaps exists in the popular Italian use of the word *casa.* We know, at any rate, that in Dauphiny *casau* means *hortus*, and that Casaubon when a young man translated his name into *Hortibonus*, and was in consequence at a later period charged with plagiary by those who did not know who Hortibonus was. Admitting this hypothesis, six jugera remain, for what comparatively speaking may be designated by the French phrase, "la grande culture." Now nothing is more marked than the tendency of agrarian measures to form themselves as far as possible (I mean of course ideally) into squares, and to be subdivided into similar smaller figures. Thus, and perhaps the change was connected with some forgotten system of four-course husbandry, the six jugera may have fallen into four agri as our acre itself has into four roods. If this be the origin of our acre, its area would be to that of the jugerum (no regard being had to the difference of the Roman and English foot) as 3 to 2, and therefore the acre should contain 4800 square yards. It *does* contain 4840, but the reason why it has thus been modified is obvious, in order that the tenth of a rood may be a square with a whole number of yards for its side, or which is the same thing, in order that the side of the square perch, which is the 160th of an acre, may be accurately expressed without any complex fraction. Had the acre remained what on my theory it ought to be, the perch would have been 30 square yards; by adding a quarter of a square yard to it, it becomes a square whose side is 5½ yards. Thus, setting aside this modification and the difference of units, the jugerum is two-thirds of our acre. The former cause reduces it by about five-sixths per cent., and the latter by about six, and we thus get an approximate rule for converting jugera into acres, namely, to take two-thirds of their number and strike off 7 per cent. from the result. How the tradition of the jugerum ever got to Ireland seems very hard to say. Perhaps it was not extinct in England in the time of Earl Strongbow, at least I am much disposed to recognize in the Irish acre, not three actus as in the English, but five, which would of course make it 8000 yards. But in order that its 160th part should be a square, it underwent rather a greater modification than the English, losing about 2 per cent. and becoming a square of 7840 yards, the linear perch being thus not 5½ yards but 7. Here again my speculation suggests an approximate rule to reduce Irish acres to English; multiply them by ten, divide by six, and diminish the result by 3 per cent. Between these two acres stands in point of magnitude the Scotch, but here we seem to come upon a tradition of the direct kind, and I mean therefore to connect it with that of the pezza from the jugerum.

7. At page 163 of the 3rd volume of Miss Winkworth's translation of Niebuhr's letters is one to Savigny, which offers several points of interest as to the connection of ancient and modern measures. He tells Savigny that at Rome there is no manual of weights and measures, nor any information to be got either from scholars or men of business, but that it having occurred to him that during the French rule tables must have been formed connecting their measures with the Roman, he had been enabled to ascertain by accurate calculation the magnitude of the pezza, and thus to show its near approach to the ancient jugerum. Consequently the rubbio being 7 pezze represents the old land allotment. The remarks on the *dépourvu* state of Rome as to information are in the querulous tone into which unhappily Niebuhr often falls. The tables he speaks of having found are doubtless those published but a very few years before, and of which Prony has given an account at the end of his work on the drainage of the Pontine Marshes. If this commission had ceased to exist in 1818, when Niebuhr wrote, its previous existence must at any rate have been matter of notoriety, and it is hard to believe either that all the commissioners had left Rome

or that any of them would have been unable to give the required information. They were certainly not all Frenchmen. Scarpellini, the only person whom Prony mentions, and whom he speaks of with respect, being, beyond all question, one of them. Was he ultimus Romanorum? the last person who knew the length in metres or French feet of a measuring chain? and might not Niebuhr have borrowed one from any village land-surveyor and measured it himself? Niebuhr's result is that the pezza is equal to 24,716 square feet and a fraction. It is not said what foot he means; most probably the French foot, which is the longest of any he was at all likely to make use of, and it is therefore worthy notice that even so his estimate is too small. I am obliged to do these things in my head, but a simple piece of mental arithmetic enables one to see that the pezza is much more than 25,000 French feet. Prony's statement, which does not differ from that of the tables, if at all, except by the correction of an almost inappreciable error (I know not if this error affected the particular result in question) is, that the pezza is equal to 2640·6257 square metres. Neglecting the last decimal, it follows that the pezza is 10 times the square of 16¼ metres. Hence if we assume the pezza to be 25,000 feet, 100 linear feet will be 32·5 metres; a result which exceeds the truth by about the 2000th part. Consequently the pezza is more by about 25 feet than 25,000 feet, and consequently Niebuhr's calculation, if he meant the French foot, is in defect by about 300. When we come to his calculations of the jugerum, another difficulty arises, namely, his not having mentioned what value he assigns to the ancient Roman foot. According to Boeckh, Niebuhr has in his history expressed a strong opinion in favour of Cagnazzi's determination. Whether he had formed it at the date of the letter of which I am speaking, I do not know. Certain it is that, if we adopt this value, which is higher than the average, Niebuhr's determination of the size of the jugerum errs in excess as that of the pezza in defect; and, the latter being the larger measure, both errors tend the same way, namely, to make the pezza and the jugerum appear more really equal than they are. The value in question is 131·325 French lines, and the result is that the jugerum is something less than 24,000 feet. Niebuhr makes it 24,310 feet and a fraction. Thus unless I am greatly mistaken, the difference between the two measures instead of being scarcely more than 400 feet is in reality over 1000. Taking the relations of the rubbio into account, and also the subdivisions of the pezza, there can be no reasonable doubt that the latter does really represent the jugerum; but in such a matter an error of Niebuhr's is worth pointing out. A little farther on he says that no one can blame him for not having sooner known that the pezza was not *any* piece of land, but one of a particular size. No one will blame him probably, and yet it is only what must have been known to a great number of persons. In 1760 Cristiani published a work which I only know by a table quoted in Diderot's *Encyclopédie* (Art. Arpent), in which in the space of a few lines the values of the pezza and the jugerum are both given (I am at least sure as to that of the former), much more accurately than by Niebuhr. The same table, or the materials from which it was compiled, have probably found their way into many other works of reference. Niebuhr tells Savigny that rubbio is also a measure of wheat of about 640 Roman pounds weight, so called because it was the usual amount of seed for the corresponding measure of land. But he has not noticed the curious circumstance that the rubbio of oats and that of wheat differ, and that the measure (for it is a measure and not a weight) is less in the case of oats, although probably in Italy as here a much larger quantity of seed is used for it than for wheat. Niebuhr has also remarked that in his opinion the stajolo, which is the smallest square measure, is derived from sextarius, so as to signify such measure of land as would require a sextarius of seed, admitting however that the stajolo must have grown smaller since it obtained its name. Greatly smaller it must have grown, for a sextarius to the stajolo would be at the rate of about 50 bushels to the English acre. It is almost incomprehensible that Niebuhr should not have seen that stajolo can be nothing else than stadiolum, whether or no the latter word occurs in dictionaries. It is moreover properly a linear measure.

Although square measures have strictly no definite form, yet one form which may be called the normal form is generally indicated by their subdivisions if not by historical evidence. In the case of the jugerum both concur in showing that it was contemplated as a double square, and probably in early times it never existed in any other form. That of the fundus of 7 jugera is not I apprehend so distinctly marked, though it can scarcely be doubted but that it formed a rectangle, as the methods of the gromatici would else have been inapplicable. The simplest hypothesis appears to be that the 6 jugera of which I have before spoken were arranged parallelly in a quadrangle, whose longer side was 480 feet and shorter 360, and that the 7th jugerum (that which I have supposed to be reserved for house and garden) was treated exceptionally, divided, that is, into 2 half jugera so as to form a strip 60 feet in width, and in length equal to and in contact with one of the longer sides of the before-mentioned rectangle. Or this strip may have been interposed among the pairs of jugera, but this is immaterial, what I am going to say depending merely on the assumption (and it is no more) of the divided jugerum. Of its two halves one I suppose occupied by house and kitchen-garden, the other by a vineyard with probably some pot-herbs growing among the vines, at least the enmity of the vine and the cabbage (depending probably on the amount of potash they both contain) could hardly have been observed where such a mode of culture had been uncommon. Half a jugerum would (speaking quite roughly) produce about 20 dozen yearly on the lowest computation admitted by Columella, twice as much according to his own estimate, and yet more according to others. Even this in a frugal family, the women and slaves drinking no wine, would have been quite enough, though of course wine was not, as with us, a luxury. My inference is that the normal form of a vineyard was in "la petite culture" a rectangle whose longer side was 4 times the shorter, and that this idea was retained when the unit of a vineyard was doubled and became in point of area a jugerum. It is thus I would account for what is plainly indicated by its divisions. That the pezza or petia (and it is to be remembered that the word is only used with reference to garden and vineyards) has to this day such a rectangle for its normal form. Perhaps the word replaced its older synonyme in consequence of this ideal change of figure. The proof that the pezza is conceived of in the form I have mentioned appears to result from the following consideration: it is divided into 4 quarters and each quarter into 40 ordini or vine-ranks. Now the length of the surveyor's chain is 10 stajoli, and therefore no length would be so naturally assigned to each ordine as this. And with this length the distance between two adjacent vine-ranks will be 1 stajolo, which is just about what one might expect to find it, being a little less than 4 feet. It is obvious that in this way the pezza will contain 1600 vines, and it is impossible to believe but that the fundamental unit in laying out a vineyard, namely, the distance between the vines, should not coincide with the unit of mensuration, namely, the stajolo. This granted, it appears probable, as I have already said, or in fact certain, that the length of the ordine is 10 stajoli or 1 fourth of the 40 stajoli, which would form the longer side of the pezza; and hence, on far the most probable supposition, namely, that the pezza was similar to its fourth part, it results that the shorter side of the former is 20, and its longer 80 stajoli. What makes all this very curious is that the length of the stajolo is of course incommensurable with the side of the old jugerum, which, for anything we see, was a perfectly convenient shape for a vineyard. I cannot help thinking therefore that there was an historical reason for the change, namely, that the form of the half jugerum was transferred to the whole one, or pezza. A change which probably occasioned a slight one in the inter-spacing of the vines. There may have been, for instance, in the primitive vineyard 4 quarters, each 30 feet in width and 120 in length, containing 40 ordini and 400 vines, the inter-spacement being only 3 feet. All this was preserved, except that the vines were put at a distance equal to 3 into the square root of 2; a change which has nothing to surprise us seeing the extreme discrepancy of opinion as to the best inter-spacement. Whether the question is settled now I do not know, but the doctrine was at one time popular that 4 or even 5 feet was always the best distance, while on the other hand it was said that even within the limits of France the distance ought to range from 3 or 4 deci-metres in the north to 2 metres in the south. Climate however is clearly only one element, and the question on the whole is probably as hard to settle as the one which relates to the proper quantity of seed corn. In the absence of definite grounds for forming an opinion men might very well have been guided by habit, and the love of the proportions which habit had made the most agreeable both to the eye and to the mind. Virgil has alluded to the vintner's love of symmetry and of the extreme delight which a man of imagination may be led to feel in the harmonious arrangement of trees.

Brown's Garden of Cyrus is a good, though perhaps an exaggerated, example. Niebuhr, on what grounds does not appear, speaks of the pezza as a square. If it was, so no doubt was its fourth part; and in that case, the actual vine-row being no doubt equal in length to the side of the square, would have been 20 stajoli; that is, just double the length of the ordine; for that the width of the latter was one stajolo does not appear to admit of question. It seems therefore to be a strong objection to his view that it would not represent the subdivisions of the pezza. What practical modifications there may have been cannot be ascertained. Every vine-dresser was of course at liberty to alter his standard or stajolo. Probably the modern practice varies less from the ancient in Italy than in France, and a good deal of information about the former may be gathered from the agricultural writers. I have forgotten what Columella says on the subject, and though the book is at hand cannot now refer to it. As Cervantes says in his last preface, "the thread is broken here but some other day I may be able to take it up;" but the other day did not come.

8. It is well known that within the boundary of a fundus there could be no usucapio for a space of 4 feet, or in other words, that up to that space there was "æterna auctoritas," a perpetual right of reclamation. It is commonly supposed that it was intended not only to prevent disputes, but to secure the existence of the numerous rights of way which form parts of the necessary evils of small property. Hugo has remarked that it does not follow that the space in question was uncultivated, and he must therefore have regarded the law as being merely a matter which concerned the owners of contiguous fundi, for if the space were cultivated at all it must have been by them. That a law leads to absurdly inconvenient consequences is not conclusive against its having existed, but on his view, if my neighbour had encroached in one place 10 feet within the boundary, he might establish a claim by usucapio to a width of six feet, while I could at any time re-establish myself in possession of the outward portion, so that what he required would be *enclavé* within land with which I could do absolutely what I pleased. It would seem therefore that the *auctoritas* lay with those whose rights of way were interfered with by the

occupation of the space in question, and not in any special manner with the two marches; and thus we come back to the common opinion that the space was really unoccupied. Now, what became of these spaces when the whole system of which they formed part was forgotten? I apprehend that the 7 jugera included only the clear cultivable space, and that consequently the strip in question came to be an addition to the area of the fundus. On the hypothesis already mentioned, that the regular form of the latter was a rectangle of which the longer side was 480 feet and the shorter 420, the whole circuit would be 1800 feet. The space gained would therefore amount to 7264 feet, or on the average jugerum, to nearly 1038 feet. Of the residual phenomenon, to use what was once a popular phrase, of the existence of 1000 Paris feet of difference between the pezza and the jugerum, my hypothesis would thus explain five-sixths and leave the difference so small as to require no explanation. It is difficult to believe that the limiting strip was counted in the allotment, because it would make the jugera for practical purposes of unequal size, a thing in itself inconvenient and liable in many cases to produce injustice, and there is therefore a strong presumption in favour of a view which at once explains what has become of it and why the pezza exceeds the jugerum.

9. The existence of a rubbio of oats of about four-fifths the size of that of wheat, while the ordinary proportion of seed corn used for oats exceeds that for wheat in about the proportion of 8 to 5, suggests the idea that oats, which were certainly but little cultivated in Italy in old times, (and probably are not now), were not usually sown over the full surface of the rubbio. There may have been a simple course of husbandry, namely, the whole crop of wheat one year, half fallow and half oats the two next years, and then wheat again, so that every part of the land was fallow at intervals of 3 years. Twenty other guesses might be made, but the fact is curious that the measure of oats would serve for about half as much land as that of wheat. As for the third rubbio, that of salt, it seems to have been formed on the principle of weighing as much as that of wheat.

At the end of Niebuhr's letter to Savigny he observes that 7 jugera must have been sufficient for one family, because he knew a vigneron who, holding 11 pezze as a metayer, supported himself and his family comfortably. The remark, if I may say so, appears to be singularly inconclusive. As the man was a vigneron, we may presume that his chief occupation was growing grapes; and how would it have been if his 11 pezze had been all under glass? and he had devoted himself to the culture of pine apples? The question is not within what area a gainful industry may be carried on, but what acreage per head of the population is necessary to secure a given degree of comfort, and whether the best possible use is made of it by dividing it into allotments. If Sancho's piece of cloth had belonged to five brothers, it might still have been better to make one wearable cap of it than one for each. The cases are not quite parallel, and there is doubtless much to be said in favour of small properties, but the question cannot be settled in the present state of things by individual instances, even if the question were not complicated by the existence of common rights of pasturage. Columella, Book III. chap. 5, speaks of rows, ordines, 240 feet in length and 3 feet apart, as the allowance for a jugerum. In each row there are 60 vines, and in the whole 2400. Thus if we divide the jugerum into 4 quarters by lines parallel to its shorter ends, we get, as in the quarter pezza, 40 rows, and each row of about the length of an ordine. The transition preserved the number of rows, but removed an objection to Columella's arrangement, namely, that the interval between trees in the same row is not equal to that between the rows, whereby Virgil's precept "to set vines square so that they may have an equal supply of nourishment all round" is transgressed. The change also increased the amount of ground for each vine from 12 to 18 feet, diminishing the number of vines from 600 to 400. I do not know whether the error in the text of the edition I have has been corrected. By it he is made to speak of 600 vines in a row of 240 feet, and of a total of 24000, numbers which are absolutely incredible. In the preceding chapter, when speaking of the number of cuttings which before the vines have grown up may be put between the rows and taken up when they have become viviradices, he allows 20000 to be planted. In this there is nothing which is not quite natural. They may then be put, I suppose, nearly as thick as asparagus, and perhaps the transcriber was misled by not seeing that Columella was speaking in the 5th chapter of the permanent arrangement of the vineyard.

That his calculation of its profitableness refers to an extent of 7 jugera as that in which it would be proper to employ one vine-dresser, is a curious instance of the existence of the idea that this extent of land is the proper allowance for one person. Here we have the old allowance passing into the modern rubbio. It is useless to look for much accuracy in such matters, but if we take Pliny's statement that the regular allowance of wheat seed was 5 modii, and assume that the amphora was 26 litres, which agrees with Cagnazzi's value of the Roman pound in French grammes, and if we also adopt the common statement that the modius is one-third of the amphora, then 35 modii (the allowance for a rubbio) will be 30 décalitres and a third, which agrees very closely with Prony's value of the rubbio of wheat. That of oats is yet more nearly equal to 4 modii. The perpetuation of the traditionary allowance of seed is in the former case more close than we could have expected. It amounts to something less than 2 bushels the acre, and is therefore under the old practice in England and in Cisalpine countries generally.

10. The divisions of the acre appear to have grown out of those of the pezza. It is divided first into 4 roods as the other into 4 quarta, and the 40 perches of the rood correspond to the 40 ordines of the quarta. But with us so small a division as the staggiolo was useless, and the perch becoming practically the smallest measure seems to have been regarded as a square. Properly it was 30 square yards, but for convenience it was increased by a quarter of a yard, so that its side became 5 yards and a half. This has been already noticed. I repeat it now because it leads to the history of the English mile, which got its name from being not much unlike in magnitude the Roman mile; though it is said not to be formally recognized in anything earlier than an act of Elizabeth. What we call a furlong is properly a firling, or 4th part. Why then is it the 8th part of a mile? The answer is, that the original unit was the *long acre*, that is, an acre consisting of perches arranged in single file. One hundred and sixty times 5 yards and a half make eight hundred and eighty, which is said to be the length of Long Acre in London. The furlong was therefore the long rood, and we have thus an instance in which a measure of length has sprung out of one of surface. Perhaps the word rood has relation to the

cross-like division of the acre into quarters, the sacred symbol replacing the cardo and decumanus of the Romans, and the name being transferred from the dividing figure to the divisions it produced. K. O. Müller's idea that mensuration was a sort of degeneration from the augurial laying out a space, seems to invert the natural order of ideas. I should be much more inclined to think that the religious ideas which in early times associated themselves with the transition from a pastoral life to tillage and individuated property in land, were earlier than any formal system of divination. The curious story how Attus Marius found the largest bunch of grapes in a vineyard by repeated subdivisions of the space and repeated taking auguries so as to ascertain the row in which it was to be found, illustrates the connexion between the two things. All this is by the way. Another incidental remark is, that the reduction known as cultellatio does not appear to be rightly understood. At least if I remember what is said on the subject, it is made to depend on the fact that if corn grows on a hill-side, the horizontal distance between the straws ought not to be greater than if they grew on level ground, and that therefore to distribute land equitably you must conceive it projected on a level surface, but a variety of other considerations ought to be taken in, and this reason has much the appearance of being invented to give a favourable colour to a practice, the real motive to which was that without it it would have been impossible to keep the boundary lines straight. A hill in the middle of a level plane would set every line wrong on one or the other side of it; on the side, namely, towards which the survey advanced.

11. It is well known that in Gaul the old native measure, the league, is never supplanted by the mile, and that a league equivalent to a mile and a half occurs in the Itineraries. Comparing this with the modern French leagues it would seem to have been a half league introduced by the Romans, in order that the intervals marked by the league stones might differ less from those to which they were accustomed. I am inclined to this view, because the short league would obviously be 5000 cubits. Now although there is a reason why 5000 feet should be the linear unit, as 5 feet make a pace, there is no reason of the same kind in favour of 5 cubits; and, considered apart from any such reason, 10,000 cubits would be a more natural unit than 5000. And this would give the league which occurs in modern times. Let it be granted, however, that the Gallo-Roman league contains as many cubits as the mile feet, it would immediately suggest itself by analogy that the unit of square measure should be a square whose side was 120 cubits, so as to correspond to the actus. Thus we are led to the Arpentum, which continued to be the unit of French surveying until the revolution, varied however in size by the difference between the French and the Roman foot. Like the actus it is described as a square, the side being of course 180 feet or 10 perches of 18 feet each. This was known as the "Arpent de Paris," but there existed another also much used called the "Arpent des Eaux et Fôrets," connected with the former by an approximately simple relation, but perhaps resulting from a tradition of the old jugerum which must have long survived in Provence. At least with the double jugerum it coincides more closely than the pezza with the jugerum. This latter arpent is a square of 220 feet, and therefore differs less than a half per cent. from being to the Arpent de Paris as 3 to 2, a circumstance which may have modified its definition. Half of it would be 24200 square feet, which would be precisely the jugerum if the Roman foot were 132 Paris lines. The calculation in Niebühr's letter to Savigny implies a greater value than this of the Roman foot, but there is no occasion to have recourse to anything doubtful in order to show how nearly the two things coincide. For if we take Gagnazzi's value already mentioned, the jugerum will not fall short of the half Arpent des Eaux et Fôrets, much more than one per cent. Two things are worthy of notice respecting this second arpent. One, that it appears to have led to a definition of the league as 60 times the side of the arpent, or 2200 toises; the other, that there can be little doubt of its being the origin of the Scotch acre, with which its coincidence in size may be called absolute. Converting this arpent into yards, according to the logarithms given in Babbage's *Constants*, I make it a fraction more than 6108 square yards, the normal Scotch acre being 6104 yards, though it is said to vary in different counties from 6084 to 6150 yards. Thus, if my fancies (for I can hardly call them more) are right, we have the pezza equal to two actus, the English acre to three, the Scotch to four, the Irish to five, and the rubbio to fourteen: all with more or less modification, but yet so as that these measures may all fairly be considered to be derived from a single unit. From the intimate connexion between France and Scotland the hypothesis of the origin of the Scotch acre does not seem improbable, at least compared with other links in my chain. The passage in the 5th book of Columella appears to show that the arepennis was regarded as a half jugerum, which of course it would be if it were considered an actus. But he speaks of two kinds of candetum, the former used in towns and being a hundred feet square, the latter agricultural and 150 feet square, that is, I suppose, 100 cubits. The transition from the latter to the arepennis or arpentum of 120 cubits square would thus be the point at which the influence of Roman ideas showed itself.

NOTES ON BOOLE'S LAWS OF THOUGHT*.

It appears to be assumed in Chapter III. Section 8, that in deriving one conception from another the mind always moves, so to speak, along the line of predicamentation, always passes from the genus to the species. No doubt everything stands in relation to something else, as the species to its genus, and consequently the symbolical language proposed is in extent perfectly general, that is, it may be applied to all the objects in the universe. But I venture to doubt whether it can express explicitly all the relations between ideas which really exist, all the threads of connexion which lead the mind from one to the other. It seems to me that the mind passes from idea to idea in accordance with various principles of suggestion, and that in correspondence with the different classes of such principles of suggestion we ought to recognize different branches of the general theory of inference. This leads me to a further doubt whether logic and the science of quantity can in any way be put in antithesis to one another. From the notion of an apple we may proceed to that of two apples, and so on in a process of aggregation, which is the foundation of the science of discrete quantity. Or again, from the notion of an apple we may proceed to that of a red apple, and this movement of the mind *in lineâ predicamentali* is the foundation of ordinary logic. But it is plain, *à priori*, that there are other principles of suggestion besides these two, and the following considerations lead me to think that there are other exercises of the reasoning faculty than those included in the two sciences here referred to.

In the first place, certain inferences not included in the ordinary processes of conversion and syllogism were recognized as exceptional cases by the old logicians. Leibnitz has mentioned some with the remark that they do not depend on the dictum

* Now first published.

de omni et nullo, but on something of equivalent evidence. The only question is whether we should be right in considering these cases as exceptions, and, if they are so, to what they owe their existence. One instance is the *inversio relationis*, e.g. Noah is Shem's father, therefore Shem is Noah's son. Here we pass from the idea of Shem to that of his father, and *vice versâ*. The movement of the mind is along a track distinct from that which it follows, either in Algebra or what we commonly call Logic. The perception of the truth of the inference depends on a recognition of the correlation of the two ideas, father and son. Again, take a similar instance: Prince Albert sat at the Emperor's right hand, therefore the Emperor sat at Prince Albert's left, &c. &c. How shall we express such inferences symbolically? Let S be Shem, N Noah, f father, s son;

$$N = fS,$$
$$sf = 1.$$

Eliminating f between these two equations, we get

$$S = sN.$$

Nothing can be simpler than this: but the symbols s, f, are of a distinct nature from those employed in the *Laws of Thought*. For fA does not denote a species of A, but an idea standing in a different relation to it. The distinction between these two kinds of symbols becomes more manifest when we reflect that f^2 is not identical with f, but denotes "father of father," or grandfather. Now I do not see how these cases of inversion of relation are to be dealt with symbolically without the introduction of such symbols. In the following examples I confine myself to the cases afforded by relationship and the succession of generations. Let A, B, C, denote three persons, s son, g grandson; then, if B is A's son and C B's, C is A's grandson, which we may express symbolically by the following equations:

$$B = sA,$$
$$C = sB,$$
$$s^2 = g.$$

Eliminating B, we get

$$C = gA.$$

It would be more accurate in these examples to introduce a symbol x or y to indicate that B is only one of the possible sons of A, an individual ranged under the species sA. I shall do that in the next example, in which the word son is replaced by the more general term descendant denoted by d. The equations will now be

$$B = xdA,$$
$$C = ydB,$$
$$d^2 = zd;$$

videlicet a descendant not of the first generation. The result of eliminating B now is

$$C = ydxdA:$$

but by a principle about to be noticed

$$dx = x'd;$$
$$\therefore\ C = yx'zdA,$$

or C is included in the class of descendants of A.

The principle just used forms one of the recognized examples of an inference not lying within the domain of Aristotelian logic. It was called *Transitio ex recto in obliquum.* Whately, though he says nothing of its nature, gives in his praxis of examples one which depends upon it. A negro is a man, therefore he who kills a negro kills a man; let this derived notion *killing* be denoted by f, which may serve to indicate a general functional dependence: then, M, N, denoting man and negro respectively, we have the following equations:

$$N = xM,$$
$$fx = x'f;$$
$$\therefore fN = x'fM,$$

or the killing of a negro is a kind of homicide. The evidence of the truth of the equation

$$fx = x'f$$

is the same as that in favour of the equation

$$xy = yx,$$

when x and y both belong to the kind of symbols used in the

Laws of Thought. I shall not stop to inquire into the limitations which it may perhaps require.

The general truth of the equations

$$x^2 = x \text{ and } xy = yx$$

appears to suffer another exception in the case of relative terms, that is, of adjectives of which the interpretation is functional of the object to which they are applied. A small St Bernard dog is not simpliciter a small dog; the word meaning that which is less than the medium size of the class of objects to which it is applied. Here neither $s^2 = s$ nor $sx = xs$. If we say that in order to save all these equations we may employ a different symbol for every application of the adjective small, how can we express the meaning which is common to them all, and in virtue of which the word small exists as an element of language?

Diffident as I am with respect to all these remarks on a method in which I find so much to admire, I am yet more so with respect to the following. But it seems to me that we cannot say that

$$x(1 - x) = 0$$

expresses *proprio vigore*, that is, in virtue of antecedent conventions, what is called the principle of contradiction.

In ordinary language we have words which, independently of this principle, express negation: we say *red*, *not red*, and the like; but in the symbols employed in the *Laws of Thought* there is no other means of expressing *not red* than by $1 - x$, x denoting *red*. Now the interpretation of this symbol $1 - x$ seems to me to be given by the principle of contradiction, and therefore I should rather say that the equation

$$x(1 - x) = 0$$

is interpreted by that principle than that it expresses it. In accordance with this view the equation

$$x^2 = x$$

would appear to be independent of the principle of contradiction.

REMARKS ON CERTAIN WORDS IN DIEZ'S ETYMOLOGISCHES WÖRTERBUCH DER ROMANISCHEN SPRACHEN (ETYMOLOGICAL DICTIONARY OF THE ROMANCE LANGUAGES)*.

Addobbare. Is it not from *adoperari?* The word occurs so early and is so much used that it can scarcely have been derived from chivalry.

Camicia. As the *b* in *cannabis* becomes *m* in Spanish, I would assume the following forms: *canamisia, can'misia, camisia;* the last being the form given by St Jerome, who speaks of it as a linen garment: this explanation is confirmed by the old French form *chainse.*

Laquais. Laquet, Naquet, Nackt, viz. Light-armed-men: the word being plainly military and German. Compare the Greek γυμνής. The same class of people were also called *pages* and *servants* or *sergeants.* Addison is right in supposing that sergeants-at-law represent the serving brothers of the Temple; the judges who called them "Brothers" being the Knights.

Albricias corresponds to the German *botenbrod,* and means what is given to a messenger if you are pleased with him, as when he brings you good news. I believe the true form is *albergias,* or something of that kind, and the literal meaning *herberge-money* or *gifts,* that is, gifts for the entertainment of the messenger. The German word in the Nibelungen Lied, and the Spanish in the Cid.

Travar. Considering the French form *entraver,* it seems more likely that the root of these words is *trabea* and not *trabs.* From *trabea* came *trabeare* and *trabeatio,* the latter used in the sense of *Incarnation.* The idea of *entraver* is simply wrapping up in clothes, so as to hinder the free motion of the limbs.

* Now first published.

Donjon. Surely from *dominatio*, the highest or most commanding part of the place. The Provençal form is *Dompnon.* The French word *donner* in the phrase 'cette fenêtre donne sur la rue' is probably not from *donare* but from *dominari*, the French word being thus a crasis of two Latin ones.

Treillis. There are two etymologies given of this word, one in the first, the other in the fourth part of the dictionary: the former from *tri-licium*, the latter from *trichila:* the second is probably the right one.

Hanter. May it not come from *hamitare* for *habitare?* The change of *b* into *m* is not very uncommon.

De balde. It does not seem necessary to refer to an Arabic source for this phrase. The filiation of ideas may have been something of this kind: In Lithuanian, *balths* means white; the "Baltic Sea" is probably a "white sea," as well as that of Archangel. Connected with this idea is that of brightness, as we see in the name Baldur, and in many other northern and German instances. Then comes the idea of doing a thing openly, boldly and *ἀναφανδόν*, balthaba, as in *Ulphilas.* Also that of brightening up, making merry, &c. as in the Provençal *Esbaudir.* Again, between whiteness and vacuity there is an obvious connexion, as in the word *blank:* and thus we get to the English bald, &c. and to the Spanish phrases, *de balde* and *en balde.*

Pantois. Paventois was perhaps an earlier form, the meaning being to breathe as a person does when frightened. Compare the Italian *Spantare.*

Assouvir. Surely from *adsopire.*

Jachère. There must have been a Latin word *jacerium* from *jacere*, in the sense of to lie neglected or fallow.

Charivari. Why the Lille Glossary translates *Chalivali larnatium* appears from the Sachsenspiegel, in which immediately after sheep and geese kasten mit upgehavenen leden, &c. are mentioned as part of the *Morgengabe.* This shows that the word was used not in a ludicrous sense, but seriously.

Zorzal seems to be simply the German *drossel.* The English form *throstle* seems to confirm this.

Massacre. The root is *macella.* In Provençal "War against the Albigeois," *Mazel* occurs in the sense of carnage. From *Macella* comes *Macellarius*, a meat man, easily condensed in French into *machecrier*, which occurs in the Roman de Rou in the sense of a butcher; and this would become *massacrier*, and thus give rise to a concrete *massacre*. At Rouen the street where meat was sold used to be called *Rue de Massacre*. Carnage, in the same way, meant originally meat-eating, as opposed to *manger maigre*.

Boucher. Surely *Boucherie* meant a place where you could get *munition de bouche*, and a *butcher* is merely a purveyor.

Piloto. The connexion of ideas between *Pilotis* and *Piloto* is illustrated by what I have seen on the Rhone, namely, two men in the bows of the vessel taking soundings with long poles, *pilotis; piloting* the vessel by means of them.

Pimiento. As the name seems to have been given to the spiced wine originally, and secondarily to the spice used in it, the idea of calling it *pigment* seems to be an allusion to the effect of spiced wine on the complexion; a ludicrous or joking name for a drink seems natural, and there are other instances of it. *Orpiment*, by the way, is *auri pigmentum*.

Fregata. What is the origin of our English word *to freight? fregata*, or frigate must be connected with this verb, as *Carrick* with *caricare* and *oneraria* with *onerare*.

Orza. I should be inclined to suggest, though doubtingly, that *a lorza* means on the left side, in consequence of the early habit of mankind, which seems to recur at all times of speaking of the cardinal points on the hypothesis of looking to the east. Thus the phrase originally meant towards the Bear or North: and secondarily to the left hand as the north is when one looks east. The earliest trace of this way of thinking is in Genesis, where it is said of Ishmael, that he shall dwell in the presence of his brethren, that is, east of them.

Aubaine. Is not this word, like *arban*, *heribannum*, as *auberge* is *herberge?* meaning the Lord's right, and especially his right against a stranger. Why it should mean this particular right, as *arban* means the *Corvée*, one cannot quite tell: but I imagine that *aubain* in the sense of a stranger was a word formed in consequence of a mistake of the meaning of the word *aubaine*.

Bis. Another instance of the use of this particle in composition to indicate something other than as it should be, is *Bistourner:* the only church at Paris not towards the east is popularly called St Benoît le Bistourné.

Trovare. In confirmation of the derivation of this word from *turbare*, it may be remarked, that *invenio* is literally to come in upon. A person comes in upon you, finds you, or disturbs you; the ideas manifestly being akin. So too the dog puts up the game, disturbs it, or finds it.

Hurepé. This word with a little modification of the spelling occurs in *La battaille de Karesme et de Charnage**, as the name of a fish; I suppose the *Barbel.*

Mora. In Dante this word is explained to mean a heap of stones. "La grave mora" occurs in one of the early cantos of the *Purgatorio.*

Estable. It seems difficult to separate *stall* from *stable:* in Marie de France (Purgatory of St Patrick) we find *estaule* in the sense of *stabilis*, and in English the *stall* was until lately spelt with a *u.*

Colmena. In the Vulgate the phrase 'corner of the house-top' is translated *angulus domatis;* and in the Vaudois translation the second word is rendered by *colme.* In the "Book of Virtues" *colme* is replaced by *meysoneta.* So that there is little doubt that *colme* was used for little dome or cupola rising above the level of the roof. The transition from this sense to that of *bee-hive* is from the form of the latter obvious, and the theme of both *colme* and *colmena* must be *culmen.*

Cafre. Diez remarks in his grammar, that of the Arabic words in Spanish hardly any refer to things of human feeling. It seems very improbable that the word for *Infidel* should have been taken from the Arabs, who did not call themselves so, and with whom the Christians were not likely to come to an understanding as to its use. Considering that the word occurs also in French, and the facility with which aspirates of different organs pass into one another, I think the word must be *Cathar*, the origin, as it is well known, of the German word *Ketzer;* though Diez has remarked, he is not aware of any case in which the Greek θ becomes *f*, but the induction must be founded on a small number of cases.

Indarno. It is so very unlikely that it should come from the Sclavonic that I think it must be simply *in danno*, with the first *n* changed into *r.* Compare the French phrase *en pure perte.*

* In *Fabliaux et Contes;* edition of Barbazan, Vol. IV. p. 85.

Trou. The word *trabucar* in Provençal seems to be a military one. It is constantly used in the "War against the Albigenses" for battering or breaching with a *trabs*, or in the Italian form *trabocco. Trabuca* is therefore probably the breach made with this instrument, and the Italian *buco* is probably a corruption from hence and not a German word.

Nebli. As the Arabic derivation seems to be a failure, why should it not come from *milvius?* The change of *m* into *n*, of *v* into *b*, and of the place of the liquid, are all matters of frequent occurrence in Spanish. Perhaps we ought to assume a diminutive form *milvillus*, the final *l* being lost with the shifting back of the accent.

Calibre. The word is principally used, speaking of the bore of a cannon, or the diameter of a column; but it has another meaning which must be the original one. The word is used by masons, carpenters and workers in metal, for the tool or model which goes round anything and enables them to see if it be of the right size; its form, &c. vary in different cases, but it is always something that embraces or clips what it is applied to. It therefore seems clear that *calibre* is simply *clipper*, the latter word being borrowed from England or Holland, for I do not know that it exists in modern German. If a Frenchman pronounced *clipper* half a dozen times it would run into something not to be distinguished from *calibre.* Compare the English word *caliper*, and *canif* from *knife;* two things of the same size are said to be of the same *calibre*, because they would fit the same; and hence the other sense of the word, which there is no occasion to derive from the Arabic.

Caviar. Καυιάρι is certainly not a Greek word. I imagine the Greeks intended the four vowels to represent the full sound of the Italian or Provençal *u*, and that the word was originally *curée* or *curata*, meaning simply *cured roe.* When it came to the West from Greece, the perplexing number of vowels caused the hardening of the *u* into *v.* Compare, for an analogous change of a travelled word, the French word *fashion* derived from *façon.*

SOME THOUGHTS ON THE FORMATION OF A CHINESE DICTIONARY, AND ON THE BEST MODE OF PRINTING CHINESE. IN A LETTER TO THE REV. J. POWER, M.A., FELLOW OF CLARE HALL, AND UNIVERSITY LIBRARIAN*.

Cantantes licet usque minus via lædit eamus.

MY DEAR SIR,

THE study of Chinese is hindered by many difficulties, the nature of which is not generally understood. The expense of printing Chinese with our kind of types is one of them. Much has been done in this matter by Breitkopf, of Leipsic, who has been employed by the American mission.

Some of the results were exhibited at the Crystal Palace, but I have not been able to learn anything of the details of the analysis to which he subjects the Chinese characters. More recently Professor Brockhaus has proposed, that in order to get a complete and inexpensive Chinese dictionary, we should have recourse to Lithography. Even so the undertaking would be a great one, and it is very desirable that such a dictionary should be arranged in the most convenient and useful manner.

Brockhaus proposes to follow the ordinary Chinese arrangement; according to which the characters are distributed under 214 radicals, remarking in favour of doing so, that it is convenient to be in accordance with the Chinese practice†.

No doubt this is true; but the objections to this mode of *proceeding are considerable.*

In the first place, it is essentially unscientific; the radicals are chosen on no definite principle. They seem to be, as Kant somewhat too boldly asserted of Aristotle's categories, *aufgerafft*, gathered up at random from all sorts of sources.

In the second place, the relation of the characters to the radicals under which they stand, is arbitrary and uncertain. In some cases the relation is one of mere resemblance, in others

* Previously printed for private circulation.

† The system of 214 radicals is not of high antiquity, and has not been always followed since its first introduction.

the radical is a component part of the character, while occasionally it is very difficult to see what the connexion between them is.

Again, there is no principle of arrangement under each radical, except according to the number of strokes of which the character is composed; and in a complete Chinese dictionary there would be on an average 200 characters under each radical, and in some cases more than 1000. The result of this is, that it is necessary to have a supplementary index of characters, of which the radical is difficult to recognize, and in this there is no principle of arrangement, except the number of strokes.

Callery, of whose improvements I am about to speak, after mentioning the way in which Chinese dictionaries are constructed, remarks, that it is not wonderful that so few persons attain to a knowledge of the language, and they, only after years of painful labour. In fact, to be able to use a dictionary, is a great part of the whole business of learning to read Chinese. The principle on which Callery proceeds, he derived from his instructor, Gonçalves, whom he speaks of as the ablest of Chinese scholars, and who has published several works, in which it is followed, though he has nowhere fully explained it.

Most of these works are in Portuguese, which is perhaps the reason why they seem not much known. Callery's own work is in a mixture of Latin and French. It was printed at Macao, and it is said that most of the copies were accidentally destroyed. It consists of two parts; the first introductory, the second, a dictionary of perhaps 13,000 characters, arranged according to his own method, in which the principle of Gonçalves is employed, in subordination to a phonetic classification. The facility with which this dictionary is used seems to make it desirable that no other should be constructed until it has been considered whether the same method, improved, if possible, ought not to be employed in it.

The principle of Gonçalves is, as he states in his *Arte China,* not wholly new. It is based on the almost necessary connexion which there is between learning to read and learning to write Chinese.

The early missionaries held that all Chinese characters consisted only of six different kinds of strokes. Gonçalves increases the number of elementary strokes to nine. These may be compared to the letters with which our words are spelled, in this respect at least, that a Chinaman, in writing a given character, will always employ them in a given order, just as when we write *man* we write the *m* first, the *a* next, and so on.

Callery affirms, that throughout China a uniform method of forming the characters obtains, though little is said by Chinese writers on the subject. Some directions, he remarks, have been given, but too vague to be of much utility. I do not know if he refers to a tract, of which Sir John Davis published an account in the *Transactions* of the Royal Society of Literature. A collection of characters arranged according to the order in which the elementary strokes are employed will, as similar cases continually recur, very soon enable any one, even without instruction, to acquire a knowledge of the Chinese method. This arrangement is of course based upon giving an order of precedence to the nine elementary strokes analogous to the order of the letters of the alphabet. But this the Chinese themselves do not seem to have done, and without this step, however definite their method of forming the characters, no principle of arrangement could be hence obtained. That they should never have introduced this improvement, is one of the many instances of the mixture of sterility and ingenuity by which they seem to be distinguished.

When we have learned the order in which the strokes are to be employed in forming characters, and have given an order to the strokes themselves, we are in possession of a principle in virtue of which all the characters of the language may be arranged, so to speak, alphabetically. In practice, however, it is desirable to combine this principle with others, and especially with that which results from the analysis of which the great majority of characters obviously admit. It is clearly more convenient to arrange the characters in the first instance according to the number of strokes, than to set out with a purely alphabetical arrangement; and this is the plan followed by both Gonçalves and Callery, in their vocabulary of primary characters. The arrangement therefore may be compared to that of a spelling-book, in which lists of short words arranged alphabetically are followed by similar lists of longer ones.

Thus far there seems no reason to deviate from their method; but the question becomes more difficult when we inquire what characters (and on what grounds) should be accounted primary. My own impression is, that as the main object to be attended to is facility of reference, we ought to be guided by the eye, and whenever a character is clearly made up of others, to treat it as a compound character, whatever the significance of its parts or the relation between them. The great majority, it has been already remarked, of characters, are in this sense compound; and no one who is at all used to Chinese writing, can have any difficulty, except in a few cases, in dividing them into their elements.

Generally speaking, one or other element is in the Chinese arrangement the radical under which the compound character is placed, and this is the part of the method of radicals which is in practice most convenient. Even so it is defective in giving the compound characters mixed up with others, which on other grounds are placed under the same radical. Still few compound characters, considering their whole number, will be found in the supplementary index.

Callery's method differs from what is now suggested in this respect; he does not avail himself* of the analysis of compound characters into simple ones, in all cases in which such an analysis is distinct and obvious, but only in a certain class, though undoubtedly the largest and most important class of cases.

More than five-sixths, according to one estimate, eleven-twelfths according to another, of all Chinese characters are not only compound, but made up on a uniform plan.

They admit of analysis into two parts, one of these simpler parts having the same sound as the whole character or a modification of it, and the other having some reference to the meaning. Thus the former indicates the pronunciation, and is therefore called by Callery 'the phonetic element.' To the other he gives the name 'classifica.' Gonçalves had called the latter Generic, and the former Differential elements†.

The names are not satisfactory, for the phonetic element may as well be made use of to constitute a class or genus as

* Or rather, he ought not.

† I do not apprehend that Gonçalves conceived it necessary that the differential element should always be phonetic. He would have given the name to whatever he found *differentiating* a generic character.

the other, and then the latter must be considered the *differentia.*

The nomenclature of both writers indicates the influence of the old view, adopted by Remusat and quoted with apparent censure by Callery, that the class of characters of which we are speaking was based on a scientific arrangement of objects, according to genus and species.

The name phonetic is unobjectionable, and the other element might be termed non-phonetic. But it would, I think, be better to call it the logical element, inasmuch as it has relation to the word as such; that is, as it has a meaning, and is not a mere sound.

Callery presents a list of 1039 characters, which enter into others as phonetic elements. These he distributes into classes according to the number of strokes, arranging those in each class on what may be called the principle of graphic analysis, namely, that of Gonçalves. Under each of these 1039 characters he places those of the same or the kindred sound, which are formed by the union of that character with one of those which he called 'classifiers,' these being arranged subject to certain modifications in the same way as the phonetic elements, under each of which we find on an average about a dozen characters.

The plan, though in many respects very instructive, is not all that one wishes for in a complete dictionary. His list of primary phonetic characters contains many themselves made up of a phonetic and a logical element.

The classifiers are for the most part phonetic elements also, but when they are not, one has to look for them in a separate list, given only in the first part of his work.

Moreover, there are certain characters, some of them common and important, and not admitting of analysis into others, which do not seem to have been used by the Chinese, in the formation of compound characters, either as logical or phonetic elements, and which therefore he does not notice at all, or if he does, only incidentally.

But the great defect is, that he cuts himself off from the advantage resulting from the analysis of characters in all cases in which neither element is phonetic. Some indeed he seems to introduce by mistake. Thus, the character for *dog* has no relation in point of sound to that for *bark,* which is compounded of *dog* and *mouth*, and is pronounced as no other character is in which *dog* is really the phonetic element. It is clear that both elements are logical, we might say quite intelligibly to *dogs-mouth*, instead of to *bark*.

It is no practical objection to Callery's arrangement, though it shows the difficulty of adhering, if we were required to do so, to a definite plan, that in some cases both elements of the characters are phonetic, though in different ways, one indicating the sound, and the other showing that the former does so*.

It is in this case a sort of signum diacriticum, and the character which is thus employed the most frequently is that for *mouth*. For instance, *five*, and one of the words for *I*, are both pronounced *eu*, and the character for the latter is the same as that for the former, with the addition of that for *mouth*.

A similar instance has given rise to one of the innumerable foolish things which are said and repeated about the Chinese.

Ho happens to mean *corn*, and *concord* or *comfort*, and the character for it, in the latter sense, is the same as in the former, with the same addition as in the preceding instance.

The *mouth* is here used diacritically, yet people have been found to say, that the grossness of the Chinese is shown by their having a character for happiness which indicates that they have no higher idea of it than mere eating.

It so happens that the same sound *ho* also means a child's crying, and in this sense also is represented by corn and a mouth differently placed in relation to one another. What authority can there be for an interpretation in the former case which is obviously inapplicable in the latter? This, however, is a less offensive error than those into which the early missionaries fell, in seeking for the doctrines of Christianity in China. I can only allude to their interpretation of the word *yang*.

If a Chinese were to say that the English are a particularly selfish people, because the same symbol denotes unity and per-

* Cases might be pointed out in which both elements are representatively phonetic, so as to form a sort of reduplication. In some cases it may be said that both elements are at once phonetic and logical. Thus assuming that the two elements of *pi*, to compare, are both *pi* spoons, the idea of comparison results from their similarity. In other cases the same element is phonetic and logical. Thus *tsien*, a small coin, consists of *tsien*, small, and *kin*, metal or coin. Compare our word 'groat.'

sonality*, he would only imitate the example of errors long gravely maintained by European scholars. No more effectual mode of getting rid of these errors presents itself than making a complete analysis of all compound characters, in order afterwards to recognize and classify the different principles which have guided their formation. Remusat's remark is perfectly just, that the Chinese characters are formed in a variety of ways, and that nothing but confusion can result from any attempt to analyse all on the same principle. Chinese etymology consists, as Humboldt has observed, of two parts, that of the characters, and that of the spoken language. Both parts involve great difficulties, and as yet neither has been treated scientifically. The former part is particularly attractive: there is no more amusing book than a Chinese dictionary. Perhaps my saying so may remind you of the painter's reflection, 'chè dolce cosa è la perspettiva.' One instance may be enough to show you the sort of interest I mean. The character for *heart*, which expresses generally all mental operations, combined with that which represents an enclosed and divided field, means to think or consider: we have here a graphic representation of the Latin 'contemplor,' formed, as there seems little reason to doubt, from 'templum,' in the sense in which the augurs used the word†.

A very interesting part of the study of the Chinese character would be the comparison of it with Egyptian hieroglyphics, not in order to revive the old notion of an historical connexion between them—between the flowery region and the lands of the lotus and the papyrus—but in order to see how similar problems have in the two cases been dealt with. It has been said that 'Thoth was wiser than Fo' (Fo-Hi), which may be true, but still the comparison is worth making.

It is curious, that while in Egypt the feet or legs seem to be the symbol of activity, so as to give a verbal signification to the symbol with which they are associated, or, in some instances, rather grotesquely joined, the hand should often serve the same purpose in China. Other points of analogy might doubtless be

* *Guesses at Truth.*

† Grimm's derivation of *templum*, from the same root as *tepeo*, making it refer to the sacrificial fire, seems open to more than one objection: in the first place, the augurial sense of the word appears to be the primary one, and this connects it with τέμνειν; and in the second, it would then be particularly strange that the house of Vesta should not have been a templum.

indicated, though the less complicated forms of the Egyptian characters can hardly admit of phenomena so various as those which are presented by the Chinese. In the latter, for instance, I believe we might trace that curious principle of language, which, for want of a better name, may be called 'the principle of intelligibility,' of which we cannot have a better instance than the conversion of 'mandragore' into 'main de gloire;' I mean, that in Chinese groups of strokes, originally forming only a part of a complex picture, have probably, in some cases from accidental suggestions, shaped themselves into the likeness of other simpler characters.

With regard to the etymology of the spoken language, such an arrangement as Callery's is of the greatest value. When we find different sounds associated together under the same phonetic element, we may, special cases being set aside, conclude that they are, with reference to the Chinese organs of speech, cognate sounds, and thus establish the laws by which our investigations are to be guided. The change of Ch into T, which we are familiar with as the peculiarity of the Fokien dialect, is one of the most obvious phenomena thus made manifest*.

To return from this digression, to the formation of an index to a Chinese dictionary. I should propose to form a list of all simple characters, and of all in which there could be any serious doubt or uncertainty as to their analysis. The latter would not be a very large addition to the number of the list. Callery's own estimate is that his 1039 phonetic elements result from about 300 primary ones.

A small number, and those capable of further analysis, of his classifying characters do not belong to his phonetic list; but if we say, that all the elements he employs cannot much exceed 300, we shall not perhaps be far from the truth. Taking account of omissions, accidental and otherwise, we may perhaps say, that 500 elements would appear in our list. I admit this seems a small number; but Callery speaks of having gone through almost all the Chinese characters, and having omitted

* Compare with this the change of *tr* into *cr*, in *craindre* from *tremere*, *veintre* (*Hymn on Eulalia*) from *vincere*, &c. The same thing is seen in the English corruption of *ask* into *ast*. Is not *triticum* formally equivalent to *κριθή*? Compare also the Greek and Latin names for Carthage; Oxmantown and Ostmantown, (Worsaae, *Danes and Northmen in England*.)

only such primary characters (he calls them by perhaps a better name, 'indivisible or fundamental characters') as were rarely found, or useless. To allow 200 for such omissions seems sufficient. However that may be, I should propose the formation of such a primary list, and its being printed in a tabular form, as a frontispiece to the index. Callery has done this with his 1039 phonetic characters, and they are all visible (and of a sufficient size for clearness) at one opening of the book. These being arranged in the manner already mentioned, the remainder of the index is to be placed under these, as keys or headings. Under each I would place, in order, all with which it combines: first, all the simple characters; then, all the binary characters, and so on. No doubt there would be a good deal of repetition in this; every compound character would be entered twice at least; those consisting of three elements, three times; and so on. But the advantage of being able to find any character you want with comparatively little trouble, as soon as you have recognized one of the elements it is composed of, seems to outweigh this disadvantage, and perhaps about 80 quarto pages would be enough for the index to a dictionary of thirty or forty thousand characters.

It is to be observed, and this I think a very important part of the plan, that I do not propose to use compound characters at all. The original 500, or whatever the number may be, would be the whole number of characters used, and therefore of types required.

You are reading, we will suppose, a Chinese book, and come to a character you do not know. Seeing that it consists of *woman*, *mouth*, and *heart*, you look for it under any one of these three characters, and in a little while find the other two grouped side by side with a number which enables you to refer to the body of the dictionary.

There is but one character in the language made up of these three elements, and therefore in order to recognize it without ambiguity, it is not necessary that you should actually see it before you in the index, or that you should even be told there what you already know, that, to speak heraldically, *woman* occupies the dexter chief, and *mouth* the sinister. Some cases undoubtedly there are—probably only few—in which the same elements, differently arranged, form different characters; but nothing can be easier than to devise diacritical signs; by which the same group of characters should be made to refer, without possibility of mistake, to the different compound characters.

Take a simple instance already noticed. Under *corn* you find *mouth*, accompanied by an arrow pointing to the right, and again a *mouth* with the arrow pointing upwards. The reference to the first would be, to *ho* in the sense of *comfort*, and to the second to *ho* in the sense of *crying;* for in the first case *mouth* stands to the right of *corn*, and in the second above it.

This principle once admitted, namely, that a character may be as clearly recognized by means of its elements alone as if a fac simile of it were given, may of course be applied much more widely than merely to forming the index of a dictionary. It seems to furnish the solution of the chief difficulty by which the study of Chinese has hitherto been impeded. For we thus get a *mezzo termine* between the unintelligibility of Chinese written with Roman characters, and the impracticable expense of a complete fount of Chinese type. Even if we had the 3000 elements*, which the ingenuity of Breitkopf has devised, where should we, here in Cambridge at least, find a compositor sufficiently learned to put them together?

The reason why so many more elements are required to imitate compound characters, than are necessary, if we content ourselves with simply representing them, is of course the variation in size and shape, requisite in order to give uniformity in these respects to the compound character. This uniformity of size and contour is a matter of Chinese taste with which in books intended for European use we need not trouble ourselves. It is, by the way, an inconvenient taste even for the Chinese, because in order to make complex characters distinct, the simpler ones must be unnecessarily large. If it be said that the Chinese would never become accustomed to characters of the proposed kind, we may answer that even if this be so, the necessity of printing in Europe books intended for Chinese use is not very obvious.

Whatever may have been the case formerly, there can henceforth be probably no difficulty to hinder the printing whatever is intended to be read in China, at presses (which, by the way,

* Am I right in thinking there are 3000 type elements? or are there only 3000 punches?

is an incorrect phrase in speaking of Chinese printing) established in the towns to which Europeans have now free access. And it must be remembered that we could scarcely hope to produce in Europe what the Chinese would account a handsome book. The softness of impressions from wood can hardly be imitated with metallic type, and Chinese paper cannot, I believe, be used in our printing presses.

A collateral advantage, resulting from what is now proposed, would be the facility of learning to read Chinese. The difficulty of analysing the characters would be removed, and when once a student was able to read a book printed in the new method, the transition to the usual characters would not cause more difficulty than Greek contractions, or than the ligatures in Sanscrit. Another advantage would be that as the characters would follow one another in regular order, accompanied only by brackets to form them into groups, and by a few simple diacritical signs, any ordinary compositor would be able to set them up. Probably it would not be found very difficult to distinguish the phonetical elements by printing them with red ink, which to beginners would be a great assistance. By similar means we might distinguish the same character, according as in any sentence it presented itself as a noun or as a verb.

Brockhaus has proposed a different way of printing Chinese for European use, namely, in Roman letters with a numerical reference under each word to its place in the dictionary he wishes to see made.

There are two or three objections to this plan. In the first place, difficult as it is to remember Chinese characters, it would be found much more difficult to remember the meaning of a number, even with the help of the pronunciation, because numbers give very little for the mind to fasten on, and can never be exclusively associated with a single class of ideas. Imagine the difficulty of remembering the plot of a story the persons in which were denoted only by numbers.

Again, in all questions relating to what I think must, by and bye, form an interesting part of comparative philology, namely, the theory of the Chinese characters, such a plan would be useless, even if we could suppose that all Chinese scholars agreed to use the same dictionary.

Lastly, this method would form no introduction to the study

of works printed in the Chinese character, a class to which the great mass of Chinese literature must always belong. A man might give years to the study of this Stratford atte Bowe Chinese without being able to read the commonest characters.

A more radical reform has sometimes been proposed, namely, simply to print Chinese in Roman letters. Why are not the absurd Chinese characters laid aside? has been asked in much the same tone as the question one occasionally hears, of why legal terms and forms should be used in conveyances? Get rid of these, it is said, and any deed might be written on a single sheet of paper. The answer in the two cases is much alike.

If it had been possible during the last six hundred years to enforce brevity in legal instruments, not only the practice but the theory of conveyancing would be very unlike what they now are. The complicated relations which have grown up amongst us, the various subtle modifications of which the idea of property has been found susceptible, could never have been developed on such a system. Not only the outward form, but that which the form represents, would have been different. Our thoughts, and the mode in which we express and record them, act and react on one another.

The influence of writing on the history of language, which has been made the subject of an interesting essay by William Humboldt, has been greater in China than anywhere else. The hand and eye have, so to speak, brought into subjection the voice and ear; the reason of which is to be sought partly in the original nature of the language, and partly in the general diffusion of education.

The language of China, especially the written language, is in many respects what it is, in virtue of the character, which we cannot now give up without introducing ambiguity and confusion*. To a certain extent the Roman letters may be used, and this has already been done, as for instance by Morison and Gonçalves. The dialogues of the latter are particularly valuable, from giving both the Mandarin and the Canton pronunciation.

* The best plan by which indeed many of the difficulties would be removed, would be to write all compound characters like fractions, I mean with the pronunciation of the whole character above, and those of its elements below a horizontal line. *Lin*, a wood, for instance, would be denoted by $\frac{\text{lin}}{\text{mou, mou}}$.

So far as the question, as to giving up the Chinese character, can be decided by authority, it seems sufficiently settled. I may refer particularly to what is said on this subject in Mr Kidd's work on China. One kind of influence exerted by the character is sufficiently peculiar to deserve mention. We have many words whose meaning has been changed in consequence of a mistake caused by accidental resemblances of sound. Such, for instance, has been, at least in popular use, the case with *demean.* Johnson even thought the secondary meaning had the authority of Shakespeare. Again, there are words whose meaning has been influenced by 'juxta-position,' by their occurring, so to speak, in contact with others. Such are *implicit,* and *buxom**.

But both kinds of influence may concur in Chinese. Not only the sound of the word, but also the way in which the sound is expressed, by bringing the character into constant association with another, may influence the meaning. Take as an instance the character already noticed, composed of *heart, woman,* and *mouth.* The two latter characters alone form a binary character, pronounced *ju,* and meaning *even as, sicut.* This binary character is the phonetic element of the ternary one, of which it forms the upper part. The latter is pronounced *shu,* and means goodness or kindness. But it is related that Confucius taught that this word is the summing up of all morality; that it means the state of mind in which a man interests himself in the happiness of others, *even as* in his own.

This development of the meaning of the word was, it is probable, merely the result of an acccidental coincidence of sound, and of the selection of the one character to be the phonetic element of the other. But, error or not, this opinion as to the meaning of the word has perpetuated itself; and what in this case is referred to the authority of Confucius, has probably happened tacitly in many others.

One of the difficulties in making a Chinese dictionary arises from the number of compound words, that is, words each of which means something separately, but when grouped together express a single idea.

Remusat went so far as to say, that the compound word was polysyllabic, and that each character merely represented a syllable. This question is scarcely worth the attention which has

* So too in German *Ehe.*

been given to it; but the important point is, that the unwary scholar frequently endeavours to give separate translations to each element of compound words. These are given in the best dictionaries; but there may be some difficulty in making a list of them complete and easy of reference. In printing it would be well, I think, to connect the elementary characters by a hyphen.

The matter is so peculiar that you will not object to my giving you an instance of the errors it is apt to produce. In Remusat's version of one of the 'Four Books,' as they are called, of Confucius, it is said that Confucius lived in accordance with the seasons and with the earth and water. The meaning of this is certainly not clear; but 'water-earth' simply means 'climate.' The habits of Confucius were not in accordance with the earth, whatever that may mean, nor with the water, but simply, which is quite intelligible, with the climate.

The instances which in the Notes to Humboldt's letter to him Remusat quotes from other languages ('horseman' is his English instance), are not quite parallel, for though no grammatical form indicates the relation between their parts, yet ideally one of them is a substantive, and the other a modifying adjective: whereas in Chinese the compound word is a new formation, of which the meaning is suggested only by those of its parts. 'Elementa *quodammodo* manent in composito*,' we cannot define the matter more precisely. We see here, as in the formation of the Chinese characters, and in the structure of the language, the tendency to merely external union. There is contact and combination, but no interpenetrating compound growth, and the whole resembles not a picture but a mosaic.

The same remark might be made as to Chinese style, which is all compact of set phrases and antitheses. I cannot enter on all the matters of detail connected with the Index, of which I have endeavoured to give you an outline. My ideas of them are of course very imperfect. Callery's merits, with respect to Chinese lexicography, are doubtless great, both in his exposition of the ultimate dissection of the characters, and in showing, more clearly and fully than had been done before, the presence of a phonetic element in the great majority of characters. His assumption, that a set of phonetic elements were deliberately

* S. Thomas Aquinas *de Principiis.*

and simultaneously invented, is unphilosophical, and seems to have led him into his principal error, that in all compound characters one element is phonetic.

This error is decidedly opposed to competent Chinese authorities, and would place us, if we adopted it, in the dilemma of either rejecting obviously correct analyses, or of setting aside the laws by which the affinities of sounds are governed.

I must here conclude these remarks. They are the result of your kindness, which has led to my seeing the works on subjects of Chinese literature recently added to the Library, and has thus recalled my thoughts to matters which my increasing illness had made me lay aside. You know the circumstances in which I write, or, to speak more accurately, dictate. *Vive et vale.*

Yours very truly,

R. L. ELLIS.

March 17, 1854.

VALUE OF ROMAN MONEY*.

GRONOVIUS'S estimate of the value of Roman money is vitiated by two principal errors: his doctrine that 100 denarii went to the pound weight of silver, a doctrine connected with his theory that the proper and direct meaning of sestertium is two pounds and a half of silver, but which is contradicted both by testimony and by the denarii, which like the bricks in Richard II. are alive to this day to witness to the contrary; and his confounding the pound Troy with the Roman pound. The errors tend to balance, one making the denarius too little in value, and the other making our currency of too small value; but his result is of course mere haphazard, to say nothing of his neglecting the question of alloy.

The basis of the calculations in the *Dictionary of Antiquities* is much more satisfactory, but the calculations themselves are wrong. The articles *Sestertius* and *Denarius* do not take into account that our shilling circulates as a counter above its intrinsic value. The value of the denarius is determined by comparing its weight of fine silver with that of the shilling. Now as our coinage since 1816 is at the rate of 66*s*. to the pound, the result is the same as if the price of silver had been taken to be 66*d*. per ounce standard, which certainly is not its real price. The rate of coinage was purposely fixed above the variations of the bullion market to prevent melting. Sixty-two pence is the price commonly assumed in calculating the par of exchange, and is rather a large average price. Taking the data given in the article Denarius, and this price of silver, the denarius of the end of the Republic is worth (not 8·6245*d*. as it is there made) but 8·099*d*., or in round numbers, not 8½*d*. but 8*d*.

The error will be nearly the same in the value of the later denarius.

* *Journal of Classical and Sacred Philology*, Vol. I. p. 92.

The value of the sestertium resulting from the value of the denarius which I have quoted is £8. 19*s.* 8*d.*, though by some error of calculation it is reduced to £8. 17*s.* 1*d.*; the real value is £8. 8*s.* 8½*d.*, so that the two mistakes, like Gronovius's, tell against one another.

It is curious that the later value of the denarius gives the sestertium £7. 7*s.* 7½*d.*, a sum in 7 as the other in 8.

In the article *Aureus*, the writer says that the sovereign contains 113·12 grains of fine gold. It really contains (neglecting the third place of decimals) neither more nor less than 113 grains. The result is that he gives the aureus as £1. 1*s.* 1*d.*, and a little more than a half-penny, instead of as nearly as possible £1. 1*s.* 2*d.*

The following is an outline of my calculation:

Required the price of 60 grains of silver, $\frac{29}{30}$ths fine, at 62*d.* per ounce, standard. (1 ounce = 480 gr.)

$$x = 60\,\frac{62}{480}\,\frac{40}{37}\,\frac{29}{30}. \quad \text{(Standard being } \tfrac{37}{40} \text{ ths fine.)}$$

Reducing,
$$x = \frac{31 \times 29}{3 \times 37},$$

$$31 \times 29 = 30^2 - 1 = 899,$$

$$3 \times 37 = 111,$$

$$x = 8{\cdot}099d. = \text{value of early denarius},$$

$$250 \text{ denarii} = 1 \text{ sestertium},$$

$$240 \text{ pence} = £1;$$

$$\therefore \text{ value of sestertium} = £\frac{809{\cdot}9}{96} = £\frac{101{\cdot}23}{12} = £8{\cdot}435,$$

$$= £8.\ 8s.\ 8d.\ 4 \text{ or } £8.\ 8s.\ 8\tfrac{1}{2}d. \text{ nearly.}$$

The later denarius is 52·5 gr. or 8·75 of the earlier, and the sestertium is in the same proportion.

THE COURSE OF MATHEMATICAL STUDIES*.

THE seventh query†, so far as it relates to the limits beyond which it is not expedient that the undergraduate course of mathematics should extend, seems naturally to form a part of a more general question, namely, how the whole time given to the study of mathematics may be most advantageously employed. In order to discuss this more general question, it is necessary to consider on what grounds the study of mathematics is made to form part of our system of education.

I. The grounds are two-fold: mathematics are studied as ancillary to natural philosophy and as a means of training and developing the mind. In the latter point of view they are chiefly valuable, because they deal with necessary and not contingent truth‡. Of every necessarily true proposition which the

* *Cambridge University Commission*, 1852. Evidence on Mathematical Studies and Examinations, p. 222.

† The seventh query is: Would you be disposed to recommend the limitation of some of the subjects included in the present range of the examinations, for instance to omit such propositions and applications of the Calculus of Variations, of the theories of Definite and Elliptic Integrals, of the Planetary and Lunar theories, of the theories of Heat, Electricity, and Magnetism, of the undulatory theory of Light, as require for their treatment a very refined and laborious analysis? Might such higher mode of treating these subjects be advantageously reserved for examination for special prizes at periods subsequent to the Degree of B.A.? Would not the concentration of the attention of Students upon a smaller number of subjects, and those restricted within narrower limits, tend to increase the accuracy and raise the character of their knowledge, and to bring their instruction more completely within the grasp of the public and recognized teaching of the University?

‡ This applies to mixed as well as to pure mathematics; the necessity of the conclusion being, however, in the latter absolute, and in the former hypothetical, —τὸ ἐξ ὑποθέσεως ἀναγκαῖον.

mind distinctly apprehends as such, the contradictory is seen to be inconceivable; this inconceivableness of the contradictory being *ex parte mentis* the criterion of necessary truth. Nevertheless, although when we think of any simple proposition in arithmetic or geometry, we perceive not merely that it is true, but that it must of necessity be so, this is nowise the case with respect to all demonstrated or demonstrable results. The intuition, so to speak, of the ablest mathematician is confined within a narrower circle than that of the truths which he can prove. He may satisfy himself of the cogency of each step of the demonstration, and yet the essence of the conclusion—the fundamental principle of its truth—remains unseen. The ὅτι is manifest, but the διότι obscure; and consequently a proposition contradictory to that to which he has been led does not appear to him an absurdity, but simply an untruth. It might, for what he sees, have been true, though he knows that actually it is not, and thus while he is aware that his conclusion is true necessarily, yet still it seems as if it were so only contingently and as a matter of fact, the demonstration appearing *assensum constringere, non rem.* In a word, his conception of the matter is still imperfect. But between this state of mind and that which is produced by the contemplation of any elementary proposition, there is no fixed or definite boundary. Every one who has really studied mathematics must remember cases in which, after long and patient thought, the reason of the truth of a proposition, with the demonstration of which he may have been acquainted for years, has seemed to dawn on him; the proposition thenceforth becoming, as it were, a part of his own mind —a matter about which he is no more capable of doubting than about the primary conceptions of form and magnitude. The mind thus brought into nearer, if not immediate, contact with necessary truth is conscious of its own development; and herein, I believe, resides the special benefit to be derived from the study of mathematics, a benefit, that is, distinct from the exercise of patience and attention which it undoubtedly requires, but which is required also in other pursuits. The study of mathematics is especially valuable, not because it gives the Student practice in ratiocination but because it enlarges the sphere of his intuition, by giving him distinct and conscious possession of truths which lay hid in his conceptions of figure, number, and the like. But in order to this kind of mental development, it is necessary not only that the Student should master the successive steps of the demonstrations set before him and retain them in his memory, but that his mind should become imbued with their spirit and essence. His real progress therefore is not to be measured simply by the extent of ground over which he has passed: it varies also according to the degree in which he has approached towards a complete intuition into the results which he is able to prove.

I believe that this principle ought to be our guide in examining the merits and defects of a course of mathematical study intended to form part of a liberal education. But the connexion of natural philosophy with mathematics must, to a greater or less extent, modify the conclusions to which it would lead us.

II. It would be impossible to trace in detail the consequences which appear to follow from this way of considering the subject. They may be classed under two heads, the choice of subjects, and the choice of methods. With respect to the former, my impression has long been that a good deal might be omitted which now enters into our course of reading, not only without impairing its utility, but with positive advantage. A more rigorous subordination of details to fundamental principles would not only save the Student's time, but would make the principles themselves be more clearly apprehended. Everything received into our course ought to justify its admission there, either by its own importance or by its connexion with something more valuable than itself. Mere exercises of industry and ingenuity, long numerical calculations, complicated processes of algebraical reduction, tricks of transformation for the evaluation of integrals and the solution of differential equations, and the like, may all be accounted comparatively useless. These things may be impressed on the memory but will hardly long remain there, and meanwhile are felt to be rather a burden than an acquisition. So in mixed mathematics,—many of the approximate formulæ in optics, the less important astronomical corrections, detailed descriptions of philosophical instruments, &c., might all be advantageously laid aside. Not that these things are not worth knowing, but that they do not properly belong to such a course of mathematics as we are considering—in which the chief end proposed is a clear insight into fundamental principles. In general it may be said that formulæ of approximation are unsuited to the end we have in view; they give little or nothing on which the mind can rest: their value resides in the practical application which is to be made of them, but which the Student never makes. It is the predominance of approximate results which renders the lunar and planetary theories unsatisfactory portions of the Student's course. They are, however, by no means to be omitted or curtailed, and with respect to the latter, the evil might be lessened, though not without some inconvenience, by giving more prominence to the general theory of the variation of parameters as we find it in the *Mécanique Analytique* and in some of Poisson's memoirs. Sir W. Hamilton's essays in the *Philosophical Transactions,* and those of Jacobi in Crelle's *Journal,* might, perhaps, give some additional materials for the formation of a course of study on this part of natural philosophy.

III. Secondly, as to the choice of methods, and especially as to the preference to be given to geometry or to analysis. Ever since the introduction of the modern analysis into University reading, there have been complaints of its having superseded the older methods and traditions of the Cambridge system. Those who favoured its progress affirmed, and most truly, that by its aid the Student advances faster, and goes farther, than he could do without it; he gains in fact more knowledge of the subjects set before him. But this argument had little weight with those who held that not the knowledge but the process of acquiring it—the training and discipline of the mind was the thing chiefly to be thought of.

It has been said that if information merely is the end in view, mathematics have less claim on our attention than many other things, and that most of the arguments in their favour cease to be applicable if geometry is discarded or disparaged. Of late these views seem to have gained ground in the University: their influence may be traced in the recent legislation on the subject of mathematical honours. The principle on which this re-action against the newer methods is chiefly based, namely, that the mind of the Student ought to be as much as possible conversant with fundamental conceptions is, I think, perfectly correct. But it does not follow that analytical methods ought to be discouraged. Demonstrations may be geometrical, and yet in a high degree artificial; and first principles may be lost sight of in a maze of triangles, no less than in a maze of equations. Though in mathematical investigations there is no royal road, yet there is a natural one, that, namely, which enables the Student, as far as possible, to grasp the natural relations which exist among the objects of his contemplation. If this route be followed, it matters but little whether the reasoning be expressed by one set or kind of symbols or by another—in plain words—in short hand—or algebraically. To change the notation is merely to translate from one language into another.

It is common to find persons in Cambridge and elsewhere who insist upon it that geometry is geometry, and analysis analysis; but it may be doubted whether this notion of an absolute separation between the two things is not the result of a want of familiarity with either. It seems to be supposed that if a mathematician treats a problem geometrically, he has to think about it for himself, whereas if he treats it symbolically, the symbols think for him. Perhaps it may be said that the fact of there being any tendency towards so childish a notion is in itself evidence of the mischief produced by the use of symbols; and certainly if symbols were never used, the notion could not exist. But neither could it exist if they were rightly used and rightly understood. The phrases which I believe may now and then be heard from some of our younger analysts, such as "the irrefragable x," and "putting it into the mill," for expressing the conditions of a problem symbolically, show perhaps that those who use them have but a half understanding of what they are doing. But this evil is not to be remedied by discouraging the use of symbols. That our methods should be geometrical is not by any means essential; they ought to be natural, and it has been too hastily supposed that they will necessarily be so if symbols are excluded: whereas it is not by precise adherence to any particular mode of expression that we are to bring the Student to a familiar apprehension of the principles of what he is engaged on. This is to be accomplished rather by a "mélange heureux de synthèse et d'analyse," to use the words of a great master in the art of which he speaks, than by imposing either on teacher or students any unnecessary restraints. Let us consider the question more generally. When the conditions of a problem have been stated, the solution may be evolved from them by innumerable sets of combinations. It does not often occur that even a practised mathematician divines the simplest and the best. His choice among the routes which he may follow is determined by an infinity of circumstances, and more especially by the way in which the conditions have been expressed. "Words shoot back on the understanding of the wisest," and so do symbols; and if the conditions of the problem, whether geometrical or mechanical, or, if we will, logical, are expressed by means of algebraical symbols, he will, in all probability, not deal with them as he would have done had they been expressed in common language: the reason of which is, that the combinations and inferences which are the most obvious when one mode of expression is employed, cease to be so when it is replaced by another. Hence and from other causes arises a variety of forms of demonstration, often, it is true, perplexing, and yet, if attentively considered, full of instruction. For as to a mind which has attained to a perfect mastery of the subject, and by which, therefore, the connexion of the data of the problem, with its solution is perceived as by intuition, all the demonstrations appear to be in their essence identical,—different modes merely of presenting the same conceptions,—so contrariwise the comparison of the different demonstrations by which a given result has been established, tends to make us recognize the grounds of their essential unity. It is not by merely fixing in the memory the successive steps of a single mode of demonstration, or even by studying several, if we allow them to remain in the mind as distinct and heterogeneous processes of thought, that we are to acquire a complete insight into the subject in hand, but by a more discursive method,—by inquiring perpetually into the grounds and reason of what we are doing,—by interpreting our symbols and following the train of geometrical or physical conceptions to which their interpretation leads, and again by retracing our steps and passing from general considerations or purely geometrical reasoning to the technical language of symbols. Every change of form should be suggestive of a new aspect of the subject, and it is thus that the simplest way of considering it is to be discovered.

In confirmation of some of the opinions which I have been endeavouring to express, I may refer to Poinsot's admirable tract on the motion of a rigid body. He has there shown, with great felicity both of thought and of expression, that the art of combining symbols is by no means the whole of mathematical analysis;—that we must join to it the art of interpretation, and that in many cases the essence and meaning of a result are scarcely more obvious in the equations which express it than in the original "mise en équation."

It did not belong to his purpose to point out that on the other hand geometrical are not necessarily natural methods of demonstration; but that there is a real distinction cannot, I think, be questioned. If it were not so,—if the Student felt that by studying a subject geometrically he acquired more real insight into it than he could else have got, geometry would be more popular in the University than it now is. In truth, the difficulty of remembering many geometrical demonstrations is in itself a proof of their artificial character; for that which the mind has once completely grasped it does not easily forget. What we want is the introduction of a freer and more liberal method*, and especially the abandonment of the notion that anything is gained by a rigorous separation of geometry and analysis. It is this which for the most part makes our geometry pedantic, and many of our analytical text books dry and sterile. If it be asked how such a change could be brought about, I am inclined to think it could only result from a change in the opinions of those on whom the character of our studies chiefly depends,—the Professors, Tutors, and Examiners. It could hardly be made the subject of direct legislation.

IV. The same remark would apply to the subject more especially suggested by the seventh query, namely, the proper limits of an undergraduate course of mathematics. In every branch of mathematics there are parts which from their abstruseness ought not to be introduced into the degree examination;

* What may be called the new geometry seems to be little studied in the University; yet the method of which it makes so much use, namely, the generation and transformation of figures by ideal motion, is more natural and philosophical than the (so to speak) rigid geometry to which our attention has been confined. It has been well said that the differential calculus is the symbolical expression of the law of continuity, and probably the principles of the calculus would be better understood if notions connected with this law were introduced at an earlier period of our course. See, on the impossibility of severing our conceptions of space from those of time and motion, Trendelenburg's *Logische Untersuchungen*.

but it would be found difficult to trace any precise and permanent line of demarcation by which these might be separated from the rest. In the progress of every science its methods tend to become simpler; and to refer especially to one of the subjects mentioned in the query, namely, Electricity, I may remind the Commissioners of the great simplification which the theory of induced electricity has recently received. Professor William Thomson's theory of electrical images has made, so to speak, elementary, problems which previously required a "very refined and laborious analysis." This theory, if electricity is to be studied at all, would now almost of necessity form a portion of the undergraduate course. It is, however, sometimes doubted whether not only electricity but also the cognate theories of heat, light, and magnetism ought not be excluded from the degree examination. I confess to being unwilling that the terminus of our mathematical studies should be made to recede, and am disposed to believe that with the changes suggested in the earlier part of these remarks, sufficient time would be found for these subjects to be, up to a certain point, satisfactorily studied. They now engage so much of the attention of scientific men that it seems particularly desirable that the highest class of Students should leave the University in a condition to follow their progress and development. A young man will not willingly forget what he learned at Cambridge if he finds that it enables him to understand the discoveries and researches which are now going on; on the contrary, he will probably always retain, at least, an interest in scientific matters. This advantage would in many cases, perhaps in most, be lost if the subjects in question were not studied until after the B.A. degree. Few even of our best Students could then be induced to devote themselves to fresh mathematical studies, notwithstanding the influence of any system of prizes either for mere proficiency or for original research. Of all such prizes* it may, I think, be said that they could not be made to form a natural element of the system of University education.

I do by no means deny, or even doubt of, their utility; but we must remember that a University may be considered in two

* The remark does not apply to the Smith's Prizes, the examination for these prizes being in effect a kind of sequel to the degree examination, and, therefore, not requiring a distinct course of reading.

points of view,—as a seat of learning and as a place of education. Much may be done by means of prizes to encourage learned men in the pursuit of the kind of knowledge to which they have especially dedicated themselves; but as things are, and perhaps as they ought to be, even a liberal education must end about the age at which the Bachelor's degree is commonly taken. And it may further be affirmed, with much show of reason, that those whom after that epoch circumstances still permit "inter silvas Academi quærere verum," may with advantage, so far as the symmetrical development of the mind is concerned, turn from mathematical to other studies.

V. With respect to the subjects mentioned in the seventh query, I have already made some remarks on the lunar and planetary theories, as well as on electricity and the kindred branches of physics. There remain, therefore, only the calculus of variations and definite and elliptic integrals. Of these subjects the first seems not unsuited for University reading. It involves important principles and admits of important applications. From its connexion with the theory of conditions of integrability it forms a natural sequel to the integral calculus; and, on the other hand, if the planetary theory were to be studied in the manner which I have suggested, some previous acquaintance with its principles would be indispensable. In favour of definite integrals there is not so much to be said, many of them depending for their evaluation on particular artifices, which must uselessly burden the Student's memory; and if in an examination he attempts to determine the value of one with which he is not already acquainted, he will often only waste time and ingenuity to no purpose. Still certain definite integrals must be known, and the theory of definite integrals of periodic functions is especially important from its connexion with Fourier's theorem and the development of discontinuous functions. The difficulties of this theory have, I think, been sufficiently removed to justify its introduction into our course; and it is not to be forgotten that no other step in the recent progress of analysis has exercised so great an influence on mathematical physics. The general theory of elliptic integrals is too extensive and too abstruse for University reading, and the study of isolated propositions almost useless. But Abel's theorem ought to be studied as the natural development of the theory of symmetrical functions, nor is there any difficulty in the demonstration by which an intelligent student can be embarrassed. Likewise Abel's method for the division of the complete elliptic function might with advantage replace Gauss's solution of the binomial equation. It includes this solution as a particular case, and from its generality is far more intelligible: Boscovich's doctrine, that the more generally a subject is considered the more easily is it understood, being for the most part true. These instances, in which portions of the theory of the comparison of transcendents serve to complete and illustrate the theory of equations, tend to show that a course of mathematical study cannot well be made to adhere precisely to any definite classification of the different branches of mathematics.

VI. One of the obstacles which hinder our mathematical studies from being quite what they ought to be is touched on in the tenth query*. It is there asked whether the number of problems proposed in the Senate-House examination is not, regard being had to the time allowed for solving them, greater than it should be. It may be answered that it is necessary to set before the candidates for high honours more problems than any one is supposed capable of solving in the given time, in order by the variety of subjects to provide sufficient employment for each, and at the same time to leave a certain freedom of choice; and, further, that if no more problems were proposed than the ablest questionists might be presumed capable of solving, the number would still be too great for those of inferior ability. All this is true, and it is therefore much less easy to point out a remedy than to perceive the evils which result from the present state of things, not only in the problem papers but throughout the examination. He who for a season can remember a great deal, and who while he remembers it can reproduce

* The tenth query is: Is it your opinion, or the contrary, that the problems proposed in the examinations bear too large a proportion to the questions derived either immediately or by a very simple deduction from the books which are commonly read? Are they for the most part so proposed, as to admit of their being readily apprehended and their solution effected by a Student who thoroughly understands the principles and applications of the branches of Mathematics upon which they depend, or are they not unfrequently involved in such a form as to require for their solution a peculiar tact distinct from accurate and philosophical knowledge? Are you not disposed to think the number of problems proposed to be solved in the time allowed (as many, generally, as seven or eight in one hour of time) is greater than the best prepared Student can be expected to complete?

it rapidly and well, generally attains to more honour than he quite deserves; and though the undue influence of mere memory is somewhat diminished by papers consisting of original problems, yet at any rate the student is trained to be quick and ready rather than wise and thoughtful. In devising problems it is difficult to avoid mere puzzles—things to be solved only by some happy guess; and if, on the other hand, examiners were to confine themselves to tolerably obvious deductions from known propositions, the problem papers would cease to be a counterpoise to the rest of the examination. If the candidates for high honours were, as in the Smith's prize examination, examined apart from the rest, it might be possible, so far as they were concerned, to diminish these difficulties by varying the modes of examination. In a select examination there could be no excessive inconvenience in giving almost unlimited time for the consideration of the questions proposed, and in these questions the candidate might be required to state accurately, and in detail, the grounds of his views on fundamental principles in analysis, geometry, or physics. Again, in such an examination books might perhaps be introduced, and if so, interesting questions might be proposed of a kind now inadmissible; to mention one class only—the candidate might be required to form an opinion on any controverted point, to examine for instance the correctness of Sir James Ivory's doctrine, that in certain cases the ordinary condition of fluid equilibrium is insufficient, and to state his reasons for adopting or rejecting it.

It must, however, be remembered that after all possible improvements the complaint that schools "lack profoundness and dwell too much on seeming," will always be more or less just. A course of study, of which the most obvious purpose is to prepare the student for an ἐπίδειξις, can never be quite what a course of study ought to be, and might be made if higher ends could always be kept in view.

CAMBRIDGE: PRINTED BY C. J. CLAY, M.A. AT THE UNIVERSITY PRESS.

CAMBRIDGE, *Nov.* 1863.

BOOKS PUBLISHED BY

DEIGHTON, BELL, & CO. CAMBRIDGE.

(Agents to the University).

Jerusalem Explored: being a description of the Ancient and Modern City, with upwards of One Hundred Illustrations, consisting of Views, Ground-plans, and Sections. By ERMETE PIEROTTI, Doctor of Mathematics, Captain of the Corps of Engineers in the Army of Sardinia, Architect-Engineer to his Excellency Soorraya Pasha of Jerusalem, and Architect of the Holy Land. [*In the Press.*

Memoir of the late Bishop Mackenzie. By the DEAN OF ELY. With portrait and illustrations. Dedicated by permission to the Lord Bishop of Oxford. [*In the Press.*

The Queen's English; being Stray Notes on Speaking and Spelling. By HENRY ALFORD, D.D., Dean of Canterbury. [*In the Press.*

Codex Bezæ Cantabrigiensis. Edited, with Prolegomena, Notes, and Facsimiles. By F. H. SCRIVENER, M.A. [*In the Press.*

A Full Collation of the Codex Sinaiticus, with the received Text of the New Testament, to which is prefixed a Critical Introduction by the Rev. F. H. SCRIVENER, M.A., Rector of St. Gerran's, Cornwall.

Kaye University Prize. The Authenticity of the Book of Daniel. By the Rev. J. M. FULLER, M.A., St. John's College, Cambridge.

Wieseler's Chronological Synopsis of the Four

GOSPELS. Translated by the Rev. E. VENABLES, M.A.

The Greek Testament: with a critically revised

Text; a Digest of Various Readings; Marginal References to Verbal and Idiomatic Usage; Prolegomena; and a Critical and Exegetical Commentary. For the use of Theological Students and Ministers. By HENRY ALFORD, D.D., Dean of Canterbury.

Vol. I. Fifth Edition, containing the Four Gospels. 1*l.* 8*s.*

Vol. II. Fourth Edition, containing the Acts of the Apostles, the Epistles to the Romans and Corinthians. 1*l.* 4*s.*

Vol. III. Third Edition, containing the Epistles to the Galatians, Ephesians, Philippians, Colossians, Thessalonians,—to Timotheus, Titus, and Philemon. 18*s.*

Vol. IV. Part I. Second Edition, containing the Epistle to the Hebrews, and the Catholic Epistle of St. James and St. Peter. 18*s.*

Vol. IV. Part II. Second Edition, containing the Epistles of St. John and St. Jude, and the Revelation. 14*s.*

Novum Testamentum Græcum, Textus Stephanici,

1500. Accedunt variæ lectiones editionum Bezæ, Elzeviri, Lachmanni, Tischendorfii, et Tregellesii. Curante F. H. SCRIVENER, A.M. 16mo. 4*s.* 6*d.*

An Edition on Writing-paper, for Notes. 4to. *half-bound.* 12*s.*

An Exact Transcript of the Codex Augiensis,

Græco-Latina Manuscript in Uncial Letters of S. Paul's Epistles, preserved in the Library of Trinity College, Cambridge. To which is added a full Collation of Fifty Manuscripts containing various portions of the Greek New Testament deposited in English Libraries: with a full Critical Introduction. By F. H. SCRIVENER, M.A. Royal 8vo. 26*s.*

The Critical Introduction *is issued separately, price* 5*s.*

The New Testament for English Readers. Con-

taining the authorized Version, with additional corrections of Readings and Renderings; Marginal References; and a Critical and Explanatory Commentary. By HENRY ALFORD, D.D., Dean of Canterbury. In two volumes.

Vol I. Part I. containining the first Three Gospels, price 12*s.*

On The Imitation of Christ. A New Translation. By the Very Rev. the DEAN of ELY. *Second Edition.* 18mo. 3*s.* 6*d.*

Fcp. 8vo. An Edition printed on LARGE PAPER, 5*s.*

Messiah as Foretold and Expected. A Course of Sermons relating to the Messiah, as interpreted before the Coming of Christ. Preached before the University of Cambridge, in the months of February and March, 1862. By the Rev. HAROLD BROWNE, B.D., Norrisian Professor of Divinity, and Canon of Exeter Cathedral. 8vo. 4*s.*

A History of the Articles of Religion. To which is added a series of Documents from A.D. 1536 to A.D. 1615. Together with illustrations from contemporary sources. By CHARLES HARDWICK, B.D., late Archdeacon of Ely. *Second Edition, corrected and enlarged.* 8vo. 12*s.*

*** A considerable amount of fresh matter has been incorporated, especially in the two Chapters which relate to the construction and revision of our present code of Articles.

Commentaries on the Gospels, intended for the English Reader, and adapted either for Domestic or Private use. By the Very Rev. H. GOODWIN, D.D., Dean of Ely. Crown 8vo.

S. MATTHEW, 12*s.* S. MARK, 7*s.* 6*d.* S. LUKE, *in the Press.*

The Psalter, or Psalms of David in English Verse. With Preface and Notes. By a Member of the University of Cambridge. Dedicated by permission to the Lord Bishop of Ely, and the Reverend the Professors of Divinity in that University. 5*s.*

A General Introduction to the Apostolic Epistles, with a Table of St. Paul's Travels, and an Essay on the State after Death. *Second Edition, enlarged.* To which are added a Few Words on the Athanasian Creed, on Justification by Faith, and on the Ninth and Seventeenth Articles of the Church of England. By A BISHOP'S CHAPLAIN. 8vo. 8*s.* 6*d.*

Bentleii Critica Sacra. Notes on the Greek and Latin Text of the New Testament, extracted from the Bentley MSS. in Trinity College Library. With the Abbé Rulotta's Collation of the Vatican MS., a specimen of Bentley's intended Edition, and an account of all his Collations. Edited, with the permission of the Master and Seniors, by the Rev. A. A. ELLIS, M.A., late Fellow of Trinity College, Cambridge. 8vo. 8*s.* 6*d.*

Three Plain Sermons, preached in the Chapel of

Trinity College, Cambridge, in the course of the year 1859. By the Rev. E. W. BLORE, Fellow of Trinity College. 8vo. 1*s*. 6*d*.

Butler's Three Sermons on Human Nature, and

Dissertation on Virtue. Edited by W. WHEWELL, D.D. With a Preface and a Syllabus of the Work. *Third Edition.* Fcp. 8vo. 3*s*. 6*d*.

A Translation of the Epistles of Clement of Rome,

Polycarp, and Ignatius; and of the Apologies of Justin Martyr and Tertullian; with an Introduction and Brief Notes illustrative of the Ecclesiastical History of the First Two Centuries. By T. CHEVALLIER, B.D. *Second Edition.* 8vo. 12*s*.

On Sacrifice, Atonement, Vicarious Oblation, and

Example of Christ, and the Punishment of Sin. Five Sermons, preached before the University of Cambridge, March, 1856. By B. M. COWIE, B.D., St. John's College. 8vo. 5*s*.

Five Sermons preached before the University of

Cambridge. By the late J. J. BLUNT, B.D., Lady Margaret Professor of Divinity. 8vo. 5*s*. 6*d*.

CONTENTS:—1. The Nature of Sin.—2. The Church of the Apostles.—3. On Uniformity of Ritual.—4. The Value of Time.—5. Reflections on the General Fast-Day (March 1847).

Five Sermons preached before the University of

Cambridge. The first Four in November, 1851, the Fifth on Thursday, March 8th, 1849, being the Hundred and Fiftieth Anniversary of the Society for Promoting Christian Knowledge. By the late Rev. J. J. BLUNT, B.D.

CONTENTS: 1. Tests of the Truth of Revelation.—2. On Unfaithfulness to the Reformation.—3. On the Union of Church and State.—4. An Apology for the Prayer-Book.—5. Means and Method of National Reform.

Two Introductory Lectures on the Study of the

Early Fathers, delivered in the University of Cambridge. By the late J. J. BLUNT, B.D. *Second Edition.* With a brief Memoir of the Author, and a Table of Lectures delivered during his Professorship. 8vo. 4*s*. 6*d*.

The Example of Christ and the Service of Christ, Considered in Three Sermons preached before the University of Cambridge, in February, 1861. To which are appended, A Few Remarks upon the Present State of Religious Feeling. By FRANCIS FRANCE, B.D., Archdeacon of Ely, and Fellow of St. John's College. Crown 8vo. 2*s*. 6*d*.

Ecclesiæ Anglicanæ Vindex Catholicus, sive Articulorum Ecclesiæ Anglicanæ cum Scriptis SS. Patrum nova collatio. Cura G. W. HARVEY, Collegii Regalis Socii. 3 vols. 8vo. Reduced to 16*s*.

The Doctrines and Difficulties of the Christian Religion contemplated from the Standing-point afforded by the Catholic Doctrine of the Being of our Lord Jesus Christ. Being the Hulsean Lectures for the Year 1855. By H. GOODWIN, D.D. 8vo. 9*s*.

'The Glory of the Only Begotten of the Father seen in the Manhood of Christ.' Being the Hulsean Lectures for the year 1856. By H. GOODWIN, D.D. 8vo. 7*s*. 6*d*.

Four Sermons preached before the University of Cambridge in the Season of Advent, 1858. By H. GOODWIN, D.D. 12mo. 3*s*. 6*d*.

Four Sermons preached before the University of Cambridge in the month of November, 1853. By H. GOODWIN, D.D. 12mo. 4*s*.

CONTENTS:—1. The Young Man cleansing his way.—2. The Young Man in Religious Difficulties.—3. The Young Man as a Churchman.—4. The Young Man called by Christ.

Christ in the Wilderness. Four Sermons preached before the University of Cambridge in the month of February, 1855. By H. GOODWIN, D.D. 12mo. 4*s*.

Parish Sermons. By H. GOODWIN, D.D. 1st Series. *Third Edition.* 12mo. 6*s*.

——— **2nd Series.** *Third Edition.* 12mo. 6*s*.

——— **3rd Series.** *Second Edition.* 12mo. 7*s*.

——— **4th Series.** 12mo. 7*s*.

——— **5th Series.** With Preface on Sermons and Sermon Writing. 12mo. 7*s*.

Short Sermons at the Celebration of the Lord's Supper. By H. GOODWIN, D.D. *New Edition.* 12mo. 4*s.*

Lectures upon the Church Catechism. By H. GOODWIN, D.D. 12mo. 4*s.*

A Guide to the Parish Church. By H. GOODWIN, D.D. Price 1*s.* sewed; 1*s.* 6*d.* cloth.

Confirmation Day. Being a Book of Instruction for Young Persons how they ought to spend that solemn day, on which they renew the Vows of their Baptism, and are confirmed by the Bishop with prayer and the laying on of hands. By H. GOODWIN, D.D. *Second Edition, 2d.*, or 25 for 3*s.* 6*d.*

Plain Thoughts concerning the meaning of Holy Baptism. By H. GOODWIN, D.D. *Second Edition.* 2*d.*, or 25 for 3*s.* 6*d.*

The Worthy Communicant; or, 'Who may come to the Supper of the Lord?' By H. GOODWIN, D.D. *Second Edition, 2d.*, or 25 for 3*s.* 6*d.*

An Historical and Explanatory Treatise on the Book of Common Prayer. By W. G. HUMPHRY, B.D., late Fellow of Trinity College, Cambridge. *Second Edition, enlarged and revised.* Post 8vo. 7*s.* 6*d.*

Liturgiæ Britannicæ, or the several Editions of the Book of Common Prayer of the Church of England, from its compilation to the last revision, together with the Liturgy set forth for the use of the Church of Scotland, arranged to shew their respective variations. By W. KEELING, B.D., late Fellow of St. John's College. *Second Edition.* 8vo. 12*s.*

The Seven Words Spoken Against the Lord Jesus:

or, an Investigation of the Motives which led His Contemporaries to reject Him. Being the Hulsean Lectures for the year 1860. By JOHN LAMB, M.A., Senior Fellow of Gonville and Caius College, and Minister of S. Edward's, Cambridge. 8vo. 5*s*. 6*d*.

Twelve Sermons preached on Various Occasions at

the Church of St. Mary, Greenwich. By R. MAIN, M.A. Radcliffe Observer at Oxford. 12mo. 5*s*.

Lectures on the Catechism. Delivered in the

Parish Church of Brasted, in the Diocese of Canterbury, By the late W. H. MILL, D.D., Regius Professor of Hebrew, Cambridge. Edited by his Son-in-Law, the Rev. B. WEBB, M.A. Fcp. 8vo. 6*s*. 6*d*.

Sermons preached in Lent 1845, and on several

former occasions, before the University of Cambridge. By W. H. MILL, D.D. 8vo. 12*s*.

Four Sermons preached before the University on

the Fifth of November, and the three Sundays preceding Advent, in the year 1848. By W. H. MILL, D.D. 8vo. 5*s*. 6*d*.

An Analysis of the Exposition of the Creed,

written by the Right Reverend Father in God, J. PEARSON, D.D., late Lord Bishop of Chester. Compiled, with some additional matter occasionally interspersed, for the use of Students of Bishop's College, Calcutta. By W. H. MILL, D.D. *Third Edition, revised and corrected.* 8vo. 5*s*.

Horae Hebraicae. Critical and Expository Ob-

servations on the Prophecy of Messiah in Isaiah, Chapter IX. and on other Passages of Holy Scripture. By W. SELWYN, B.D., Lady Margaret's Reader in Theology. Revised Edition, with Continuation. 8*s*.

THE CONTINUATION, separately. 3*s*.

Excerpta ex Reliquiis Versionum, Aquilæ, Sym-

machi, Theodotionis, a Montefalconia aliisque collectis. GENESIS. Edidit GUL. SELWYN, S.T.B. 8vo. 1*s*.

Notæ Criticæ in Versionem Septuagintaviralem. EXODUS, Cap. I.—XXIV. Curante GUL. SELWYN, S.T.B. 8vo. 3*s*. 6*d*.

Notæ Criticæ in Versionem Septuagintaviralem. Liber NUMERORUM. Curante GUL. SELWYN, S.T.B. 8vo. 4*s*. 6*d*.

Notæ Criticæ in Versionem Septuagintaviralem. Liber DEUTERONOMII. Curante GUL. SELWYN, S.T.B. 8vo. 4*s*. 6*d*.

Origenis Contra Celsum. Liber I. Curante GUL. SELWYN, S.T.B. 8vo. 3*s*. 6*d*.

Testimonia Patrum in Veteres Interpretes, Septuaginta, Aquilam, Symmachum, Theodotionem, a Montefalconio aliisque collecta paucis Additis. Edidit GUL. SELWYN, S.T.B. 8vo. 6*d*.

The Will Divine and Human. By T. SOLLY, B.D., late of Caius College, Cambridge. 8vo. 10*s*. 6*d*.

Tertulliani Liber Apologeticus. The Apology of Tertullian. With English Notes and a Preface, intended as an Introduction to the Study of Patristical and Ecclesiastical Latinity. By H. A. WOODHAM, LL.D. *Second Edition.* 8vo. 8*s*. 6*d*.

Rational Godliness. After the Mind of Christ and the Written Voices of the Church. By ROWLAND WILLIAMS, D.D., Professor of Hebrew at Lampeter. Crown 8vo. 10*s*. 6*d*.

Paraméswara-jnyána-goshthi. A Dialogue of the Knowledge of the Supreme Lord, in which are compared the claims of Christianity and Hinduism, and various questions of Indian Religion and Literature fairly discussed. By ROWLAND WILLIAMS, D.D. 8vo. 12*s*.

MISCELLANEOUS.

Verses and Translations. By C. S. C. *Second Edition.* Fcp. 8vo. 5*s*.

A Lecture on Sculpture delivered in the Guildhall, Cambridge, before the Cambridge School of Art March 17, 1863. By RICHARD WESTMACOTT, R.A., F.R.S., Professor of Sculpture in the Royal Academy. Price 1*s*.

The Student's Guide to the University of Cambridge. Fcap. 8vo. 5*s*. 6*d*.

Contents: INTRODUCTION, by J. R. SEELEY, M.A.—ON UNIVERSITY EXPENSES, by the Rev. H. LATHAM, M.A.—ON THE CHOICE OF A COLLEGE, by J. R. SEELEY, M.A.—ON THE COURSE OF READING FOR THE CLASSICAL TRIPOS, by the Rev. R. BURN, M.A.—ON THE COURSE OF READING FOR THE MATHEMATICAL TRIPOS, by the Rev. W. M. CAMPION, B.D.—ON THE COURSE OF READING FOR THE MORAL SCIENCES TRIPOS, by the Rev. J. B. MAYOR, M.A.—ON THE COURSE OF READING FOR THE NATURAL SCIENCES TRIPOS, by Professor LIVEING, M.A.—On LAW STUDIES AND LAW DEGREES, by Professor J. T. ABDY, LL.D.— MEDICAL STUDY AND DEGREES, by G. M. HUMPHRY, M.D.—ON THEOLOGICAL EXAMINATIONS, by Professor E. HAROLD BROWNE, B.D.—EXAMINATIONS FOR THE CIVIL SERVICE OF INDIA, by the Rev. H. LATHAM, M.A.—LOCAL EXAMINATIONS OF THE UNIVERSITY, by H. J. ROBY, M.A.—DIPLOMATIC SERVICE.—DETAILED ACCOUNT OF THE SEVERAL COLLEGES.

"Partly with the view of assisting parents, guardians, schoolmasters, and students intending to enter their names at the University—partly also for the benefit of undergraduates themselves—a very complete, though concise, volume has just been issued, which leaves little or nothing to be desired. For lucid arrangement, and a rigid adherence to what is positively useful, we know of few manuals that could compete with this Student's Guide. It reflects no little credit on the University to which it supplies an unpretending, but complete, introduction."—SATURDAY REVIEW.

The Mathematical and other Writings of ROBERT LESLIE ELLIS, M.A., late Fellow of Trinity College, Cambridge. Edited by WILLIAM WALTON, M.A., Trinity College, with a Biographical Memoir by the Very Reverend HARVEY GOODWIN, D.D., Dean of Ely. [*In the Press.*

Lectures on the History of Moral Philosophy in England. By the Rev. W. WHEWELL, D.D., Master of Trinity College, Cambridge. New and Improved Edition, with Additional Lectures. Crown 8vo. 8*s.*

The Additional Lectures are printed separately in Octavo for the convenience of those who have purchased the former Edition. Price 3*s.* 6*d.*

Athenae Cantabrigienses. By C. H. Cooper, F.S.A. and THOMPSON COOPER, F.S.A.

This work, in illustration of the biography of notable and eminent men who have been members of the University of Cambridge, comprehends, notices of: 1. Authors. 2. Cardinals, archbishops, bishops, abbats, heads of religious houses and other Church dignitaries. 3. Statesmen, diplomatists, military and naval commanders. 4. Judges and eminent practitioners of the civil or common law. 5. Sufferers for religious and political opinions. 6. Persons distinguished for success in tuition. 7. Eminent physicians and medical practitioners. 8. Artists, musicians, and heralds. 9. Heads of Colleges, professors, and principal officers of the university. 10. Benefactors to the university and colleges or to the public at large.

Volume I. 1500—1585. 8vo. *cloth.* 18*s.* Volume II. 1586—1609. 18*s.*

Volume III. *Preparing.*

Cairo to Sinai and Sinai to Cairo. Being an Account of a Journey in the Desert of Arabia, November and December, 1860. By W. J. BEAMONT, M.A., Fellow of Trinity College, Cambridge. With Maps and Illustrations. Fcp. 8vo. 5*s.*

A Concise Grammar of the Arabic Language. Revised by SHEIKH ALI NADY EL BARRANY. By W. J. BEAMONT, M.A., Fellow of Trinity College, Cambridge, and Incumbent of St. Michael's' Cambridge, sometime Principal of the English College, Jerusalem. Price 7*s.*

Livingstone's Cambridge Lectures. With a Prefatory Letter by the Rev. Professor SEDGWICK, M.A., F.R.S., &c., Vice-Master of Trinity College, Cambridge. Edited, with Introduction, Life of Dr. LIVINGSTONE, Notes and Appendix, by the Rev. W. MONK, M.A., F.R.A.S., &c., of St. John's College, Cambridge. With a Portrait and Map, also a larger Map, by Arrowsmith, granted especially for this work by the President and Council of the Royal Geographical Society of London. Crown 8vo. 6*s*. 6*d*.

This Edition contains a New Introduction, an Account of Dr. Livingstone's New Expedition, a Series of Extracts from the Traveller's Letters received since he left this country, and a History of the Oxford and Cambridge Mission to Central Africa.

Newton (Sir Isaac) and Professor Cotes, Correspondence of, including Letters of other Eminent Men, now first published from the originals in the Library of Trinity College, Cambridge; together with an Appendix containing other unpublished Letters and Papers by Newton; with Notes, Synoptical View of the Philosopher's Life, and a variety of details illustrative of his history. Edited by the Rev. J. EDLESTON, M.A., Fellow of Trinity College. 8vo. 10*s*.

A Manual of the Roman Civil Law, arranged according to the Syllabus of Dr. HALLIFAX. By G. LEAPINGWELL, LL.D. Designed for the use of Students in the Universities and Inns of Court. 8vo. 12*s*.

The Study of the English Language an Essential Part of a University Course: An Extension of a Lecture delivered at the Royal Institution of Great Britain, February 1, 1861. With Coloured Language-Maps of the British Isles and Europe. By ALEXANDER J. D. D'ORSEY, B.D., English Lecturer at Corpus Christi College, Cambridge, late Head Master of the English Department in the High School of Glasgow. Crown 8vo. cloth. 2*s*. 6*d*.

Graduati Cantabrigienses: sive Catalogus exhibens nomina eorum quos ab anno academico admissionum 1760 usque ad decimum diem Octr. 1856. Gradu quocunque ornavit Academia Cantabrigienses, e libris subscriptionum desumptus. Cura J. ROMILLY, A.M., Coll. SS. Trin. Socii atque Academica Registrarii. 8vo. 10*s*.

JONATHAN PALMER, PRINTER, SIDNEY STREET, CAMBRIDGE.

www.ingramcontent.com/pod-product-compliance
Lightning Source LLC
LaVergne TN
LVHW020551110826
845149LV00002B/227

* 9 7 8 1 4 1 8 1 8 4 5 3 7 *